T0143253

Programming for Hybrid Multi/Manycore MPP Systems

Chapman & Hall/CRC Computational Science Series

SERIES EDITOR

Horst Simon
Deputy Director
Lawrence Berkeley National Laboratory
Berkeley, California, U.S.A.

PUBLISHED TITLES

COMBINATORIAL SCIENTIFIC COMPUTING
Edited by Uwe Naumann and Olaf Schenk

CONTEMPORARY HIGH PERFORMANCE COMPUTING: FROM PETASCALE TOWARD EXASCALE
Edited by Jeffrey S. Vetter

CONTEMPORARY HIGH PERFORMANCE COMPUTING: FROM PETASCALE TOWARD EXASCALE, VOLUME TWO
Edited by Jeffrey S. Vetter

DATA-INTENSIVE SCIENCE
Edited by Terence Critchlow and Kerstin Kleese van Dam

ELEMENTS OF PARALLEL COMPUTING
Eric Aubanel

THE END OF ERROR: UNUM COMPUTING
John L. Gustafson

EXASCALE SCIENTIFIC APPLICATIONS: SCALABILITY AND PERFORMANCE PORTABILITY
Edited by Tjerk P. Straatsma, Katerina B. Antypas, and Timothy J. Williams

FROM ACTION SYSTEMS TO DISTRIBUTED SYSTEMS: THE REFINEMENT APPROACH
Edited by Luigia Petre and Emil Sekerinski

FUNDAMENTALS OF MULTICORE SOFTWARE DEVELOPMENT
Edited by Victor Pankratius, Ali-Reza Adl-Tabatabai, and Walter Tichy

FUNDAMENTALS OF PARALLEL MULTICORE ARCHITECTURE
Yan Solihin

THE GREEN COMPUTING BOOK: TACKLING ENERGY EFFICIENCY AT LARGE SCALE
Edited by Wu-chun Feng

GRID COMPUTING: TECHNIQUES AND APPLICATIONS
Barry Wilkinson

HIGH PERFORMANCE COMPUTING: PROGRAMMING AND APPLICATIONS
John Levesque with Gene Wagenbreth

PUBLISHED TITLES CONTINUED

HIGH PERFORMANCE PARALLEL I/O
Prabhat and Quincey Koziol

HIGH PERFORMANCE VISUALIZATION:
ENABLING EXTREME-SCALE SCIENTIFIC INSIGHT
Edited by E. Wes Bethel, Hank Childs, and Charles Hansen

INDUSTRIAL APPLICATIONS OF HIGH-PERFORMANCE COMPUTING:
BEST GLOBAL PRACTICES
Edited by Anwar Osseyran and Merle Giles

INTRODUCTION TO COMPUTATIONAL MODELING USING C AND
OPEN-SOURCE TOOLS
José M Garrido

INTRODUCTION TO CONCURRENCY IN PROGRAMMING LANGUAGES
Matthew J. Sottile, Timothy G. Mattson, and Craig E Rasmussen

INTRODUCTION TO ELEMENTARY COMPUTATIONAL MODELING: ESSENTIAL
CONCEPTS, PRINCIPLES, AND PROBLEM SOLVING
José M. Garrido

INTRODUCTION TO HIGH PERFORMANCE COMPUTING FOR SCIENTISTS
AND ENGINEERS
Georg Hager and Gerhard Wellein

INTRODUCTION TO MODELING AND SIMULATION WITH MATLAB® AND PYTHON
Steven I. Gordon and Brian Guilfoos

INTRODUCTION TO REVERSIBLE COMPUTING
Kalyan S. Perumalla

INTRODUCTION TO SCHEDULING
Yves Robert and Frédéric Vivien

INTRODUCTION TO THE SIMULATION OF DYNAMICS USING SIMULINK®
Michael A. Gray

PEER-TO-PEER COMPUTING: APPLICATIONS, ARCHITECTURE, PROTOCOLS,
AND CHALLENGES
Yu-Kwong Ricky Kwok

PERFORMANCE TUNING OF SCIENTIFIC APPLICATIONS
Edited by David Bailey, Robert Lucas, and Samuel Williams

PETASCALE COMPUTING: ALGORITHMS AND APPLICATIONS
Edited by David A. Bader

PROCESS ALGEBRA FOR PARALLEL AND DISTRIBUTED PROCESSING
Edited by Michael Alexander and William Gardner

PROGRAMMING FOR HYBRID MULTI/MANYCORE MPP SYSTEMS
John Levesque and Aaron Vose

PUBLISHED TITLES CONTINUED

SCIENTIFIC DATA MANAGEMENT: CHALLENGES, TECHNOLOGY, AND DEPLOYMENT
Edited by Arie Shoshani and Doron Rotem

SOFTWARE ENGINEERING FOR SCIENCE
Edited by Jeffrey C. Carver, Neil P. Chue Hong, and George K. Thiruvathukal

Programming for Hybrid Multi/Manycore MPP Systems

John Levesque
Aaron Vose

CRC Press is an imprint of the
Taylor & Francis Group, an **informa** business
A CHAPMAN & HALL BOOK

CRC Press
Taylor & Francis Group
6000 Broken Sound Parkway NW, Suite 300
Boca Raton, FL 33487-2742

Printed on acid-free paper
Version Date: 20170825

International Standard Book Number-13: 978-1-4398-7371-7 (Hardback)

Library of Congress Cataloging-in-Publication Data

Names: Levesque, John M. author. | Vose, Aaron, author.
Title: Programming for hybrid multi/manycore MPP systems / John Levesque, Aaron Vose.
Description: Boca Raton : CRC Press, Taylor & Francis, 2017. | Series: Chapman & Hall/CRC computational science | Includes index.
Identifiers: LCCN 2017018319 | ISBN 9781439873717 (hardback : alk. paper)
Subjects: LCSH: Parallel programming (Computer science) | Multiprocessors--Programming. | Coprocessors--Programming.
Classification: LCC QA76.642 .L+475 2017 | DDC 005.2/75--dc23
LC record available at https://lccn.loc.gov/2017018319

Visit the Taylor & Francis Web site at
http://www.taylorandfrancis.com

and the CRC Press Web site at
http://www.crcpress.com

Printed and bound in the United States of America by
Edwards Brothers Malloy on sustainably sourced paper

To my life's love, my wife of 53 years, who has had to put up with bad moods, good moods, and absent days for me to pursue another of my life's loves – my work.

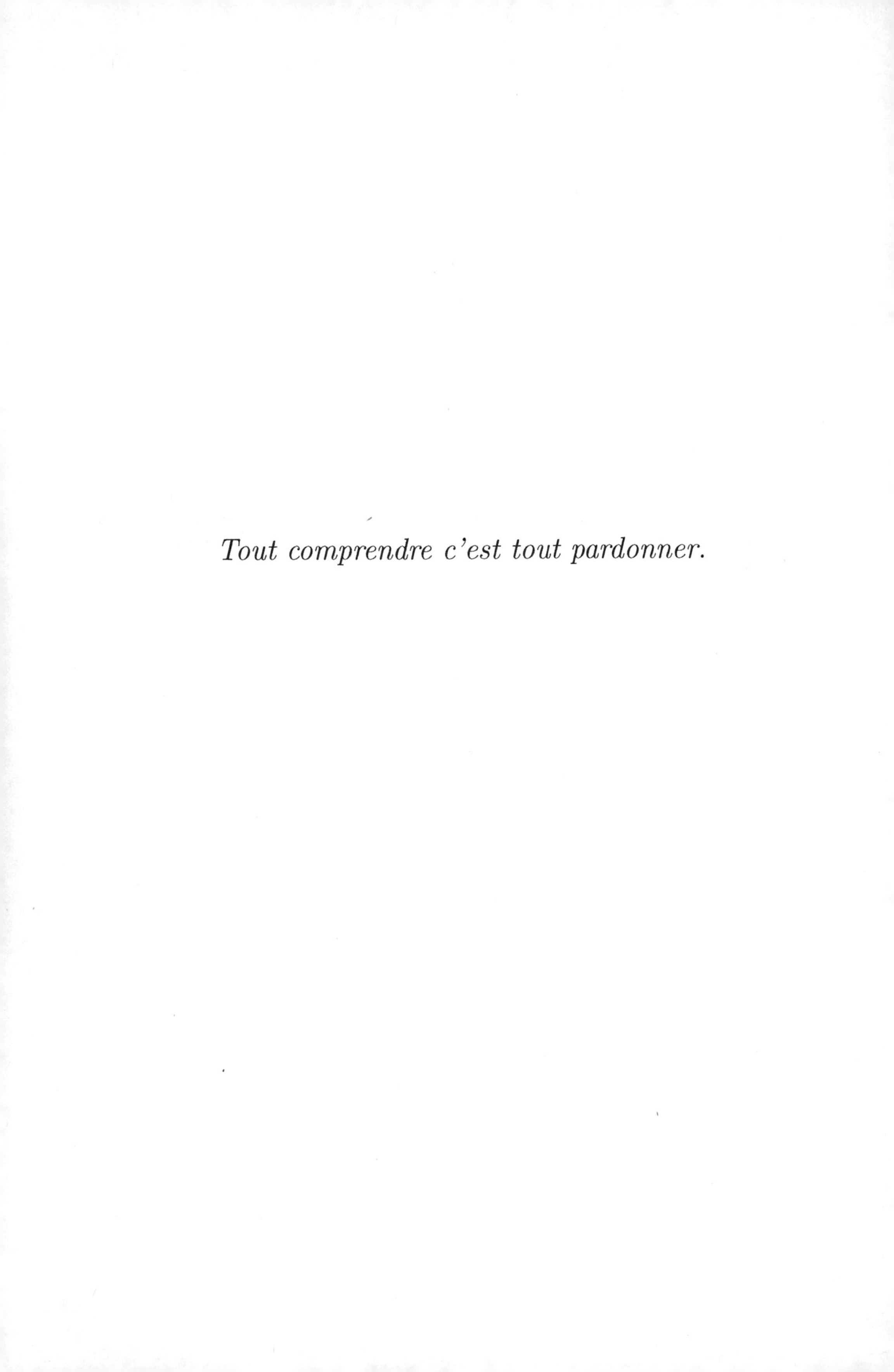

Tout comprendre c'est tout pardonner.

Contents

Preface

For the past 20 years, high performance computing has benefited from a significant reduction in the clock cycle time of the basic processor. Going forward, trends indicate the clock rate of the most powerful processors in the world may stay the same or decrease slightly. When the clock rate decreases, the chip runs at a slower speed. At the same time, the amount of physical space that a computing core occupies is still trending downward. This means more processing cores can be contained within the chip.

With this paradigm shift in chip technology, caused by the amount of electrical power required to run the device, additional performance is being delivered by increasing the number of processors on the chip and (re)introducing SIMD/vector processing. The goal is to deliver more floating-point operations per second per watt. Interestingly, these evolving chip technologies are being used on scientific systems as small as a single workstation and as large as the systems on the Top 500 list.

Within this book are techniques for effectively utilizing these new node architectures. Efficient threading on the node, vectorization to utilize the powerful SIMD units, and effective memory management will be covered along with examples to allow the typical application developer to apply them to their programs. Performance portable techniques will be shown that will run efficiently on all HPC nodes.

The principal target systems will be the latest multicore Intel Xeon processor, the latest Intel Knight's Landing (KNL) chip with discussion/comparison to the latest hybrid/accelerated systems using the NVIDIA Pascal accelerator.

The following QR code points to `www.hybridmulticore.com`, the book's companion website, which will contain solutions to the exercises in the book:

About the Authors

John Levesque works in the Chief Technology Office at Cray Inc. where he is responsible for application performance on Cray's HPC systems. He is also the director of Cray's Supercomputing Center of Excellence for the Trinity System installed at Los Alamos Scientific Laboratory at the end of 2016. Prior to Trinity, he was director of the Center of Excellence at the Oak Ridge National Laboratory (ORNL). ORNL installed a 27 Petaflop Cray XK6 system, Titan, which was the fastest computer in the world according to the Top 500 list in 2012; and a 2.7 Petaflop Cray XT4, Jaguar, which was number one in 2009. For the past 50 years, Mr. Levesque has optimized scientific application programs for successful HPC systems. He is an expert in application tuning and compiler analysis of scientific applications. He has written two previous books, on optimization for the Cray 1 in 1989 [20] and on optimization for multicore MPP systems in 2010 [19].

Aaron Vose is an HPC software engineer who spent two years at Cray's Supercomputing Center of Excellence at Oak Ridge National Laboratory. Mr. Vose helped domain scientists at ORNL port and optimize scientific software to achieve maximum scalability and performance on world-class, high-performance computing resources, such as the Titan supercomputer. Mr. Vose now works for Cray Inc. as a software engineer helping R&D to design next-generation computer systems. Prior to joining Cray, he spent time at the National Institute for Computational Sciences (NICS) as well as the Joint Institute for Computational Sciences (JICS). There, he worked on scaling and porting bioinformatics software to the Kraken supercomputer. Mr. Vose holds a Master's degree in computer science from the University of Tennessee at Knoxville.

List of Figures

List of Tables

List of Excerpts

CHAPTER 1

Introduction

CONTENTS

1.1 INTRODUCTION

The next decade of computationally intense computing lies with more powerful multi/manycore nodes where processors share a large memory space. These nodes will be the building block for systems that range from a single node workstation up to systems approaching the exaflop regime. The node itself will consist of 10's to 100's of MIMD (multiple instruction, multiple data) processing units with SIMD (single instruction, multiple data) parallel instructions. Since a standard, affordable memory architecture will not be able to supply the bandwidth required by these cores, new memory organizations will be introduced. These new node architectures will represent a significant challenge to application developers.

We are seeing these new architectures because they can supply more performance for less power. For the past 20 years, Moore's Law has prevailed with more and more logic placed on the chip to perform functions important to the performance of applications. Ten years ago, the higher density resulted in higher frequency clock cycles, which resulted in better performance. Then the computer industry hit the power wall where the energy to run the chip at the higher frequencies drove the cost of running the chip to an extreme level. Now the higher density is being used for replicating the processors on the chip and expanding SIMD instructions. Improvements over the last 10 years do not deliver an automatic performance boost. In fact, the single core is running slower, and the application developer must employ more parallelism on the node and ensure that the compiler vectorizes their application to achieve a performance gain. Moore's Law no longer guarantees increased computational performance of the chip as it has in the past decade. For example, current x86 architectures are optimized for singe thread performance. Wide superscalar functional units, speculative execution, and special hardware to support legacy operations use more power than the arithmetic units. Today, chip developers need to design computing devices that can be used in phones, tablets, and laptops as well as components of large web servers and devices used for gaming. In all of these markets, it is important to minimize the power used by the chip and memory. For phones and tablets, and to some extent laptops, the goal is to preserve battery life. For larger servers it comes down to minimizing operational cost. Moving forward, the increased chip density promised by Moore's Law will be used for added computation capability with a decreased power requirement. Added computational capability will be delivered by increasing the number of cores on the chip, which may come with a reduction in the complexity and clock rate of the core. Further, more capable vector functional units supplying added computational power per watt will be introduced. These changes in the chip architecture will require application developers to refactor programs to effectively utilize the new hardware components that supply more computational power per watt.

Over the next five years, application developers requiring more computational power to solve scientific problems will be faced with a significant challenge to effectively utilize the additional computational resources. For the past five years, we have seen high performance computing getting walled in. First, we hit the memory wall and try to understand how to restructure our programs to exhibit locality to take advantage of existing memory hierarchies. Then we hit the power wall where the cost of the electricity to operate large scientific computers is more expensive than the computer itself. The Department of Energy (DOE) mandated an exaflop system should not use more than 20 MW of power. Today's traditional 20 petaflop supercomputing systems utilizing multicore nodes typically utilize 5 to 10 MW of power. How can a factor of 50 increase in performance be delivered with only a factor of 2 to 4 increase in power? It will take radical new designs with well-structured programs to achieve increased performance on less power-hungry hardware. Since the radical new designs are made at the node level, everyone who programs a computer to perform complex operations will be challenged to redesign algorithms and refactor existing applications to realize higher performance from future systems.

For example, consider the general purpose, graphics processing units (GPU) being supplied by NVIDA. These chips have special purpose functional units that are good at performing operations required for displaying high speed graphics for gaming systems. Only recently have these systems been extended to higher precision with error checking memory, not because they are targeting HPC, but because game developers want to include more realistic affects within games. Given the increased floating-point computation required by game developers, these systems are now potential candidates for the high performance computing (HPC) community.

While these systems have many similarities, they are also quite different. Whereas NVIDA's systems utilize GPUs unlike existing x86 processors, Intel's Xeon Phi processors are more like x86, but with more processors sharing memory. Table 1.1.1 provides the relative MIMD and SIMD dimensions of chips from NVIDA and Intel. The lack of sufficient cache close to the processor is a huge disadvantage for the NVIDA GPUs and the Intel KNL systems.

TABLE 1.1.1 Key MIMD and SIMD attributes of current hardware.

Arch	MIMD Procs	SIMD DP Width	Close KB/core	HBM GB	HBM GB/sec	Far GB	Far GB/sec
K20X	14	32	90	6	250	-	-
K40	15	32	96	12	288	-	-
K80(/2)	15	32	96	12	240	-	-
Pascal	56	32	73	16	720	-	-
Ivybridge	16	8	2887	-	-	128	100
Haswell	32	8	2887	-	-	256	120
KNL	60	16	557	16	480	384	80

Moving forward, each vendor has a different path: NVIDA is considering attached accelerators with a single address space across the host and accelerator, and Intel is considering no host at all – with their chip handling all host and accelerator duties. Further, each of these vendors are supplying different programming approaches for their systems: NVIDA has CUDA and OpenACC, and Intel suggests its users use pthreads and OpenMP 4.5. There is a trend to OpenMP 4.5 from the newer users of NVIDA systems. However, OpenACC is easier to employ and more performant on the GPU. Thus, the picture facing application developers is quite bleak. Which of these vendors will become dominant? Which programming approach works best?

This book attempts to briefly describe the current state-of-the-art in programming these systems and proposes an approach for developing a performance-portable application that can effectively utilize all of these systems from a single application. Does this sound too good to be true? A leap of faith must be made. At some point the final translation from the user code to the executable code running on the target system must be performed by a compiler. Compilers can be very good at generating code for these systems, if and only if the input program is structured in a way the compiler understands, as a sequence of required operations in a context that allows those operations to be mapped to a framework consisting of a MIMD parallel collection of SIMD instructions.

1.2 CHAPTER OVERVIEWS

Chapter 2 proposes a strategy for optimizing an application for multi/manycore architectures. The strategy starts with a good understanding of the target hardware. In this case, there are several potential architectures we may want to move to. First, we have the attached accelerator, with the host orchestrating the execution and the accelerator or GPU performing the compute/memory-intensive parts of the application. Second, we have the self-hosted manycore architecture like the Xeon Phi. Last, we have the traditional multicore Xeon and ARM systems. An extremely important aspect of the porting and optimization process is selection of the input problem, or more importantly, asking the question "what do we want to compute on the target architecture?" Problem size will dictate how one formulates the parallel and vector strategies. Unfortunately, the optimization approach for a given application depends upon the problem being executed. If a significant amount of restructuring is to be performed, it is important that the person doing the restructuring is addressing the correct problem set. While possible and desirable, optimization of an application while targeting a wide range of problem sizes is extremely difficult. Also in this chapter we will address acceptance of the restructured application by the principal developers. The application teams must be involved in the process so they understand where changes are being made and how future enhancements of the physics might be implemented within the new code structure.

Chapter 3 looks at the three typical architectures, all with advantages and disadvantages. Developing an efficient application that is performance portable across all three will be a challenge. We hope to point out the areas where choices may have to be made to target one architecture over the other. The chapter discusses a high level view of hardware supplied by NVIDA and Intel. While this discussion will not be as extensive as one might get from a paper on the hardware design, the intent is to supply enough information so the reader understands issues including memory organization, how MIMD units work and communicate, and the restrictions on utilization of SIMD units. While the book is targeted to the effective utilization of the node, utilization of multiple nodes for an application is extremely important. However, application developers have spent the last 20 years perfecting the utilization of MPI.

Chapter 4 looks at the other important component of the target, the compiler. The compiler will ultimately convert the input language to executable code on the target. From Chapter 3 we have an idea of what we want to take advantage of and what we want to avoid. In this chapter we will learn how to make the compiler do what we want. Additionally, this chapter addresses the application developer's relationship with the compiler. So often the developer expects the compiler to generate efficient code from their application and are very surprised when it doesn't. Often the developer expects far too much from the compiler – they must learn to work with the compiler to generate the best code from a given application.

> "Ask not what your compiler can do for you, ask what you can do for your compiler."
>
> John Levesque
> Director of Cray's Supercomputing Centers of Excellence

Chapter 5 talks about gathering runtime statistics from running the application on the important problem sets discussed in Chapter 4. We must find the bottlenecks at various processor levels. Is the application memory bandwidth limited? Is it compute bound? Can we find high level looping structures where threading may be employed? We must have an idea of how much memory we will require, as this will undoubtedly drive the number of nodes that must be employed.

Chapter 6 talks about how best to utilize available memory bandwidth. This is the most important optimization for a given application. If the application is memory bound and cannot be refactored to reduce the reliance on memory then vectorization and parallelization will not be as helpful. Some algorithms are naturally memory bound and there is little that can be done. However, most of the time the application is not using the memory subsystem in the most optimal manner. Using statistics from Chapter 5, we will be able to determine if the application is memory bound, and if there are typical bottlenecks that can be eliminated. Memory alignment, the TLB (Table Lookaside Buffer), caches, and high-bandwidth memory play a role in the efficient utilization of memory bandwidth. In this chapter we will also investigate the use of the new stacked memory technology being used on CPUs such as the Intel Phi Knight's Landing processor.

Vectorization is investigated in Chapter 7. Twenty years ago, everyone knew how to work with the compiler to vectorize important sections of the application. In this chapter we will reintroduce vectorization. In the 1970s, vectorization was introduced to enhance a super-scalar processor with more powerful instructions that could produce more results per clock cycle than a scalar instruction. This gave rise to SIMD systems. Initially, SIMD instructions were used to take advantage of vector function units on the CDC Star 100 and Texas Instruments ASC. Then came SIMD parallel systems like the Illiac IV and DAP. In both of these systems, a single instruction could generate a large number of results. The principal requirement is that all the results had to be produced with the same operation. With the advent of the register-to-register vector processors from Cray, vectors were much more successful than parallel SIMD systems. However, they disappeared with the "attack of the killer micros" in the 1990s. Over the past 20 years, systems still employed vector instructions. However, the benefit from using them rarely exceeded more than a factor of two. Today vector or SIMD or SIMT instructions are reappearing due to their ability to generate more results per watt. Going forward, SIMD instructions will undoubtedly supply factors of 10 to 30 in performance improvement with the added benefit of using less power. Today very few people really understand what the compiler needs to safely generate vector instructions for a given looping structure. What are the trade-offs when restructuring to attain vectorization?

In Chapter 8 we will look at hybridization of a program. In this context we refer to a hybrid multi/manycore system that contains nodes consisting of many shared memory processors. Such a system may or may not have an accelerator. The approach consists of identifying enough parallelism within the program, and structuring it in the right way to expose the parallelism to the compiler. Hybridization of a program is a two-step procedure which will be evaluated in this chapter. First the programmer must consider refactoring the application to use MPI between units of the multi/manycore system, whether it is a die, socket, or node and OpenMP within the unit. For example, on a 32-core dual socket Haswell node, MPI could be between the 32 cores on the node without any OpenMP, between the two 16-core sockets using 16 OpenMP threads, or between the four 8-core groups each using 4 OpenMP threads. Hybridizing the application is extremely important on the GPUs, since multiple MPI tasks cannot be employed on the GPU. This section will be recommending significant refactoring of the application to achieve high granularity OpenMP, no automatic parallelization is allowed.

In Chapter 9, we look at several major applications, particularly looking at approaches that can be used to make them more amenable to vectorization and/or parallelization. These applications may be smaller than many large production codes. However, they have problems whose solutions can be applied to their larger counterparts. This section will concentrate on performance portable implementations where possible. Due to differences in compilers' capabilities, one rewrite may not be best for all compilers.

Finally, Chapter 10 examines future hardware advancements, and how the application developer may prepare for those advancements. Three future hardware trends are uncovered during this examination: (1) more cores per SoC (system-on-chip) and more SMT threads per core, (2) longer vector lengths with more flexible SIMD instructions, and (3) increasingly complex memory hierarchies.

CHAPTER 2

Determining an Exaflop Strategy

CONTENTS

2.1 FOREWORD BY JOHN LEVESQUE

In 1965 I delivered bread for Mead's Bakery and was working on finishing my degree at the University of New Mexico. A year later I was hired at Sandia National Laboratory in Albuquerque, New Mexico where I worked for researchers analyzing data from nuclear weapon tests. In 1968, I went to work for Air Force Weapons Laboratory and started optimizing a finite difference code called "Toody". The project looked at ground motions from nuclear explosions. All I did was step through Toody and eliminate all the code that was not needed. Toody was a large application with multi-materials, slip lines, and lots of other options. Well, the result was a factor of 20 speedup on the problems of interest to the project. That is how I got the bug to optimize applications, and I still have not lost the excitement of making important applications run as fast as possible on the target hardware. The target hardware at that time was a Control Data 6600. I worked closely with Air Force Captain John Thompson and we had one of two 6600s from midnight to 6:00 AM most nights. John and I even used interactive graphics to examine the applications progress. We could set a sense switch on the CDC 6600 and the program would dump the velocity vector field to a one-inch tape at the end of the next timestep. We then took the tape to a Calcomp plotter to examine the computation. If things looked good, we would go over to the Officers club and have a couple beers and shoot pool. The operator would give us a call if anything crashed the system.

There was another Captain "X" (who will remain anonymous) who was looking at the material properties we would use in our runs. There are two funny stories about Captain X. First, this was in the days of card readers, and Captain X would take two trays of cards to the input room when submitting his job. My operator friend, who was rewarded yearly with a half gallon of Seagram's VO at Christmas time, asked if I could talk to Captain X. I told the operator to call me the next time Captain X submitted his job, and I would come down and copy the data on the cards to a tape that could be accessed, instead of the two trays of cards. The next time he submitted the job, X and I would be watching from the window looking into the computer room. I told the operator to drop the cards while we were looking so X would understand the hazards of using trays of cards. Well, it worked – X got the shock of his life and wouldn't speak to me for some time, but he did start using tapes.

The other story is that X made a little mistake. The sites were numbered with Roman numerals, and X confused IV with VI. He ended up giving us the wrong data for a large number of runs.

Some Captains, like John Thompson, were brilliant, while others, like X, were less so. John and I had an opportunity to demonstrate the accuracy of our computations. A high explosive test was scheduled for observing the ground motions from the test. We had to postmark our results prior to the test. We worked long hours for several weeks, trying our best to predict the outcome of the test. The day before the test we mailed in our results, and I finally was able to go home for dinner with the family. The next day the test was cancelled and the test was never conducted. So much for experimental justification of your results.

2.2 INTRODUCTION

It is very important to understand what architectures you expect to see moving forward. There are several trends in the high performance computing industry that point to a very different system architecture than we have seen in the past 30 years. Due to the limitations placed on power utilization and the new memory designs, we will be seeing nodes that look a lot like IBM's Blue Gene on a chip and several of these chips on a node sharing memory. Like IBM's Blue Gene, the amount of memory will be much less than what is desirable (primarily due to cost and energy consumption).

Looking at the system as a collection of nodes communicating across an interconnect, there will remain the need to have the application communicate effectively – the same as the last 20 years. The real challenges will be on the node. How will the application be able to take advantage of thousands of degrees of parallelism on the node? Some of the parallelism will be in the form of a MIMD collection of processors, and the rest will be more powerful SIMD instructions that the processor can employ to generate more flops per clock cycle. On systems like the NVIDIA GPU, many more active threads will be required to amortize the latency to memory. While some threads are waiting for operands to reach the registers, other threads can utilize the functional units. Given the lack of registers and cache on the Nvidia GPU and Intel KNL, latency to memory is more critical since the reuse of operands within the cache is less likely. Xeon systems do not have as much of an issue as they have more cache. The recent Xeon and KNL systems also have hyper-threads or hardware threads – also called simultaneous multithreading (SMT). These threads share the processing unit and the hardware can context switch between the hyper-threads in a single clock cycle. Hyper-threads are very useful for hiding latency associated with the fetching of operands.

While the NVIDIA GPU uses less than 20 MIMD processors, one wants thousands of threads to be scheduled to utilize those processors. More than ever before, the application must take advantage of the MIMD parallel units, not only with MPI tasks on the node, but also with shared memory threads. On the NVIDIA GPU there should be thousands of shared memory threads, and on the Xeon Phi there should be hundreds of threads.

The SIMD length also becomes an issue since an order of magnitude of performance can be lost if the application cannot utilize the SIMD (or vector if you wish) instructions. On the NVIDIA GPU, the SIMD length is 32 eight-byte words, and on the Xeon Phi it is 8 eight-byte words. Even on the new generations of Xeons (starting with Skylake), it is 8 eight-byte words.

Thus, there are three important dimensions of the parallelism: (1) message passing between the nodes, (2) message passing and threading within the node, and (3) vectorization to utilize the SIMD instructions.

Going forward the cost of moving data is much more expensive than the cost of doing computation. Application developers should strive to avoid data motion as much as possible. A portion of minimizing the data movement is to attempt to utilize the caches associated with the processor as much as possible. Designing the application to minimize data motion is the most important issue when moving to the next generation of supercomputers. For this reason we have dedicated a large portion of the book to this topic. Chapters 3 and 6, as well as Appendix A will discuss the cache architectures of the leading supercomputers in detail and how best to utilize them.

2.3 LOOKING AT THE APPLICATION

So here we have this three-headed beast which we need to ride, and the application is our saddle. How are we going to redesign our application to deliver a good ride? Well, we need a lot of parallelism to get to an exaflop; we are going to need to utilize close to a billion degrees of parallelism. Even for less ambitious goals we are going to need a million degrees of parallelism. So the first question is: "do we have enough parallelism in the problem to be solved?" While there are three distinct levels of parallelism, both the message passing and the threading should be at a high level. Both have relatively high overhead, so granularity must be large enough to overcome the overhead of the parallel region and benefit from the parallelization. The low level parallel structures could take advantage of the SIMD instructions and hyper-threading.

If the application works on a 3D grid there may be several ways of dividing the grid across the distributed nodes on the inter-connect. Recently, application developers have seen the advantage of dividing the grid into cubes rather than planes to increase the locality within the nodes and reduce the surface area of the portion of the grid that resides on the node. Communication off the node is directly proportional to the surface area of the domain contained on the node. Within the node, the subdomain may be subdivided into tiles, once again to increase the locality and attempt to utilize the caches as well as possible.

A very well designed major application is the Weather Research and Forecasting (WRF) Model, a next-generation mesoscale numerical weather prediction system designed for both atmospheric research and operational forecasting needs. Recent modifications for the optimization of the WRF application are covered in a presentation by John Michalakes given at Los Alamos Scientific Laboratories in June, 2016 [22]. These techniques were further refined for running on KNL, which will be covered in Chapter 8. The salient points are:

1. MPI decomposition is performed over two-dimensional patches.
2. Within the MPI task the patch is divided into two-dimensional tiles, each having a two-dimensional surface and depth which represents the third dimension.
3. The dimensions of the tile were adjustable to most effectively utilize the cache.

When identifying parallelism within the application, care must be taken to avoid moving data from the local caches of one processor to the local caches of another processor. For the past 15 years, using MPI across all the cores in a distributed multicore system has out-performed employing OpenMP on the node with MPI only between nodes. The principal advantage that MPI had was that it forced data locality within a multicore's NUMA memory hierarchy. On the other hand, OpenMP has no notion of data locality. On all Intel architectures the level-1 and level-2 caches are local to the processor. However, the level-3 cache tends to be distributed around the socket, and then there tends to be multiple sockets on a node (2 to 4). If the data being accessed with the OpenMP region extends to level-3 cache, then there will be cache interference between the threads. Examples will illustrate this in Chapter 6. Additionally, some looping structures may be parallelized with a different decomposition than other looping structures. Such an approach would necessitate moving data from one thread's low-level caches to another thread's caches. To get the most out of the supercomputers of the future, one must minimize this data motion. Recently, several applications have employed OpenMP successfully in a SPMD style like MPI. While this approach, discussed in Chapter 8, is more difficult to implement, it performs extremely well on multi/manycore NUMA architectures.

Since most large applications generate a lot of data, we should design the application to perform efficient I/O. The decomposition of the problem across the nodes and within the node can be done to allow asynchronous parallel I/O. See Appendix E for an I/O discussion.

Your strategy depends upon the state of the existing application. When writing an application from scratch, one has complete flexibility in designing an optimized application. When a legacy application is the starting point, the task can be significantly more difficult. In this book we will concentrate on the legacy application.

Your strategy depends upon the target problem, which hopefully is large enough to utilize the million degrees of parallelism on the new system. There are two approaches for scaling a problem to larger processor counts. First, there is the idea of weak scaling where the problem size grows as the number of processors grow. Consequently, each processor has a constant amount of work. This is a great way to scale the problem size if the increased size represents good science. For example, in S3D, a combustion and computational fluid dynamics application, the increase in problem size gives finer resolution and therefore better science. S3D will be discussed further in Chapter 9. On the other hand, many weather/climate codes at some point cannot take advantage of finer resolution, since the input data is sparse. If the finer grid cannot be initialized, then often times the increased refinement is of no help. For the weather/climate application, strong scaling is employed. With strong scaling, the problem size is fixed, and as the number of processors are increased, the amount of work performed by each processor is reduced. Strong scaling problem sets are the most difficult to make effectively utilize large parallel systems. Strong scaling does have a "sweet spot" when the mesh size within each MPI task is small enough to fit into the low level cache. A super-linear performance boost will be seen when this point is reached. This is particularly true on KNL when the problem size can fit within the MCDRAM, allowing the application to run completely out of the high bandwidth memory. However, exercise caution with this line of thinking. Once the problem is using enough nodes to run within MCDRAM, one should determine if the increase in system resources required to obtain the faster results comes with a similar decrease in compute time. For example, if a strong scaling application obtains a factor of 1.5 decrease in wall clock time when scaled from 120 to 240 nodes, the decrease does not account for the factor of two increase in system resources. Of course some credit needs to be given to decreased turn-around which results in more productivity.

Given the problem size, the most important task is to find a parallel decomposition that divides the problem up across the processors in a way that not only has equal work on each processor, but also minimizes the communication between the processors. As mentioned earlier, the hope is to achieve locality within a domain for both MPI domains as well as threaded domains. With multidimensional problems, the surface area of the MPI domain should be minimized, which is achieved by using cube domains as depicted in Figure 2.3.1. For threaded regions, locality is achieved by threading across tiles. The size of the tile should allow the computation within the tile to be conducted within the cache structure of the processor. To achieve this, the working set (the total amount of memory utilized) should be less than the size of the level-2 cache. If multiple cores are sharing level-2 cache, then the working set per core must be reduced to account for the sharing.

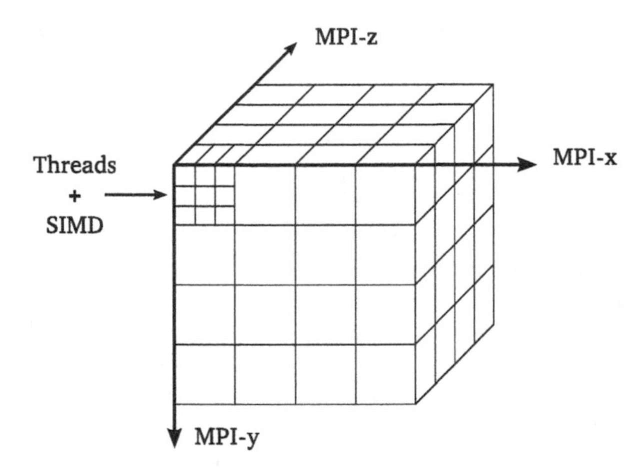

FIGURE 2.3.1 3D grid decomposition minimizing MPI surface area.

In large, multiphysics, irregular, unstructured grid codes, decomposing the computational grid can be a challenge. Poor decompositions can introduce excessive load imbalance and excessive communication. Many three-dimensional decomposition packages exist for addressing this problem, and the decomposition itself can take a significant amount of time.

In the decomposition across MPI tasks and threads, the processor characteristics must be taken into account. The state-of-the-art Xeon tends to have a modest number of cores within the node and few NUMA regions, so the number of MPI and threading domains are relatively small compared to what is required to run efficiently on a GPU. GPUs tend to have a similar number of symmetric processors (SM), and the SIMD units have a longer length (32 eight-byte words). However, the GPU requires thousands of independent threads for amortizing the latency to memory. Knight's Landing is closer to the Xeon, in that it has 60 to 70 cores on the node with a SIMD length of 8 for double-precision. KNL does not need as many independent threads since it only has four hyper-threads per core. This is a significant difference between the GPU and other systems; the GPU requires significantly more shared memory threading than the other systems, and it could require a significantly different strategy for the threaded domains. While the Xeon and KNL can benefit from having multiple MPI tasks running within the node, within the GPU, all of the threads must be shared memory threads as MPI cannot be employed within the GPU.

The decomposition is also coupled with the data layout utilized in the program. While the MPI domain is contained within the MPI task's memory space, the application developer has some flexibility in organizing the data within the MPI task. Once again, irregular, unstructured grids tend to introduce unavoidable indirect addressing. Indirect addressing is extremely difficult and inefficient in today's architectures. Not only does operand fetching require fetching of the address prior to the operand, cache utilization can be destroyed by randomly accessing a large amount of memory with the indirect addressing. The previously discussed tile structure can be helpful, if and only if the memory is allocated so that the indirect addresses are accessing data within the caches. If the data is stored without consideration of the tile structure, then cache thrashing can result.

Without a doubt, the most important aspect of devising a strategy for moving an existing application to these new architectures is to design a memory layout that supplies addressing flexibility without destroying the locality required to effectively utilize the cache architecture.

2.4 DEGREE OF HYBRIDIZATION REQUIRED

For the past 15 years we have seen the number of cores on the node grow from 1 to 2 to 4 and now on to 40 to 70. Many believed employing threading on the node and MPI between nodes was the best approach for using such a system, and as the number of cores on the node grew that belief grew stronger. With the advent of the GPU, a hybrid combination of MPI and threading was a necessity. However, performance of all-MPI (no threading) on the Xeon and now even the KNL has been surprisingly good, and the need for adding threading on those systems has become less urgent.

Figure 2.4.1 shows the performance of S3D on 16 KNL nodes running a problem set that fits into KNL's high bandwidth memory. In this chart the lines represent the number of MPI tasks run across the 16 nodes. The ordinate is the number of OpenMP threads employed under each MPI task. The performance is measured in timesteps per second, and higher is better. The graph illustrates that using 64 MPI tasks on the node for a total of 1024 MPI tasks gives the best performance and in this case two hyper threads per MPI task increases the performance and four hyper threads decreases the performance. S3D is very well threaded; however, it does suffer from not using a consistent threading strategy across all of the computational loops. More will be discussed about S3D in Chapter 9.

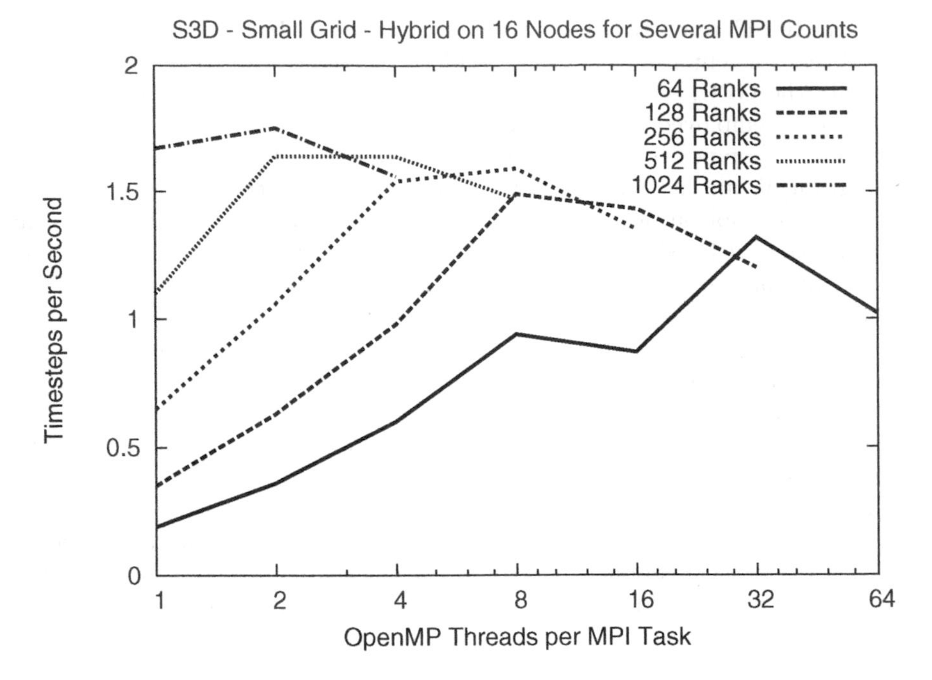

FIGURE 2.4.1 Performance of S3D on KNL with different MPI rank counts.

The two primary reasons for the superior performance of MPI on these systems are the locality forced by MPI and the fact that MPI allows the tasks to run asynchronously, which allows for better utilization of available memory bandwidth. When MPI is run across all the cores on a node, the MPI task is restricted to using the closest memory and cache structure to its cores. On the other hand, threading across cores allows the threads to access memory that may be further away, and multiple threads running across multiple cores have a chance to interfere with each other's caches. On KNL, running MPI across all the cores and threading for employing the hyper-threads seems to be a good starting point in a hybrid MPI+OpenMP approach. The performance of a hybrid application is directly proportional to the quality of the threading. In fact, as seen in the OpenMP chapter, the SPMD threading approach, which mimics the operation of MPI tasks, performs very well. OpenMP has one tremendous advantage over MPI: it can redistribute work within a group of threads more efficiently than MPI can, since OpenMP does not have to move the data. There is and will always be a place for well-written OpenMP threading. Of course, threading is a necessity on the GPU; one cannot run separate MPI tasks on each of the symmetric processors within the GPU.

2.5 DECOMPOSITION AND I/O

The incorporation of a good decomposition and efficient parallel I/O go hand in hand. While we do not talk about efficient parallel I/O in great detail in this book, some discussion is provided in Appendix E, and it should always be considered when developing a good strategy for moving to new systems.

2.6 PARALLEL AND VECTOR LENGTHS

When the application has to scale to millions of degrees of parallelism, the division of the dimensions of the problem across the MPI ranks, threads, and SIMD units is an important design criteria. In Chapter 7, we will see that vector performance on the Xeon and KNL systems top out around 80 to 100 iterations, even though the vector unit is much smaller. Additionally, we want the vector dimension to be aligned on cache boundaries and contiguous in memory – non-unit striding will significantly degrade performance. On the GPU, the vectors are executed in chunks of 32 contiguous operands, and since the GPU wants a lot of parallel work units, one way is by having very long vector lengths.

Beyond the vector length, the next question is how to divide parallelism between MPI and threading. As was discussed earlier, the Xeon and KNL seem to do well with a lot of MPI. Running MPI across 64 cores within each node and across 10,000 nodes gives 640,000 MPI tasks and, in this case, on the order of 4 to 8 threads with hyper-threads. On the GPU system, one would want the number of MPI tasks to be equal to the number of GPUs on each node, and one would want thousands of threads on the node. Much more threaded parallelism is required to effectively utilize the GPU system.

2.7 PRODUCTIVITY AND PERFORMANCE PORTABILITY

An important consideration when moving to the next generation of multi/manycore systems is striving to create a refactored application that can run well on available systems. There are numerous similarities between the multi/manycore systems of today. Today's systems have very powerful nodes, and the application must exploit a significant amount of parallelism on the node, which is a mixture of MIMD (multiple instruction, multiple data) and SIMD (single instruction, multiple data). The principal difference is how the application utilizes the MIMD parallelism on the node. Multicore Xeons and manycore Intel Phi systems can handle a significant amount of MPI on the node, whereas GPU systems cannot. There is also a difference in the size of the SIMD unit. The CPU vector unit length is less than 10, and the GPU is 32. Since longer vectors on the multi/manycore systems do run better than shorter vectors this is less of an issue. All systems must have good vectorized code to run well on the target systems. However, there is a problem.

"Software is getting slower more rapidly than hardware becomes faster."

Niklaus Wirth
Chief Designer of Pascal, 1984 Turing Award Winner

Prior to the advent of GPU systems and KNL, application developers had a free ride just using MPI and scalar processing, benefiting from the increased number of cores on the node. Today we have a situation similar to the movement to distributed memory programming which required the incorporation of message passing. Now, to best utilize all the hardware threads (including hyper-threads) applications have to be threaded and to harness the vector processing capability the applications must vectorize. Given the tremendous movement to C++, away from the traditional HPC languages Fortan and C, the modifications to achieve vectorization and threading are a tremendous challenge. Unless the developers really accept the challenge and refactor their codes to utilize threading and vectorization they will remain in the gigaflop performance realm and realize little improvement on the new HPC systems. The cited reference to the COSMOS weather code is an example of how that challenge can be realized with some hard work.

The "other p" – there is a trend in the industry that goes against creating a performant portable application: the attempt to utilize high-level abstractions intended to increase the productivity of the application developers. The movement to C++ over the past 10 to 15 years has created applications that achieve a lower and lower percentage of peak performance on today's supercomputers. Recent extensions to both C++ and Fortran to address productivity have significantly contributed to this movement. Even the developer of C++, Bjarne Stroustrup, has indicated that C++ can lure the application developer into writing inefficient code.

"C makes it easy to shoot yourself in the foot; C++ makes it harder, but when you do it blows your whole leg off."

"Within C++, there is a much smaller and cleaner language struggling to get out."

Bjarne Stroustrup
Chief Designer of C++

The productivity argument is that the cost of talented labor is greater than the cost of the high performance computing system being utilized, and it is too time consuming to improve the performance of the application. On the other hand, time spent optimizing application performance not only makes better use of expensive machines, it also reduces operational costs as energy costs continue to rise for future systems.

Several years ago, a team lead by Thomas Schulthess and Oliver Fuhrer of ETH Zurich refactored the production version of COSMOS, the climate modeling application used by MetroSwiss, and found that not only did the application run significantly faster on their current GPU system, the cost of the effort would be more than repaid by the savings in energy costs over a couple of years [12]. Figure 2.7.1 shows the performance increase and Figure 2.7.2 shows the power consumption decrease for the refactored application.

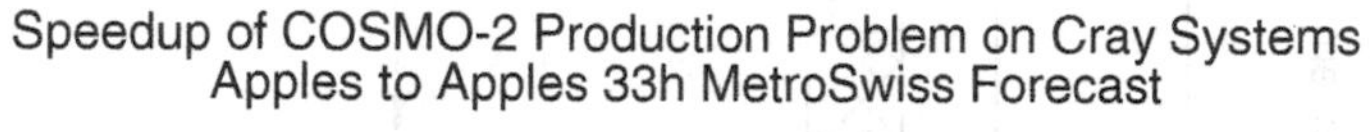

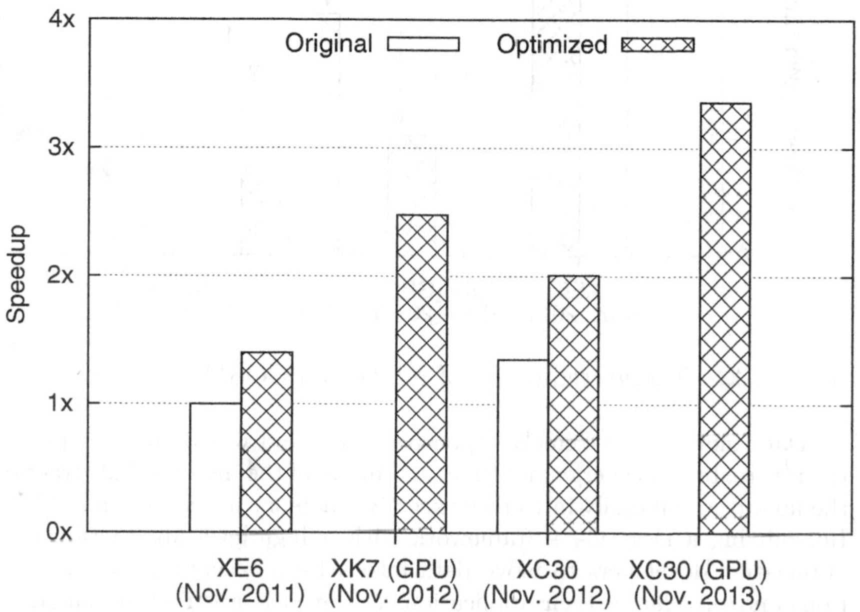

FIGURE 2.7.1 Performance increase of refactored COSMOS code.

The work on COSMOS employed C++ meta-programming templates for the time-consuming dynamical core, which resulted in the instantiation of CUDA kernels on the GPU and optimized assembly on the x86 systems. This work is an example of the developers' understanding the architecture and restructuring the application to utilize its features with the high-level C++ abstractions.

The bulk of the code – the physics – was Fortran, and OpenACC was used for the port to the accelerator. This is an excellent example that shows how an investment of several person-years of effort can result in an optimized application that more than pays for the investment in the development cost.

This work does show that a well-planned design can benefit from C++ high-level abstraction. However, there has to be significant thought put into the performance of the generated code.

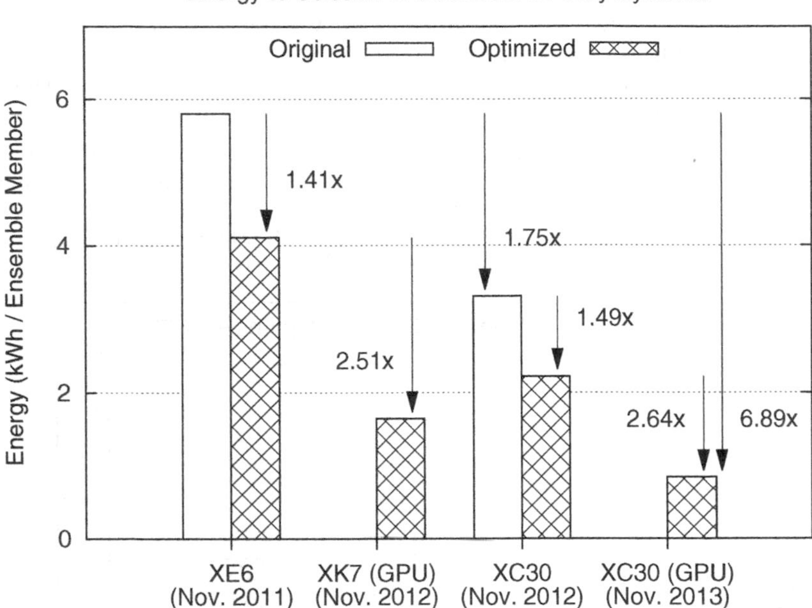

FIGURE 2.7.2 Energy reduction of refactored COSMOS code.

Data motion is extremely expensive today and will be more expensive in the future, both in energy and time, and many of the high-level abstractions in the language can easily introduce excessive memory motion in an application. Retrofitting a large C++ framework with a high-level abstraction requires application developers to move data into a form acceptable to the abstractions and/or refactor their applications to have the abstraction manage their data structures. The first approach introduces too much data motion, and the second approach often requires a significant rewrite. Once such a rewrite has been performed, the application is dependent upon those interfaces making efficient use of the underlying architecture. Additionally, most complex multi-physics applications are a combination of computations that flow from one set of operations to another, and breaking that flow up into calls to the low-level abstractions could result in poor cache utilization and increased data motion.

Much of the blame of this productivity movement has to be placed on the language standards committees that introduces semantics that make the application developer more productive without thinking about compilation issues or the efficiencies of executing the compiled code on the target system. Considering the recent additions, both Fortran and C++, it seems that the committee really does not care about how difficult it might be to optimize the language extensions and that their principal goal is to make programmers more productive. When users see an interesting new feature in the language, they assume that the feature will be efficient; after all, why would the language committee put the feature in the language if it wouldn't run efficiently on the target systems?

At some point this trend to productivity at the expense of performance has to stop. Most, if not all of the large applications that have taken the productivity lane have implemented MPI and messaging outside of the abstraction, and they have obtained an increase in performance from the parallelism across nodes. Increased parallelism must be obtained on the node in the form of threading and/or vectorization, with special attention paid to minimizing the movement of data within the memory hierarchy of the node. At some point, application developers have to put in extra work to ensure that data motion on the node is minimized and that threading and vectorization are being well utilized.

2.8 CONCLUSION

The primary strategy for designing a well-performing application for these systems is to first get the memory decomposition correct. We must have good locality to achieve a decent performance on the target system. This memory layout imposes a threading decomposition on the node, which could also be employed for the MPI decomposition on the node. Ideally, a legacy code may already have found a good decomposition for MPI usage across all the nodes. In review, we would like the vector dimension to be contiguous, and we would like the threads to operate on a working set that fits into the cache structure for a single core. Then we want to ensure that the tasks do not interfere with each other, by either doing OpenMP right or by using MPI.

2.9 EXERCISES

2.1 Compare and contrast weak and strong scaling.

2.2 When choosing a parallel decomposition which equalizes computational load, what other aspect of the decomposition is of critical importance?

2.3 What is a good decomposition strategy to use with multidimensional (e.g., 3D) problems to minimize communication between MPI domains?

2.4 *Construct a case study*: Consider an application and a target system with respect to the amount of parallelism available (the critical issue of data motion will be covered in more detail in later chapters).

   a. Select a target system and characterize it in terms of:

      i. Number of nodes.

ii. Number of MIMD processors per node.
iii. Number of SIMD elements per processor.

b. Select an application for further examination:
   i. Identify three levels of parallelism in the application (e.g., grid decomposition, nested loops, independent tasks).
   ii. Map the three levels of parallelism identified in the application to the three levels available on the target system: MPI, threads, and vectors.

c. For each level of parallelism identified in the application and mapped to the target system, compare the amount of parallelism available in the application at that level to the amount of parallelism available on the system:
   i. Compare the number of MPI ranks the application could utilize to the number of nodes available on the system as well as the number of MIMD processors per node.
   ii. Compare the number of threads per MPI rank the application could utilize to the number of MIMD processors on a node as well as the number of hardware threads per processor.
   iii. Compare the trip count of the chosen vectorizable loops to the length and number of the system's SIMD units.

   Do these comparisons show a good fit between the parallelism identified in the application and the target system? If not, try to identify alternative parallelization strategies.

2.5 Consider the execution of HPL on the KNL system and assume with full optimization it sustains 80% of peak performance on the system:

a. What would the percentage of peak be if we only used one of the 68 processors on the node?

b. What would the percentage of peak be if we did not employ SIMD instructions?

c. What would the percentage of peak be if we did not employ SIMD instructions and we only used one of the 68 processors on the node?

CHAPTER 3

Target Hybrid Multi/Manycore System

CONTENTS

3.1 FOREWORD BY JOHN LEVESQUE

In 1972 I moved the family to Los Angeles to work for a defense contractor, R&D Associates, actually doing the same work I was doing at Air Force Weapons Laboratory. However, this time on a beautiful machine, the Control Data 7600, and started vectorizing the application. While the 7600 did not have vector hardware instructions, it had segmented functional units. This was the first machine I worked on that was designed by Seymour Cray and it was absolutely a pleasure to work on. We were working on a system at Lawrence Berkeley Laboratories. Given the work we were doing on the CDC 7600, I got a contract with DARPA to monitor ILLAC IV code development efforts. The ILLIAC IV had 64 SIMD parallel processors and what a challenge, not only due to the modifications that had to be made to the applications to get them running, but the system was up and down like a yo-yo.

There's an interesting story about the fragility of this system. A picture of the ILLIAC IV on the Internet shows it with one of the doors open. In reality, if you closed the door, the machine would crash. At one DARPA review, they were demonstrating the ILLIAC IV to a group of DARPA higher-ups and one of them closed the door causing the machine to crash. Not sure they were ever able to find out the reason for the crash.

NASA/AMES would always make three runs of their computations, each on 21 processors. At each timestep they would check answers and if they all agreed, great. If two of the three agreed, they would continue by replicating the good answers and resuming the computation. If all three disagreed they would stop the run. We may have to do something like that as current day computers have zillions of components, each of which has a finite probability that they will fail.

One of the ILLIAC IV projects was to develop an automatic parallelizer which was the first example of a preprocessor that translated standard Fortran to parallel code for running on the ILLIAC IV. This product was appropriately named the PARALYZER – oh without the Y.

At RDA I also became the head of the computer group. We had a remote batch terminal where cards could be read, jobs submitted to Berkeley, and printouts would be returned. Several members of the group started a newsletter called the BiWeekly Abort. In addition to having great articles about submitting jobs, optimizing applications, etc., it had several characters. There was always a column by Phineas Boltzmann, Cora Dump, Io Baud and Chip Store. The mass store device at Berkeley was a system that saved files on chips that were placed within small boxes. At this time the author of the Phineas Boltzmann column developed an optimized vector processing library for the CDC 7600, which we called RDALIB under contract to NASA/AMES.

3.2 UNDERSTANDING THE ARCHITECTURE

It is very important to understand the architecture that you are targeting. Although the targets can be different in the number of processors on the node, the width of the SIMD unit, and the memory hierarchy, at a high level all the current supercomputing architectures are very similar. For example, for the next 5 years we will see more powerful nodes with complex memory hierarchies. The number of processors on the node will range from 16 up to 100. If you then consider hyper-threading, the amount of parallelism available on the node is enormous. Of course, this parallelism can be used by some combination of MPI and threading on the node. We will likely never be able to employ MPI across all of the hyper-threads on the node and certainly not across the symmetric multiprocessors on the GPU. Additionally, we will never want to use OpenMP across all of the processors on the node. Part of the task of porting an application to a hybrid multicore system is determining what mixture of MPI and threading is optimal. Table 3.2.1 illustrates some of the principal differences between the architectures that were available in 2016.

The most alarming difference between Xeon and the "accelerated" systems such as NVIDIA's GPUs and Intel's Phi is the total amount of cache and registers. The Xeon cores have several orders of magnitude more cache per core than either of the other accelerated systems. Xeons have always had a significant amount of cache for addressing the latency to memory. With the advent of the accelerated systems, a faster (albeit high latency in many cases) memory has been substituted for the higher levels of cache. This has a significant impact on performance of large computational kernels that cannot be blocked to reside in level-1 and level-2 caches.

TABLE 3.2.1 Key architectural attributes of current hardware.

Arch	MIMD Procs	SIMD DP Width	Close KB/core	HBM GB	HBM GB/sec	Far GB	Far GB/sec
K20X	14	32	90	6	250	-	-
K40	15	32	96	12	288	-	-
K80(/2)	15	32	96	12	240	-	-
Pascal	56	32	73	16	720	-	-
Ivybridge	16	8	2887	-	-	128	100
Haswell	32	8	2887	-	-	256	120
KNL	60	16	557	16	480	384	80

On the Xeon HPC systems, we see approximately 12 to 20 cores per socket. On these high-performance computing systems, we will see two to four sockets per node. In addition, most systems have hyper-threading, which allows several threads to share the floating-point unit of a core. An important additional characteristic of such a node is the memory architecture. All nodes used for HPC have a non-uniform memory architecture (NUMA). All the cores within a NUMA region have good latency and memory bandwidth to its memory. However, when a core accesses memory in a different region, the latency is longer and the memory bandwidth is lower. One of the difficulties of using OpenMP on such a system is that, in the major computational parts of the program, a thread may be operating on memory that is not in its NUMA region. This will cause significant inefficiencies and degrade the scaling of OpenMP. A very successful approach to deal with NUMA affects is to use an MPI task for each NUMA region, and then only thread within the NUMA region. This is an obvious initial separation between MPI and OpenMP.

3.3 CACHE ARCHITECTURES

Appendix A has a detailed description of how caches work. We will assume in this chapter that the reader has a good understanding of Appendix A. Additionally, the translation look-aside buffer (TLB) is discussed in Appendix B. Any access of an operand from memory must include the translation of the virtual address (program's viewpoint of memory) to the physical address in main memory. Following is a review to acquaint readers with the process of fetching an operand to memory (for Intel Haswell).

1. All memory is organized in physical pages within memory.
   (a) Within a physical page all operands are stored contiguously.
   (b) Two adjacent virtual pages may not be adjacent in memory.
   (c) Default page size is 4096 bytes and can be set larger.
   (d) The TLB is 4-way associative with 64 entries.
2. All data transfers from memory are performed in cache line increments of 64 bytes.
3. When the processor receives a fetch instruction,
   (a) It must first check to see if the operand is already in its cache or in another processor's cache on the node. This is referred to as cache coherency.
   (b) If the cache line is not in another cache, it checks to see if the translation entry is in the TLB. If it isn't, it must fetch the TLB entry from memory which takes ~100 clock cycles.
   (c) The cache line is then fetched to its level-1 cache.
4. When the operand arrives it will be moved to a register and placed in level-1 cache.
   (a) It usually takes >100 clock cycles to receive the operand from memory.
   (b) It usually takes < 4 clock cycles to receive the operand from level-1 cache.
   (c) It usually takes < 12 clock cycles to receive the operand from level-2 cache.
   (d) It usually takes < 36 clock cycles to receive the operand from level-3 cache.
5. When level-1 cache is full and/or there is an associativity hit (8-way), the least recently used (LRU) cache line from level-1 cache will be moved to level-2 cache.

3.3.1 Xeon Cache

Over the years, Intel has perfected cache organization. There are typically three caches. Each core has a level-1 data cache that is 32 KB and 8-way associative, a level-2 cache that is 256 KB and 8-way associative, and finally a last-level cache (level-3 cache) that is 40 MB per socket and 16-way associative on newer Xeons. As you move from inner level-1 cache to the outer caches, the latency gets longer and the bandwidth is lower, with the last level cache having lower latency and higher bandwidth than main memory. All transfers are performed in units of complete cache lines, except for level-1 cache to registers.

3.3.2 NVIDIA GPU Cache

NVIDIA's GPUs do not have level-3 cache, and more importantly, when the processor must spill intermediate results from the registers, the values spill to memory and not to level-1 cache. This results in a significant decrease in performance. When programming for the NVIDIA GPU, the application developer must pay very close attention to register utilization. This is the principal reason why CUDA is the dominate language for writing performant applications for the GPU. CUDA gives the programmer much more flexibility in utilizing the registers, caches, and other close memories in the system which support graphics.

3.4 MEMORY HIERARCHY

The most important characteristic of a node is its memory hierarchy. The memory hierarchy consists of the cache structure and the various levels of memory. The simplest example may be a standard Xeon node with a small number of sockets, each with their own CPU caches and NUMA domain with DDR memory. However, newer architectures have a more complex memory hierarchy. For example, on the new Intel Knight's Landing (KNL) processor, there is a large "capacity" memory and a smaller high-bandwidth memory. This is not unlike recent GPU systems, with a GDDR or high-bandwidth memory close to the GPU in addition to the larger standard DDR connected to the CPUs. In all cases, the best performance is obtained when the major computation is operating on arrays that are allocated in high-bandwidth memory. In the case of the KNL processor, the bandwidth of the high-bandwidth memory is five times higher than the bandwidth to the larger DDR memory. The high-bandwidth memory is also high-latency memory with a latency that is slightly higher than that from main memory.

On the GPU systems, there is a high-bandwidth GDDR or HBM memory available for the processing units in the GPU, and on some systems the GPU can access memory on the host. Accessing memory on the host or on another GPU in a multi-GPU system has a much longer latency and significantly lower bandwidth than the memory on the local GPU. NVIDIA is introducing a new GPU interconnect called NVLink to help address the latency and bandwidth issues traditionally seen with PCIe. NVLink supports up to 40 GB/s of bi-directional bandwidth per link, with initial Pascal GPUs supporting up to 4 links for a total bi-directional bandwidth of 160 GB/s. NVLink seems promising when compared to a standard PCIe 3.0 x16 link at 15.76 GB/s.

The new Pascal GPU architecture from NVIDIA complicates the typical NUMA situation, as GPUs and CPUs can be connected by PCIe, by NVLink, or by both at the same time. Additionally, these connections can be asymmetric in the sense that latency and bandwidth seen by GPU "A" when accessing memory on GPU "B" can be different than if "A" were instead accessing memory on GPU "C". If one can run with a single MPI rank per GPU, these

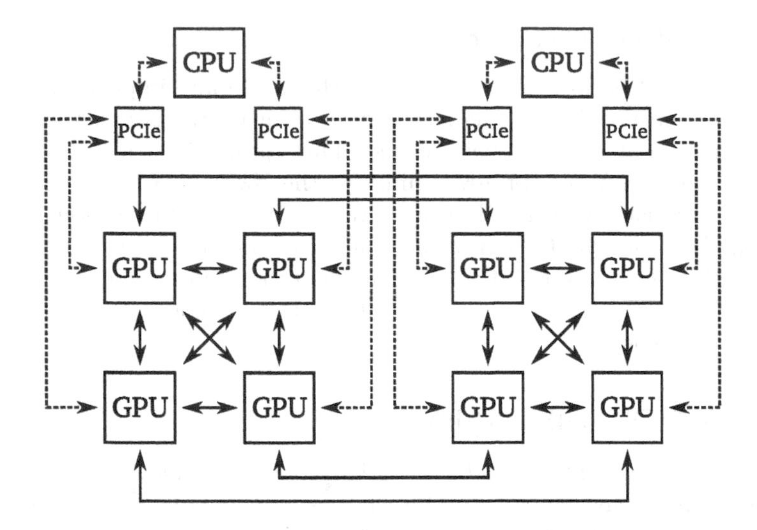

FIGURE 3.4.1 NVLink (solid) and PCIe (dashed) connecting two CPUs and eight GPUs.

complications may not matter much. However, in the case where memory is shared between multiple GPUs, attention to the NVLink topology will start to matter more. Figure 3.4.1 shows eight GPUs and two CPUs connected with a combination of PCIe and NVLink. Note that this figure presents an example where the connectivity between the GPUs is asymmetric, with some GPUs seperated by two NVLink hops and others seperated by only a single hop. This 8-GPU case is also fairly large for a compute node. It might be more typical to see nodes with one or two GPUs as depicted in Figure 3.4.2. In these cases, the NVLink topology is symmetric.

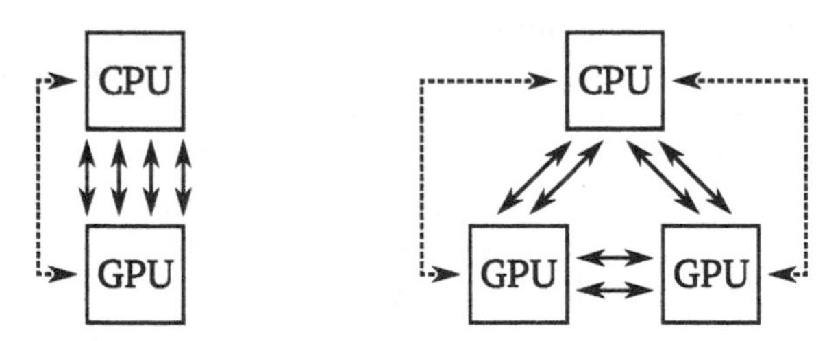

FIGURE 3.4.2 NVLink (solid) and PCIe (dashed) connecting one CPU and one to two GPUs.

On all of these systems there is an over-abundance of floating-point performance. It is a rare situation where an algorithm is hindered by the floating-point performance. Most of the time, the effective memory bandwidth is the limiting factor. We see that effective utilization of the computational units requires that the operands reside in the high-bandwidth memory. The size of the high-bandwidth memory is thus very important when targeting the efficient utilization of the node.

3.4.1 Knight's Landing Cache

Knight's Landing does not have a traditional level-3 cache. The high-bandwidth MCDRAM is meant to replace level-3 cache and must be utilized efficiently to achieve good performance on that processor. The MCDRAM can be used in three different ways. First, it can be used as a level-3 cache. When utilized in this mode, the cache is direct-mapped. That is, it is one-way set-associative. This could potentially cause significant thrashing in the cache. An additional problem is that the latency to MCDRAM when used as a cache is quite long for a cache, around 240 clock cycles. This is longer than the latency to the main memory, 212 clock cycles. However, the bandwidth of the MCDRAM in cache mode is around five times faster than DDR. The easiest approach is when the entire application and memory fits within the 16 GBytes of MCDRAM on the node. This method would not use any main memory and the bandwidth is significantly faster; however, much of the node resource, the main memory is not used. The third approach is to use the high-bandwidth memory as a separate memory to store cache-unfriendly arrays. This puts a tremendous burden on the user for identifying which arrays should be placed in the high-bandwidth memory. One can choose to split the MCDRAM 25-75% or 50-50% cache and separate address space. Since memory bandwidth is typically the bottleneck on applications, the intelligent use of high-bandwith memory is extremely important for overall performance. Chapter 6 covers more of the KNL memory hierarchy.

Core_0		Core_1		Core_{n-1}		Core_n	
32KB L1D	32KB L1I	L1D	L1I	L1D	L1I	L1D	L1I
1MB Shared L2				L2			
16GB Shared MCDRAM							

FIGURE 3.4.3 KNL cache hierarchy.

3.5 KNL CLUSTERING MODES

TABLE 3.5.1 Latency and bandwidth of KNL's cache and memory.

Memory	(mode)	Latency: ns	(clocks)	STREAM Triad GB/s
Level-1		2.9	(4)	
Level-2		13.6	(19)	
MCDRAM	(cache)	173.5	(243)	329
MCDRAM	(flat)	174.2	(244)	486
DDR	(cache)			59
DDR	(flat)	151.3	(212)	90

In this section, we will concentrate on the Knight's Landing system with its large, relatively slow memory bandwidth and small, high-bandwidth MCDRAM. In addition, KNL has a wimpy cache structure consisting of level-1 and level-2 caches. Figure 3.4.3 depicts the memory hierarchy for KNL. The original KNL was designed to have 36 tiles each containing two cores, two level-1 caches and a level-2 cache which is shared between the two cores.

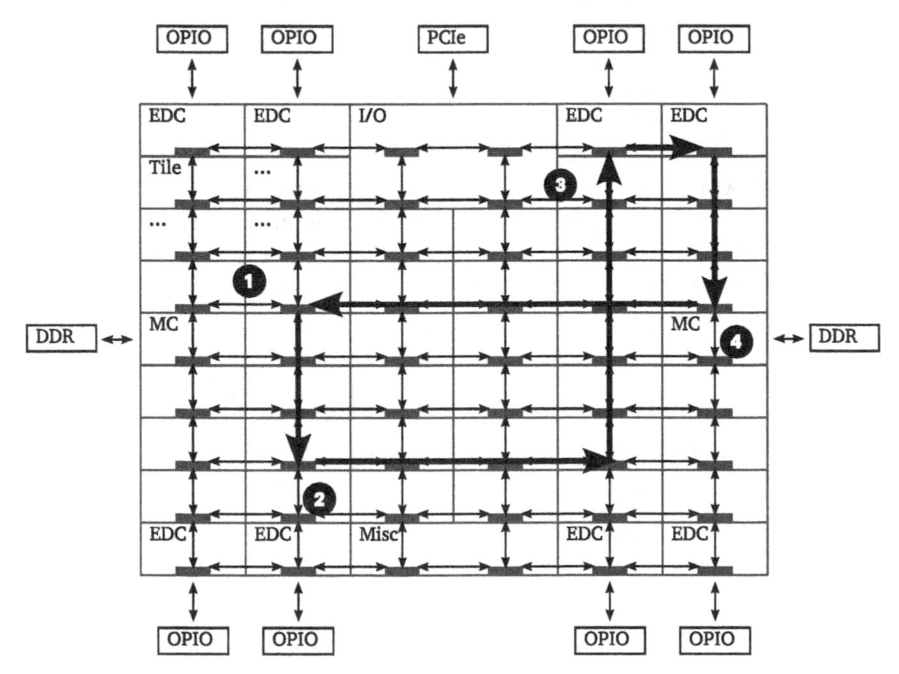

FIGURE 3.5.1 All-to-all clustering mode on KNL: (1) level-2 miss, (2) directory access, (3) memory access, and (4) data return.

Also contained on the tile is a cache homing agent (CHA) that communicates with other tiles via a 2D communication grid. The CHAs are responsible for maintaining cache coherency across the shared memory system. Each CHA has ownership of a portion of main memory (DDR) and a portion of the high-bandwidth MCDRAM. When one of the cores requires an operand and the operand does not reside within its level-1 and level-2 caches, it notifies the CHA owning the section of memory that contains the cache line that contains the operand. That CHA then determines if another tile has the cache line or if it needs to be fetched from memory. If another tile has the cache line, the cache line is transferred from that tile to the requesting tile. If the cache line is not contained within any tile, then the cache line is fetched from memory and shipped to the requesting tile. This sounds like a great deal of communication and testing to gain access to a particular cache line, and it results in a relatively long latency from both DDR and MCDRAM. Table 3.5.1 gives latency and bandwidth numbers for KNL.

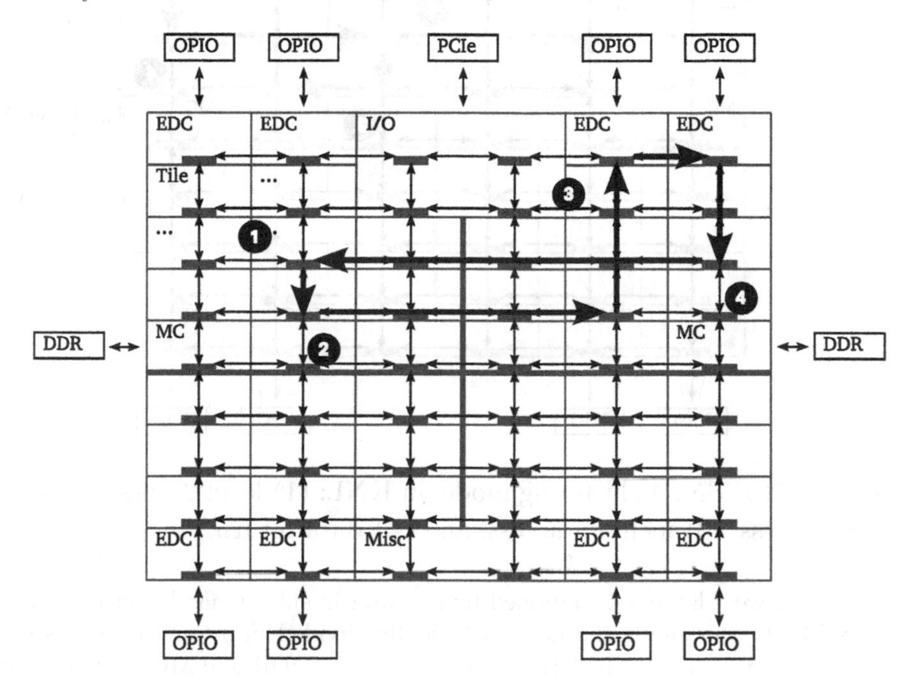

FIGURE 3.5.2 Quandrant clustering mode on KNL: (1) level-2 miss, (2) directory access, (3) memory access, and (4) data return.

There are several clustering modes that the KNL can be booted into that address some of the locality issues. What was discussed is true for all clustering modes. However, some clustering modes restrict the distance that the requesting tile would have to communicate to obtain the cache line. The first and least efficient clustering mode is all-to-all, where the requesting tile and the tile that owns memory can reside anywhere on the node. Figure 3.5.1 shows how the operation would proceed with this clustering mode. A second mode which is somewhat faster is quadrant. In this mode, the CHA always has ownership of DDR and MCDRAM space that is located closest to the quadrant of the node that contains the CHA. In this mode, the CHA can be anywhere on the node. However, it does not have far to go for accessing the memory it is responsible for. This does reduce the latency and improves the bandwidth.

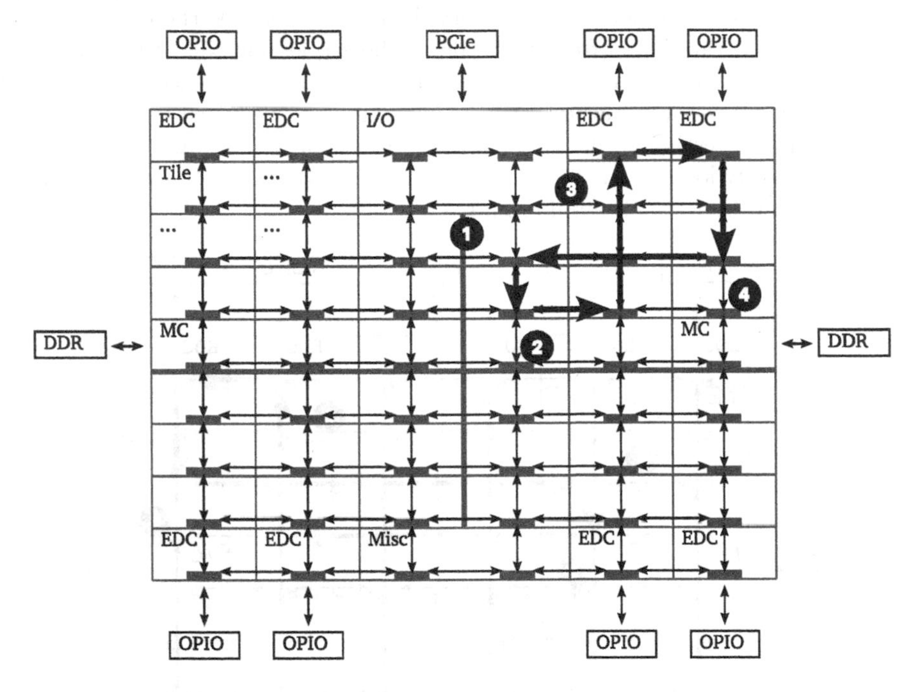

FIGURE 3.5.3 SNC4 clustering mode on KNL: (1) level-2 miss, (2) directory access, (3) memory access, and (4) data return.

The lowest latency is obtained by running in sub-numa clustering mode 4 (SNC4). In this mode the node is logically divided into four sockets, so the requesting tile and all the CHAs that control the DDR and MCDRAM closest to the quadrant are located within the same quadrant. While this gives the lowest latency, a collection of cores within the quadrant can only get 1/4 of the total bandwidth of DDR and 1/4 of the total bandwidth of MCDRAM. Additionally, latency is improved only for the memory closest to the quadrant. Latency and bandwidth to a different quadrant would be longer and slower, respectively. Figure 3.5.2 and Figure 3.5.3 show the difference between these two modes. There is also a hemisphere and a SNC2 mode which divides the node up into two sections, with hemisphere being similar to quadrant and SNC2 being similar to SNC4. Table 3.5.2 gives latencies and bandwidths for the different clustering modes.

The first test we will look at for clustering modes is the Himeno benchmark which is extremely memory bandwidth limited. Figure 3.5.4 shows performance on 128, 256, and 512 nodes using 32-256 MPI tasks/node. The higher MPI tasks/node employed hyper-threads. This test did not employ OpenMP threads. All these tests used MCDRAM as cache.

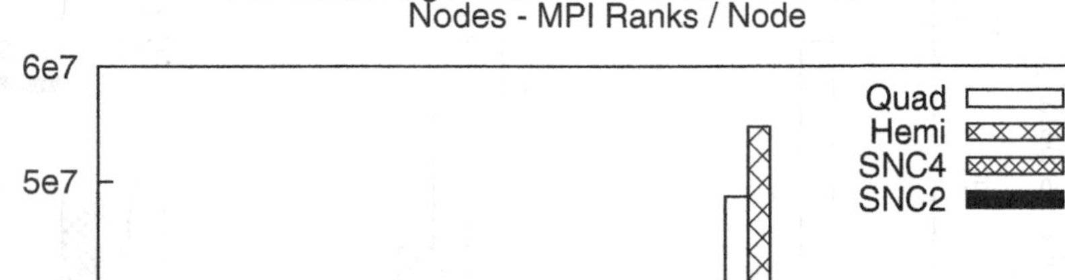

FIGURE 3.5.4 KNL clustering modes with Himeno and 1 thread per rank.

Notice that the best performance is obtained using either Hemi or Quad mode. Himeno is extremely memory bandwidth limited and benefits from the collection of MPI tasks getting to all the memory bandwidth. These two modes allow a collection of MPI tasks within a region (hemisphere or quadrant) to get all of the bandwidth from MCDRAM. At this node count all of the data resides in the MCDRAM cache. Clearly the all MPI version of Himeno prefers hemi or quad. We now add threads and see how the performance varies with clustering modes when a hybrid MPI/OpenMP code is used. Figure 3.5.5 illustrates using 4 OpenMP threads with each MPI task.

Still, hemi and quad performs best. However, the SNC modes are doing somewhat better. The MPI counts above 64 MPI tasks per node are employing hyper-threads which help somewhat. It is interesting to note that the Himeno run on 256 nodes using 32 MPI tasks per node and 4 threads, which utilizes 2 hyper-threads, is running faster than the run on 512 nodes using 32 MPI task per node and 4 threads, which uses 4 hyper-threads. The conclusion with this test of the Himeno memory bandwidth limited application is that all-MPI across 512 nodes using Hemi or Quad clustering is the highest performing run.

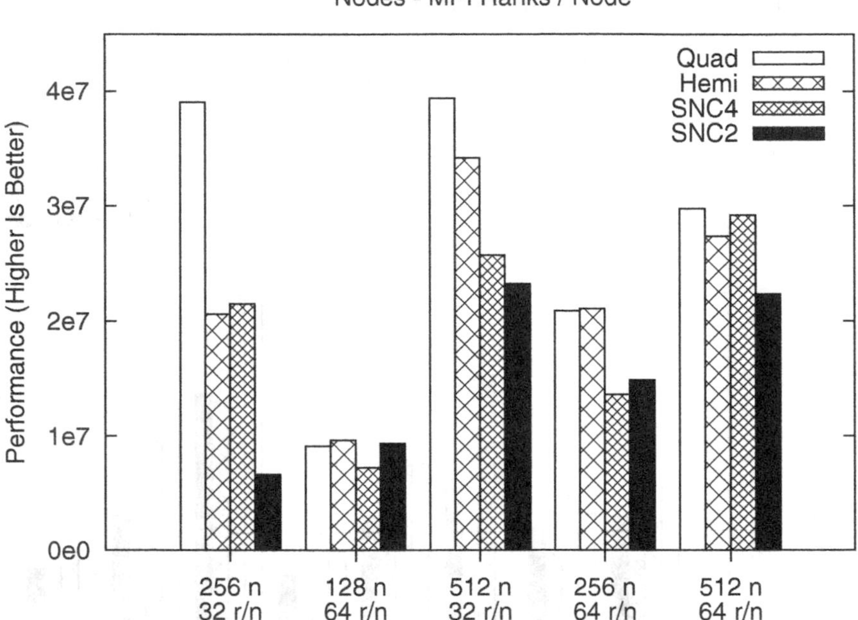

FIGURE 3.5.5 KNL clustering modes with Himeno and 4 threads per rank.

Our next test is S3D which is well-structured with complete vectorization, high-level threading, and good cache utilization. Figure 3.5.6 illustrates the performance of the different modes running S3D. In these charts the measure of performance is wallclock time, taken for each time step, lower is better. There is not much difference between the various modes at higher node counts, and once again Hemi and Quad usually result in the fastest times. Not shown here is the chart for four threads per MPI task which shows very little difference between all four clustering modes.

In conclusion, the best clustering mode to use for these two applications, and generally most of the applications one will look at, is Quadrant. Since re-provisioning the system to change modes requires a reboot, the default should be Quadrant.

Given the discussion of the clustering strategy, the KNL node has an extremely non-uniform memory architecture (NUMA). It will be difficult to thread across many tiles well unless the NUMA affects are taken into account within the application. Threading and NUMA will be discussed along with OpenMP in Chapter 8.

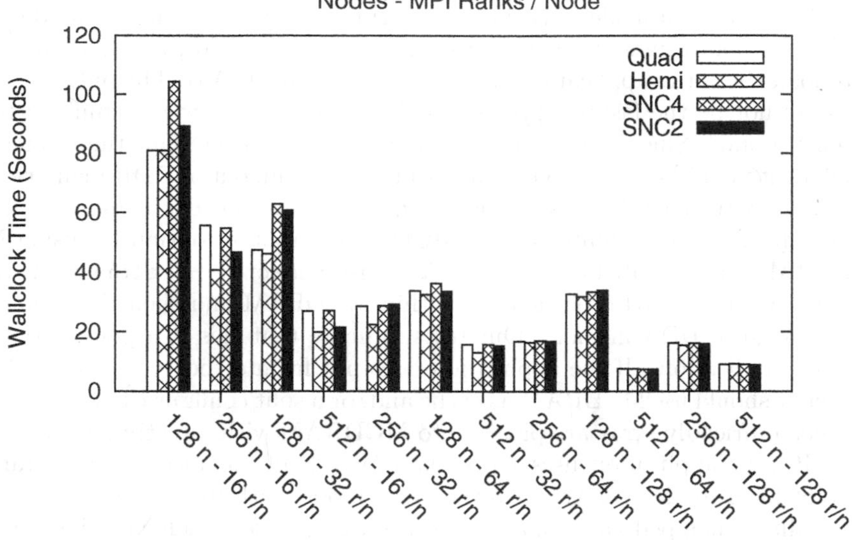

FIGURE 3.5.6 KNL clustering modes with S3D and 1 thread per rank.

TABLE 3.5.2 Latency and bandwidth with KNL clustering modes.

Memory	(mode)	Latency: ns	(clocks)	STREAM Triad GB/s
MCDRAM	(cache)	174	(243)	315
DDR	(cache)			83
MCDRAM	(flat)	174	(244)	448
DDR	(flat)	151	(212)	83
MCDRAM	(SNC2)	170-178	(239-250)	448
DDR	(SNC2)	146-172	(205-241)	83
MCDRAM	(SNC4)	167	(234)	455
DDR	(SNC4)	144-159	(203-223)	83

3.6 KNL MCDRAM MODES

There are four different ways that MCDRAM can be configured. First, it can be configured as cache as discussed above. Second, it can be configured as flat. That is, it is a separate address space. In this mode there is a method to run an entire executable and data within the flat MCDRAM if it does not require more than 16 GB of memory. Then there is an equal mode where 8 GB are configured as cache and 8 GB as separate address space. Finally, there is a split mode where MCDRAM can be configured as 4 GB cache and 12 GB as a separate address space.

An obvious choice for those applications that require less than 16 GB per node is to use flat mode. It is also easy to run the application in this mode. Simply incorporate `numactl --membind=1` into your runscript, which will attempt to load the entire application and its data into MCDRAM. The only time this would not be the fastest approach is if the application were significantly latency bound. Since the latency on MCDRAM is longer than the latency on DDR, it might be better to run in flat mode and run out of DDR memory. Additionally the more latency bound an application is, the less of an increase can be expected when running out of MCDRAM. Another option for using MCDRAM as a separate address space is to use `numactl --preferred=1` which puts up to 16 GBytes of memory within MCDRAM and then the remaining arrays go to DDR memory. Our tests indicate that this is a poor option for most applications. If the application does not fit into the 16 GB MCDRAM, then it should use MCDRAM as cache and/or a split configuration where some cache unfriendly array are placed into MCDRAM with directives or pragmas.

If your application uses more than 16 GB of memory, a good starting point for using MCDRAM is to try cache mode, with the understanding that the direct mapped cache may cause a scaling issue. On KNL, the memory bandwidth to DDR is one-fifth the bandwidth to MCDRAM. For memory bandwidth limited applications, application developers are encouraged to use MCDRAM as much as possible. There are several ways of employing MCDRAM. It can be used as a level-3 cache – while it does have a relatively high latency for a cache, many applications have shown good performance using MCDRAM as a cache. The other issue using MCDRAM as a cache is that it is direct mapped. Consider the Appendix A discussion where there is only one associativity level. Since MCDRAM is relatively large (16 GB) and most KNL chips come with 96 or 128 GB of main memory, a cache line in MCDRAM (64 bytes) must service 6 cache lines for 96 GB of main memory and 8 cache lines for 128 GB of main memory, as depicted in Figure 3.6.1. If two cache lines within the 1.5 to 2.0 GB of memory are needed at the same time, then there would be a cache eviction, and the likelihood of cache thrashing would be high.

Most of the KNL systems being delivered are running some version of Linux. When a system is initially brought up, the physical pages have a high probability of being contiguous in physical memory. However, as applications are run on a node, the physical memory can become fragmented, and the likelihood of the physical pages being contiguous becomes very low. Since the different nodes probably had different applications running on them, the physical page layout for the nodes will not be the same. When an application runs on a large number of nodes (>100) the likelihood of cache thrashing with the MCDRAM being used as cache increases. Some nodes may have the application's arrays allocated in such a way that cache conflicts in MCDRAM would happen. When a cache conflict happens on one of the nodes in a large parallel run, load imbalance would be introduced and the overall performance of the parallel run would suffer – overall application performance is often limited by the speed of the slowest node. One way to investigate whether an application has such an issue would be to run a scaling study up to hundreds of nodes using MCDRAM as cache and then run the same scaling study totally ignoring MCDRAM. That is, running in flat mode without allocating all arrays in MCDRAM. The DDR run should run slower. However, if it scales significantly better than the run using MCDRAM as cache, then the MCDRAM run is probably incurring cache conflicts.

128 GB of Main Memory							
Mem	Mem	Mem	Mem	Mem	Mem	···	Mem
Mem	Mem	Mem	Mem	Mem	Mem	···	Mem
Mem	Mem	Mem	Mem	Mem	Mem	···	Mem
Mem	Mem	Mem	Mem	Mem	Mem	···	Mem
Mem	Mem	Mem	Mem	Mem	Mem	···	Mem
Mem	Mem	Mem	Mem	Mem	Mem	···	Mem
Mem	Mem	Mem	Mem	Mem	Mem	···	Mem
Mem	Mem	Mem	Mem	Mem	Mem	···	Mem
↓	↓	↓	↓	↓	↓	···	↓

16 GB of MCDRAM as Direct-Mapped Cache							
64B Cache Line	64B	64B	64B	64B	64B	···	64B

FIGURE 3.6.1 MCDRAM as a direct-mapped cache for 128GB of DDR.

Once again the performance one gets from using MCDRAM as cache versus as a separate address space is application dependent. Considering the HIMENO benchmark using only MPI, we see a significant improvement using MCDRAM as separate address space on low node counts and then little improvement as we scale up to 512 nodes, as shown in Figure 3.6.2. Since the problem being computed is a fixed size for all node counts, as the number of nodes increases, the amount of data required by each node decreases. At 128 nodes, the data fits into MCDRAM. However, it uses a high percentage of MCDRAM. In cache mode, there are issues that impact the performance, such as the direct-mapped aspect of the cache. As the number of nodes increases, the likelihood of cache conflict decreases since less and less of MCDRAM is being used. Notice in Table 3.6.2 that the largest difference between using MCDRAM as cache or flat when the entire application fits into MCDRAM is the bandwidth. As cache the bandwidth is 329 GB/s and as flat it is 486 GB/s. Latencies are pretty much identical. This is important to understand since the differences in running with MCDRAM configured as flat may not run much faster than running with MCDRAM configured as cache.

If your application uses more than 16 GB and using MCDRAM as cache is not giving good performance, the next approach would be to identify arrays that can be stored into MCDRAM. The immediate question that comes up is which arrays should be placed in MCDRAM? Since KNL does have a level-1 and level-2 cache structure, some arrays may effectively utilize that structure, the so-called cache-friendly arrays. Other arrays which may use non-temporal

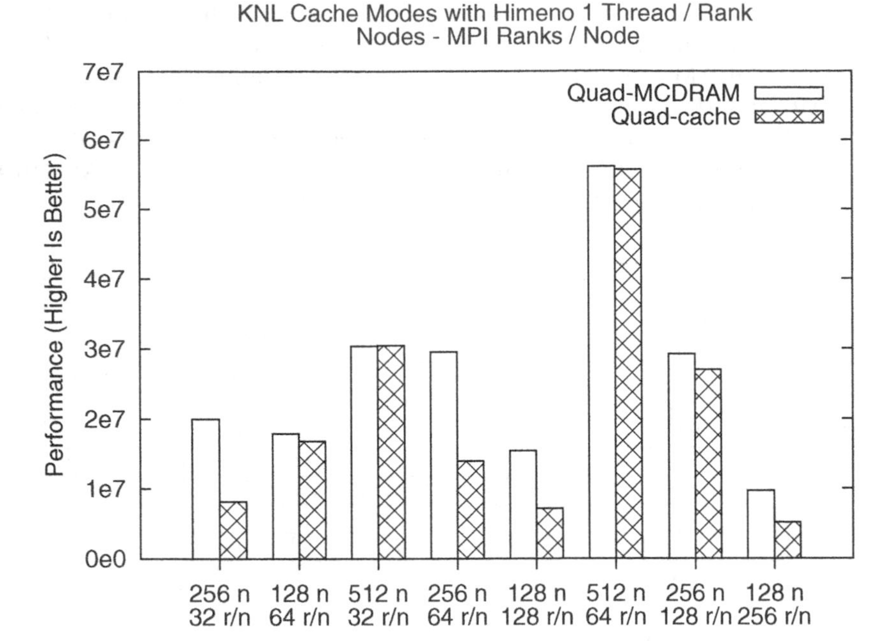

FIGURE 3.6.2 KNL cache modes with Himeno and 1 thread per rank.

reads are very cache-unfriendly, and they would be good candidates for residence in MCDRAM. Non-temporal arrays are arrays that have no cache reuse. Identifying the principal arrays to place into MCDRAM is non-trivial and both Cray and Intel have developed tools for identifying MCDRAM candidates in a running program. The analysis that is performed is to examine hardware counters during the execution of the program that give the arrays that are encountering the most level-2 cache misses. When an array has a lot of level-2 cache misses, it must then go to memory to obtain the required cache line. Undoubtedly there are a lot of arrays with that characteristic, and the tools then need to tabulate the ones which have the most level-2 cache misses and indicate after the run which arrays may be good candidates for storing into MCDRAM.

Once the application developer has a good understanding of which arrays should be stored into MCDRAM, the question is whether to use all of MCDRAM as flat. That is, should all of MCDRAM be a separate address space, or should some MCDRAM be configured as cache. If all of the arrays are not stored into MCDRAM, they may benefit from using some of MCDRAM as cache. The combination of flat and cache is something that should be experimented with. Table 3.6.1 is the output from the memory analysis tool supplied by Cray Inc. This is the output from running a large problem with S3D using 8192 MPI tasks on 128 nodes. As indicated in the report, the problem uses

1.6 GBytes/MPI task. At 64 MPI tasks/node the application requires 102 GBytes/node. The top two arrays are local automatic arrays utilized in the chemical species computation. **yspecies** is a global array, and **diffflux** is a local array in the **rhsf** routine. If we employ MCDRAM as a separate address space, then we can only place arrays that use 250 Mbytes of memory/MPI task in MCDRAM. For example, putting **rb**, **rf**, **yspecies**, and **diffflux** into MCDRAM with directives would use 176.052 Mbytes.

TABLE 3.6.1 Output from Cray memory analysis tool showing candidate MCDRAM objects in S3D.

Table 1: HBM Objects Sorted by Weighted Sample

Object Sample Weight%	Object Sample Count%	Object Sample Count	Object Max Active	Object Max Size (MBytes)	Object Location PE=HIDE
100.0%	100.0%	42,088.7	1,958.4	1,611.734	Total
15.6%	28.3%	11,894.3	1.0	0.276	$$_rf@local@<unknown>@line.0
13.1%	23.6%	9,931.3	1.0	0.276	$$_rb@local@<unknown>@line.0
11.4%	5.3%	2,234.8	1.0	43.875	yspecies@alloc@init_field.f90@line.83
9.8%	5.2%	2,198.0	1.0	131.625	diffflux@local@rhsf.f90@line.101
6.9%	3.2%	1,359.3	1.0	141.750	q@alloc@init_field.f90@line.83
6.6%	1.9%	782.7	1.0	141.750	tmmp@local@rhsf.f90@line.89
5.6%	8.9%	3,766.1	1.0	131.625	grad_ys@local@rhsf.f90@line.101
5.0%	2.9%	1,201.1	1.0	0.844	phi@alloc@mixfrac_m.f90@line.166
3.8%	6.0%	2,545.1	1.0	43.875	h_spec@local@rhsf.f90@line.111
3.3%	0.3%	146.7	1.0	2.531	u@alloc@init_field.f90@line.83
2.0%	3.5%	1,487.7	1.0	0.051	$$_dif@local@<unknown>@line.0
1.8%	3.3%	1,375.1	1.0	0.051	$$_c@local@<unknown>@line.0
1.7%	0.1%	34.5	1.0	0.051	$$_xs@local@<unknown>@line.0
1.6%	0.6%	267.7	1.0	43.875	tmmp2n@local@rhsf.f90@line.91
1.3%	0.5%	199.6	1.0	45.703	neg_f_y_buf@alloc@deriv_m.f90@line.334
1.0%	1.7%	704.8	1.0	43.875	diffusion@local@rhsf.f90@line.147
1.0%	0.4%	177.1	1.0	47.250	q_err@alloc@erk_m.f90@line.463
1.0%	1.5%	614.8	1.0	43.875	ds_mxvg@local@integr_erk.f90@line.60

The memory tool uses heuristics to identify cache unfriendly arrays that are potential candidates for MCDRAM. The first column gives the percentage of the event-based weight as to the importance of the array being placed into MCDRAM. The second column gives the percentage of samples gathered. The percentages in column one and two are different due to the weighting factors. The fifth column gives the amount of data allocated for the arrays and the last column gives the location within the program where the data is allocated. The tool cannot always identify the location of local arrays.

When putting arrays within MCDRAM using either 100%, 75%, or 50% separate address space, that amount of MCDRAM used for separate address space is taken away from the ability of using MCDRAM as cache. When an application like S3D is very cache friendly, using any amount of MCDRAM as separate address space will degrade the performance the application is achieving from using MCDRAM as cache. Even with the assistance from the memory tools, trying to place cache unfriendly arrays into MCDRAM runs slower than just using all of MCDRAM as cache. Excerpt 3.6.1 shows an example of using Cray's directives for placing `yspecies` in MCDRAM.

```
! primitive variables
  real, allocatable :: yspecies(:,:,:,:)      !mass fractions for ALL species
!dir$ memory(bandwidth)yspecies
  real, allocatable :: u(:,:,:,:)             !velocity vector (non-dimensional)
  real, allocatable :: volum(:,:,:)           !inverse of density (non-dimensional)
  real, allocatable :: pressure(:,:,:)        !pressure (non-dimensional)
  real, allocatable :: temp(:,:,:)            !temperature (non-dimensional)

! and RF and RB
      real*8 RF(maxvl,283),RB(maxvl,283)
!dir$ memory(bandwidth) RF, RB
```

Excerpt 3.6.1 Code showing use of Cray's HBM memory placement directives.

The results are given in Table 3.6.2. Once again, taking cache away by using it for specific variables hurts the performance, so the time you gain by having cache unfriendly arrays in MCDRAM is degraded by the lack of cache for the other variables. Times are in seconds per time step – lower is better.

TABLE 3.6.2 Performance of S3D with different MCDRAM cache ratios.

	`rf, rb, yspecies`	`rf, rb, diff`	`rf, rb, diff, yspecies`
100% Cache	32.52	32.52	32.52
50% Cache	33.87	out of memory	out of memory
25% Cache	35.17	51.75	51.38
0% Cache	47.80	61.57	61.11

3.7 IMPORTANCE OF VECTORIZATION

The next characteristic of the node important to consider is the width of the SIMD unit. For double-precision, this will range from 4 on some multicore processors up to 32 on the GPU. This width is related to the performance one can obtain when the inner loops of the computation can be vectorized. There are advantages and disadvantages to having shorter SIMD widths or longer SIMD widths. In a vector operation, the performance that one obtains is related to the number of iterations of the loop being vectorized. That length is then chunked up into units that are the SIMD length. On the GPU it is important to have either vectors a multiple of the SIMD length, called WARPS (i.e., 32 elements), or have very long vector lengths.

Figure 3.7.1 illustrates performance of the SIMD unit as a function of the length of the vector. Notice that the efficiency drops significantly when the vector length goes from 32 to 33. That is because two WARPS are required to handle the 33rd element. As the vector length gets longer, the efficiency drop is less severe. On Xeon and KNL the performance of a vector operation grows until we reach about 150 to 200 elements. This depends upon the loop being vectorized. However, longer is better most of the time. So on both systems we see that long vectors are good. There is more of a degradation on the GPU system when we have short <10 length vectors.

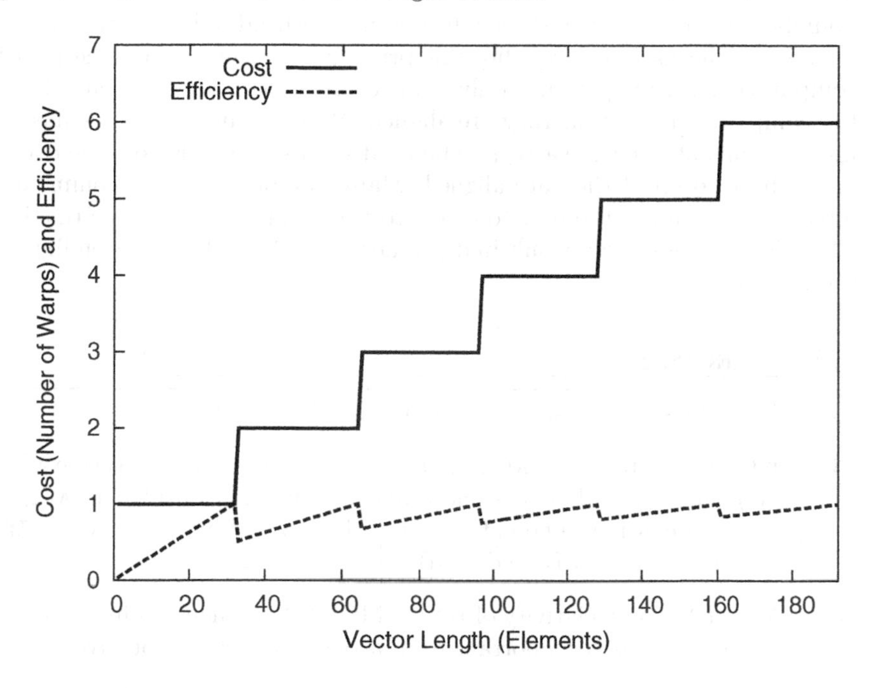

FIGURE 3.7.1 Efficiency of vectorization on GPU with increasing vector lengths.

The length of the SIMD unit also indicates the performance you can achieve by using a vector instruction. On the GPU, one can gain close to a factor of 30 over scalar when using SIMD instructions. On the KNL, the length of the SIMD unit is the length of a cache line or 8 eight-byte words. On the Xeons prior to the Skylake processor, the SIMD length was four. With Skylake the SIMD width is the same as KNL: 8 eight-byte words.

If an application is not vectorizable, the GPU will perform very poorly, whereas the Xeon will show a small difference, and the Intel Phi loses about a factor of ten in performance. Also, an algorithm may be vectorizable with a significant rewrite for the GPU, and that rewrite may not be performant on the Xeon or Intel Phi due to the overhead introduced to achieve vectorization.

For example, gathering and scattering of operands to vectorize a very complex decision process may result in more overhead, and thus the performance gain will be less than expected. Much more will be discussed about vectorization in Chapter 7.

3.8 ALIGNMENT FOR VECTORIZATION

All of the systems we are discussing need to have the vectors aligned on multiple cache line boundarys. If the operands are not aligned on such a boundary, the compiler must introduce some overhead to handle the unaligned instances. The user can expedite this process by ensuring that the principal computational arrays are not only aligned on cache boundary, but also that the compiler can tell that they are aligned. When a subroutine is called with many arguments that are arrays to be used within a vector loop, the compiler must check to see if they are aligned. Many compilers supply comment line directives that allow the user to assert to the compiler that they are aligned. The alignment issue can result in degradations of 10 to 100% depending upon the architecture.

3.9 EXERCISES

3.1 What does "TLB" stand for, and what does a TLB do?

3.2 In the array statement `A(IA(:)) = B(IA(:)) + C(:)`, describe what (if any) cache or TLB issues one would encounter. Consider if `IA(I) = I` and then consider different domains of `IA(:)`. What about when `IA(:)` accesses 1 KB, 1 MB, and 1 GB?

3.3 What is the associativity of KNL's MCDRAM when run in cache mode, and why is this so important from a performance perspective?

3.4 What does the `numactl --preferred=1` command do? What does `numactl --membind=1` do?

3.5 What compiler directive can be used to tell the compiler to place an array in high-bandwidth memory?

3.6 Which KNL clustering mode is frequently found the best and is thus a reasonable default?

3.7 What is the bandwidth of NVLink? How does this compare to PCIe?

3.8 *Construct a case study*: Revisit the application and target system selected in the exercise section at the end of Chapter 2 – this time with respect to data motion.

   a. Characterize the target system in terms of:

   i. Number of NUMA nodes per system node.
   ii. Size, bandwidth, and latency of memory per NUMA node.
   iii. Size, bandwidth, and latency of each cache level (including per-cycle load/store bandwidth between the level-1 cache and the core).
   iv. Draw a tree/graph of the memory hierarchy/topology of the target system's node architecture including each core, cache, NUMA node, and memory, as well as their logical connections.

b. For the selected application:
   i. Identify three levels of parallelism in the application (e.g., grid decomposition, nested loops, independent tasks).
   ii. Map the three levels of parallelism identified in the application to the three levels available on the target system: MPI, threads, and vectors.

c. For each level of parallelism identified in the application and mapped to the target system, compare the working set of the application at that level to the amount of memory and bandwidth available on the target system:
   i. Compare the working set size of the chosen vectorizable loops to the size and bandwidth of each cache level.
   ii. Compare the working set size of each thread to the total amount of (private) cache available to a single core.
   iii. Compare the working set size of an MPI rank to the size and bandwidth of memory on a system node. If HBM is available, can it hold the working sets of all the ranks on the node?

   Do these comparisons show a good fit between the parallelism identified in the application and the target system? If not, try to identify alternative parallelization strategies.

3.9 In Table 3.6.2, why doesn't the placement of important arrays into MCDRAM result in better performance?

3.10 Consider Table 3.6.2. Can you describe a case where the placement of the important arrays might give a better performance gain?

3.11 Describe the different NUMA regions for KNL for each clustering mode. Start with the first region being the tile containing two processors, two level one caches, and one level two cache.

3.12 Give application characteristics where Quadrant-Cache would not be the best configuration on a KNL node. How much of an improvement over Quadrant-Cache would be expected for each clustering/MCDRAM configuration?

CHAPTER 4

How Compilers Optimize Programs

CONTENTS

4.1 FOREWORD BY JOHN LEVESQUE

At the end of the 1970s, vectorization was an emerging technology. Cray Research had released a vectorizing compiler and small software firms were developing preprocessors that would convert the input program into either vectorizable Fortran or calls to vector processing libraries. While the PARALZER was marginally successful on the ILLIAC IV, it had potential, so I left R&D Associates and started at the west coast office of Massachusetts Computer Associates, the developer of the PARALZER tool. They extended

PARALZER into VECTORIZER and had code generators for the CDC 7600, which converted vectorizable Fortran loops into calls to RDALIB (mentioned previously). RDALIB was developed to perform well when called with short vector lengths. At the time, Lawrence Livermore National Laboratory (LLNL) had a similar product call STACKLIB which required longer vector lengths to gain performance. We joked that RDALIB simulated the Cray 1, and STACKLIB simulated the Star 100 – a very difficult CDC Vector machine that Seymour Cray did not work on. Given our work on the ILLIAC IV (now had a group of about 10 people) we received a contract to port/optimize an important application from Los Alamos National Laboratory (LANL) for the new Cray 1 – Seymour's first real vector machine. This was June 1976 and there was no vectorizing compiler, so we cross-compiled on a Data General Ellipse and used Cray Vector Primitives to access the vector instructions, which the VECTORIZER generated. What a wonderful machine. While initially porting to use vector instructions was difficult, scalar performance on the Cray 1 was better, with a 12.5 nanosecond clock compared to the CDC 7600's 27.5 nanosecond clock. Then when vector instructions were used, another factor of 5 to 6 was obtained. It was at this time I met a young graduate student from University of New Mexico, Jack Dongarra.

Well, things didn't go so well with Massachusetts Computer Associates, and the whole group including our administrative assistant moved to Pacific Sierra Research where we struck oil. First, we built our own vectorizer from scratch and began working for several oil companies who started to become interested in the Cray 1. In particular, John Killough of ARCO gave a talk at a conference claiming that his solution technique, Strongly Implicit Procedure (SIP), could not be vectorized. Well, I dusted off some ILLIAC IV documentation and found a paper by Leslie Lamport, who worked at Massachusetts Computer Associates at the time and found a way to vectorize Killough's solution technique using diagonals in two dimensions and hyperplanes in three dimensions. Interestingly, Lamport also worked on the development of LaTeXwhich is being used for the layout of this book. Killough then came to us and said, "We need to determine how many oil drilling rigs we should send up through the Bering strait before it freezes over and we have to port/optimize our reservoir model to the Cray". The next 10 years really were the golden age of supercomputing, and Pacific Sierra Research was so happy they hired the 10 members of the west coast office of Massachusetts Computer Associates. I started giving optimization workshops for the Cray customers and Cray themselves – teaching them how to use their own machine. Since we had the most experience on the first Cray 1 delivered to Los Alamos, we had the best material. After many years of workshops, sometimes doing 2 to 3 a month, I published my first book *Guidebook to Fortran on Supercomputers* in 1989 [20].

4.2 INTRODUCTION

To many application developers, compilation of their code is a "black art". Often times, a programmer will question why the compiler cannot "automatically" generate the most optimal code from their application. Choice of language is one of the first things that determines how well the compiler will be able to understand how to optimize critical parts of the application. One of the reasons Fortran has been so successful for the past 50 years is because it limits the use of pointers which prevents the compiler from doing many of the optimizations one expects from it. For C and C++ there are ways of restricting the impact of pointers, and many C programmers have learned how to write "Ctran" which is a name used for C written like Fortran. While most of the examples we investigate in this book will be Fortran, many of the same general principles apply to C/C++ as well. Hopefully the discussion of the Fortran examples will give the C and C++ programmers ideas of how one might improve their applications so that the compiler can optimize them. Additionally, some C/C++ specific issues, such as aliasing, are covered in more detail.

4.3 MEMORY ALLOCATION

As was discussed in the introduction, power consumption is one of the limiting issues facing hardware architects and within current hardware designs moving data uses more power than the computational units. Data bandwidth within a node design is much less than what is required for most computational algorithms. While many application developers may try to minimize data motion within their programs, they do not understand that the compiler may introduce memory movement without their knowledge. For example, Fortran allows the programmer to pass non-contiguous array sections into a subroutine. While this may be convenient for the programmer, the compiler may have to copy the non-contiguous array section into a contiguous array section prior to passing it to the subroutine.

Memory organizations going forward will become more complex and more difficult to utilize effectively. In the previous chapter, performance issues using memory were discussed. When the application program is compiled, the compiler does not know how best to allocate the arrays. The efficiency of an application depends heavily upon the way arrays are allocated. One important aspect of the allocation of arrays is the alignment of the array to other arrays as well as how a multidimensional array is allocated. This alignment can impact the TLB, the cache associativity as well as efficient transfers and alignment for using the SIMD instructions. It is unlikely that the compiler will be able to guarantee the best alignment for an application. The users can assist by making sure that the major computational arrays are always aligned to a page boundary (4096 bytes or 512 8-byte words). In this way, the starting of an array will always be on a cache boundary and most SIMD instructions can work with aligned data. Once again we want to reduce memory transfers as much as possible.

The programmer has several different ways to allocate the data referenced in a program. They can be allocated explicitly by giving the compiler the dimensions of the arrays at compile time; however, most modern applications dynamically allocate their arrays after reading the input data and then only allocate the amount of data required. The most often used syntax to do this is the typical malloc in C and/or using `ALLOCATABLE` arrays in Fortran. A more dynamic way to allocate data is by using automatic arrays. This is achieved in Fortran by passing the shape of the computation as arguments to a subroutine; this shape is then used to dynamically allocate the arrays within the subroutine within a dynamic scratch memory called the stack. When arguments are passed to a routine that are aligned, then comment line directives should be employed to alert the compiler to that fact. The application developer should strive to allocate their arrays so that they are aligned as discussed.

Remember the address space that the programmer deals with is not the same as the physical memory of the device. All data is allocated in pages and those pages are not guaranteed to be contiguous in memory. Check out Appendix B where we talk about the Translation Look-Aside Buffer (TLB). Interestingly the best situation for an application is right after a system is booted. The first application to run will likely get contiguous pages from the operating system. When an array is allocated on contiguous pages the hardware prefetcher can work more efficiently than if the array is allocated on disjoint pages. How can the programmer get contiguous pages when they are not running on a freshly booted node? They can help by allocating their data in larger chunks. When an array is allocated within a larger allocation, and that allocation is only performed once, the likelihood that the array is allocated on contiguous physical pages is higher. When arrays are frequently allocated and deallocated, the likelihood that the array is allocated on contiguous physical pages will be very low. There is a significant benefit to having an array allocated on contiguous physical pages. Compilers can only do so much when the applications dynamically allocate and deallocate memory. When all the major work arrays are allocated together on subsequent `ALLOCATE` statements, the compiler will usually allocate a large chunk of memory and then sub-allocate the individual arrays, this is good. When a user writes their own memory allocation routine and calls it from numerous places within the application to allocate arrays, particularly if they are small, the compiler cannot help and the data may not be allocated in a contiguous form. Allocating and deallocating arrays will increase compute time to perform garbage collection. Garbage collection is the term used to describe the process of releasing unnecessary memory areas and combining them into available memory for future allocations. Additionally the frequent allocation and deallocation of arrays will undoubtedly lead to memory fragmentation.

Another approach to improve contiguous physical pages is to use larger page sizes. The default page size is 4096 bytes; however, page sizes from 2 Mbytes to 512 Mbytes can be employed. It is easy to test to see if different page sizes can give one an improvement. Typically one needs to rebuild and rerun the application with the desired page size selected.

4.4 MEMORY ALIGNMENT

As discussed in Section 4.3, the way program memory is allocated impacts runtime performance. How are arrays, that are often used together, aligned and does that alignment facilitate the use of SIMD instructions, the cache and the TLB? The compiler will try to align arrays to utilize cache effectively; however, the semantics of the language do not always allow the compiler to move the location of one array relative to another. The following Fortran structures will inhibit a compiler from padding and/or otherwise aligning arrays:

1. Fortran `COMMON` block.
2. Fortran `MODULE`.
3. Fortran `EQUIVALENCE`.
4. Passing arrays as arguments to a subroutine.
5. Any usage of `POINTER`.
6. `ALLOCATABLE` arrays.

When the application contains any of these structures there cannot be an implicit understanding of the location of one array in relation to another. Fortran has very strict storage and sequence association rules, which must be obeyed when compiling code. While common blocks are being replaced by modules in later versions of Fortran, application developers have come to expect the strict memory alignment imposed by common blocks. For example, when performing I/O on a set of arrays, application developers will frequently pack the arrays into a contiguous chunk of logical memory and then write out the entire chunk with a single write. For example, consider the common block in Excerpt 4.4.1.

```
COMMON A(100,100), B(100,100), C(100,100)

WRITE (10) (A(I), I =1, 30000)
```

Excerpt 4.4.1 Example Fortran common block.

Using this technique results in a single I/O operation, and in this case a write outputting all of A, B, and C arrays in one large block, which is a good strategy for efficient I/O. If an application employs this type of coding, the compiler cannot perform padding on any of the arrays A, B, or C. Consider the use of these arrays in the call to a subroutine CRUNCH as seen in Excerpt 4.4.2.

```
CALL CRUNCH (A(1,10), B(1,1), C(5,10))

SUBROUTINE CRUNCH (D,E,F)
DIMENSION D(100), E(10000), F(1000)
```

Excerpt 4.4.2 Subroutine call passing variables from common block.

This is legal Fortran and it will certainly keep the compiler from moving A, B, and C around. Unfortunately since Fortran does not prohibit this type of coding, the alignment and padding of arrays by compilers is very limited.

A compiler can pad and modify the alignment of arrays when they are allocated as automatic or local data. In this case the compiler will allocate memory and add padding if necessary to properly align the arrays for efficient access. Unfortunately, the inhibitors to alignment and padding far outnumber these cases. The application developer should accept responsibility for allocating their arrays in such a way that the alignment is conducive to effective memory utilization. The ultimate solution is for the user to handle all data allocations in such a way that alignment is adequate for vectorization and parallelization. This will be covered in more detail in later chapters.

4.5 COMMENT-LINE DIRECTIVE

Back in 1976 when Cray Research was delivering the Cray 1, they had an opportunity to sell a system to National Center for Atmospheric Research (NCAR). Unfortunately, the major NCAR model allocated a large array and equivalenced different variables within that array. Their loops might look something like that shown in Excerpt 4.5.1.

```
DO 100 I = 1, NCELLS
A(IA(IFORCE)+I) = A(IA(IMASS)+I) * A(IA(IACCEL)+I)
 100    CONTINE
```

Excerpt 4.5.1 Example loop structure representing that in the NCAR model from 1976.

Well the Cray compiler would not vectorize their loops since it felt that there was a data dependency, which would be the case if the array sections overlapped. So Cray told NCAR that they had to rewrite their application so the compiler would vectorize them. Well NCAR told Cray that they would not buy a machine without a less drastic solution. A bright compiler engineer at Cray Research – it might have actually been Dick Hendrickson – came up with the idea to introduce a comment line directive in the form `CDIR$ IVDEP`, which when placed prior to the loop, would tell the compiler to ignore potential data dependencies in the loop. Also, any compiler that did not understand the directive would simply ignore the comment line. It worked, NCAR bought a Cray, and Cray Research's stock went through the roof. A simple comment line directive jump-started Cray Research into the Golden Age of Supercomputing.

Since that time, Cray and other companies have embraced comment line directives and of course pragmas in C and C++ as a useful way to relay information to the compilers. Unfortunately, the companies have not standardized on the directive names. Fortunately all compilers recognize `IVDEP`. There are many other useful directives that will be given and compared in Appendix C.

4.6 INTERPROCEDURAL ANALYSIS

If only the compiler knew more about the subroutine it is trying to optimize, it could do a better job generating optimal code. Even though Fortran 90 introduced interface blocks into the language for the user to provide useful inter-procedural information to the compiler, few programmers use the syntax, leaving the compiler to do its own analysis. Inter-procedural analysis (IPA) requires the compiler to look at a large portion of the application as a single compilation unit. At the basic level, the compiler needs to retain information about the arguments that are passed from the caller to the called routine. The compiler can use that information to perform deeper optimizations of the called routine, especially when literal constants are being passed for a variable in a call. The most drastic inter-procedural optimization is to completely inline all of the called routines that can be inlined. While this could increase the compile time, significant performance gains frequently can be achieved by eliminating costly subprogram call/returns. This is particularly true for C++ programs that employ templates.

For example, Intel's compiler, like many others, has an option for performing inter-procedural analysis (`-ipo`) and the Cray compiler uses (`-ipa(n)`) that generates additional data files during compilation which are used for optimization analysis. In one particular case, the compile time went from 4 to 5 minutes to 20 to 30 minutes, and the performance gain was 10 to 15% of the computation time. If this application is infrequently compiled and the executable is used for thousands of hours of computation, this is a good tradeoff.

4.7 COMPILER SWITCHES

All of the compilers' optimization options can be controlled through the use of command line option flags and comment line directives. When fine tuning an application, the programmer can use these options across the entire code or on just a portion by using comment line directives. There are options that control compiler optimization functions. Each compiler has its own set of options, but they all provide control over the level of optimization the compiler will use. It's important to familiarize yourself with the options available with the compiler you are using. Some that are very important are

1. Complex vectorization is typically not done by default. While straight forward vectorization is done by default, some compilers need flags to vectorize complex decision processes and gather/scatter in the loop.

2. Inter-procedural analysis is not always performed. This is very important for applications that call numerous small routines and in particular for C and C++ applications.

3. Unrolling, cache blocking, prefetching, etc. These optimizations are usually all combined into a general optimization flag like `-O3`; however, many cases may benefit by selectively utilizing these individually on specific loops using compile line directives.

4. Always get information about what optimization the compiler performed. The Intel compiler has a `-qopt-report[=n]` to show what analysis was performed. The Cray compilers use `-hlist=a` that shows what optimizations took place in the compilation of a routine and then gives reasons why some optimizations, like vectorization, could not be performed.

5. Don't use automatic shared memory parallelization. Automatic shared memory parallelization will not give good OpenMP performance. In some cases one part of an application might get a good speedup while another portion gives a slow down. Blindly using automatic parallelization is not recommended, unless runtime statistics are gathered to show which routines are improved and which are degraded.

6. The Intel compiler does not recognize OpenMP directives as a default. Use the -qopenmp flag to enable this capability.

Appendix C has a table of important compiler switches for the Intel compiler and the Cray Compilation Environment.

4.8 FORTRAN 2003 AND INEFFICIENCIES

With the development of Fortran 90, 95, 2003, and now 2008, new semantics have been introduced into the language that are difficult or even impossible to compile efficiently. As a result, programmers are frequently disappointed with the poor performance when these features are used.

Following are a few of the newer Fortran features that will cause most compilers to generate inefficient code and should be avoided:

1. Array syntax.
2. Calling standard Fortran functions not linked to optimized libraries.
3. Passing array sections.
4. Using modules for local data.
5. Derived types: struct of arrays versus array of structs.

4.8.1 Array Syntax

Array syntax was first designed for the legacy memory-to-memory vector processors like the Star 100 from Control Data Corporation. The intent of the syntax was to give the compiler a form they could convert directly into a memory-to-memory vector operation. When these vector machines retired, array syntax was kept alive by Thinking Machine's Connection Machine. Here the compiler could generate a SIMD parallel machine instruction that would be executed by all the processors in a lock-step parallel fashion.

Unfortunately, array syntax is still with us and while many programmers feel it is easier to use than the standard old Fortran DO loop, the real issue is that most Fortran compilers cannot generate good cache efficient code from a series of array assignment statements.

Consider the following sequence of array syntax from the POP ocean model as seen in Excerpt 4.8.1 [27]. In this example, all of the variables that are in capitalized letters are arrays. In the test we are running, the sizes of the arrays are (500,500,40). Each array assignment will completely overflow the TLB and all levels of cache and when the next array assignment is performed, the data will have to be reloaded from memory.

```
! DP_1/DT
    WORK3 = mwjfnums0t1 + TQ * (c2*mwjfnums0t2 +     &
      c3*mwjfnums0t3 * TQ) + mwjfnums1t1 * SQ
! DP_2/DT
    WORK4 = mwjfdens0t1 + SQ * mwjfdens1t1 +         &
      TQ * (c2*(mwjfdens0t2 + SQ*SQR*mwjfdensqt2) +  &
      TQ * (c3*(mwjfdens0t3 + SQ * mwjfdens1t3) +    &
      TQ *  c4*mwjfdens0t4))
    DRHODT = (WORK3 - WORK1*DENOMK*WORK4)*DENOMK
```

Excerpt 4.8.1 Array syntax from the POP ocean model.

Most of the variables shown are multidimensioned arrays. Today's compilers will generate three looping structures around each of the statements. When the arrays used in the first array assignment are larger than what can be fit in level-1 cache, those variables used in both the first and second array assignment will have to be fetched from level-2 cache and if the arrays are larger than what can be held in level-2 cache, they will have to be retrieved from level-3 cache or memory. On the other hand, writing the three statements in a single looping structure as seen in Excerpt 4.8.2 results in better cache utilization.

```
! DP_1/DT
DO K=1,NZBLOCK
DO J=1,NYBLOCK
DO I=1,NXBLOCK
  WORK3(I,J,K) =  mwjfnums0t1 + TQ(I,J,K) *                       &
    (c2*mwjfnums0t2 + c3*mwjfnums0t3 * TQ(I,J,K)) +               &
    mwjfnums1t1 * SQ(I,J,K)
  WORK4(I,J,K) =  mwjfdens0t1 + SQ(I,J,K) *                       &
    mwjfdens1t1 +                                                 &
    TQ(I,J,K) *                                                   &
    (c2*(mwjfdens0t2 + SQ(I,J,K)*SQR(I,J,K)*mwjfdensqt2) +  &
    TQ(I,J,K) *                                                   &
    (c3*(mwjfdens0t3 + SQ(I,J,K)*mwjfdens1t3) +                   &
    TQ(I,J,K) *  c4*mwjfdens0t4))
  DRHODT(I,J,K) = (WORK3(I,J,K) - WORK1(I,J,K) *                  &
    DENOMK(I,J,K) * WORK4(I,J,K)) * DENOMK(I,J,K)
ENDDO ; ENDDO ; ENDDO
```

Excerpt 4.8.2 Explicit DO loops improve cache utilization in POP.

Now the compiler will generate very efficient cache-friendly code. The variables used in the three statements are only fetched once from memory and subsequent uses will come from level-1 cache. Each sweep through the I loop will fetch up cache lines which can then be reused in subsequent statements in the DO loop. Additionally, we get much better utilization of the TLB since we are uniformly accessing each array. We get the following impressive speed-up as seen in Table 4.8.1.

TABLE 4.8.1 POP ocean model DO loop versus array syntax speedup.

Implementation	Time (seconds)	TLB refs/miss	L1 Cache Hit %
Original Array Syntax	2.02	28	69
Fortran DO Loop	.459	499	84

The POP code is an extremely well-written code and a majority of the major computational kernels are written in DO loops. However, this sample code from the equation of state is written in Fortran 90 array syntax and it can definitely be improved. Interestingly, when the number of processors is increased using strong scaling of a given grid size, the size of the arrays on each processor becomes smaller. At some point the restructuring shown above would not be needed since the size of all the arrays would fit into level 2 cache. At this point, POP exhibits super-linear scaling. The super-linear scaling comes about because the array assignments are working on arrays that do fit in level-1 and level-2 cache. When control passes from one statement to the next, the arrays are still in cache and the restructuring shown above would be unnecessary.

4.8.2 Use Optimized Libraries

This is simply a matter of understanding what the compiler does with calls such as `MATMUL` and `DOT_PRODUCT`. Whenever an application uses the more complex calls, make sure that the compiler uses the appropriate routine from the optimized libraries that the vendor supplies. The safest approach is to call the BLAS and LAPACK routines directly rather than trust the Fortran 2000 intrinsics. Some compilers have command line options to automatically recognize standard library calls and link them to the appropriate optimized versions. This is very compiler dependent and not obvious to the user unless a particular Fortran intrinsic like `MATMUL` takes a lot more time than it should.

4.8.3 Passing Array Sections

Whenever you find a call to a small subroutine that is using a significant amount of time, it's probably because an array section is being passed to the called subroutine. Consider the code in Excerpt 4.8.3 from the combustion code S3D. This call invokes a massive amount of memory movement to copy `grad_Ys` into contiguous temporary array to pass to `compScalarGrad`. Most compilers will notice that `yspecies` is already a contiguous section of memory and not perform any data movement on it. The `grad_Ys` array will be copied to a completely separate portion of memory that is passed to the routine. Then on return from the routine the result will be copied back into `grad_Ys`. Typically, this copy is performed with an optimized memory copy routine. If you see an unusual amount of time being spent in a system copy, it could be due to this sort of array section passing.

```
do n=1,n_spec
  call compScalarGrad(yspecies(:,:,:,n),grad_Ys(:,:,:,n,:))
enddo
```

Excerpt 4.8.3 Subroutine call using array sections from S3D.

The best method of passing arrays to a routine is by passing an address. If the reference were to `grad_Ys(1,1,1,n,1)`, that would be an address and no data motion would be performed. But that is not the same as what was used in the original call. The programmer would have to reorganize `grad_Ys` into a structure where `n` is on the outer-most subscript to achieve the improvement. This example is typical of applications that use Fortran 2003, and the programmer may be unaware of the inefficiencies of moving the data.

4.8.4 Using Modules for Local Variables

When OpenMP directives are used at a high level in the call chain, care must be taken to assure that the only variables referenced in a module are shared. This is due to the inability of the compiler to generate thread private copies of module data. Prior to modules, common blocks had the same problem and a special kind of common block (THREADPRIVATE) was included in the OpenMP standard to handle this situation. This will be discussed in much more detail in Chapter 8.

4.8.5 Derived Types

Derived types can have a dramatic impact on efficient memory usage. Consider the code Excerpt 4.8.4 from one of the SPEC OMP benchmarks as well as a summary of hardware counters for this loop in Table 4.8.2.

```
!$OMP PARALLEL DO DEFAULT(SHARED) PRIVATE(N)
DO N = 1,NUMRT
  MOTION(N)%Ax = NODE(N)%Minv * (FORCE(N)%Xext-FORCE(N)%Xint)
  MOTION(N)%Ay = NODE(N)%Minv * (FORCE(N)%Yext-FORCE(N)%Yint)
  MOTION(N)%Az = NODE(N)%Minv * (FORCE(N)%Zext-FORCE(N)%Zint)
ENDDO
!$OMP END PARALLEL DO
```

Excerpt 4.8.4 Loop from SPEC OMP benchmark using derived types.

Notice the poor TLB utilization; any TLB reference/miss below 512 is not good. While the loop appears as if it is contiguous in memory, since we are accessing elements of a derived type, there is a stride of the number of elements within the derived type. Rather than the arrays being dimensioned within the derived type, the derived type is dimensioned as seen in Excerpt 4.8.5. This results in each of the arrays having a stride of 12, which is hurting both TLB and cache utilization.

```
TYPE :: motion_type
        REAL(KIND(ODO))  Px, Py, Pz  ! Initial position
        REAL(KIND(ODO))  Ux, Uy, Uz  ! Displacement
        REAL(KIND(ODO))  Vx, Vy, Vz  ! Velocity
        REAL(KIND(ODO))  Ax, Ay, Az  ! Acceleration
END TYPE
TYPE (motion_type), DIMENSION(:), ALLOCATABLE :: MOTION
```

Excerpt 4.8.5 Derived type from SPEC OMP benchmark.

TABLE 4.8.2 SPEC OMP hardware performance counters.

```
USER / solve_.LOOP@li.329
------------------------------------------------------------------------
 Time%                                               4.5%
 Time                                           12.197115 secs
 Imb.Time                                        0.092292 secs
 Imb.Time%                                           1.0%
 Calls                         42.9 /sec            523.0 calls
 PAPI_L1_DCM                13.700M/sec         167144470 misses
 PAPI_TLB_DM                 0.448M/sec           5460907 misses
 PAPI_L1_DCA                89.596M/sec        1093124368 refs
 PAPI_FP_OPS                52.777M/sec         643917600 ops
 User time (approx)         12.201 secs       32941756956 cycles  100.0%Time
 Average Time per Call                           0.023321 sec
 CrayPat Overhead : Time      0.0%
 HW FP Ops / User time      52.777M/sec         643917600 ops    0.5%peak(DP)
 HW FP Ops / WCT            52.777M/sec
 Computational intensity      0.02 ops/cycle      0.59 ops/ref
 MFLOPS (aggregate)          52.78M/sec
 TLB utilization            200.17 refs/miss     0.391 avg uses
 D1 cache hit,miss ratios    84.7% hits          15.3% misses
 D1 cache utilization (M)     6.54 refs/miss     0.817 avg uses
```

4.9 C/C++ AND INEFFICIENCIES

While Fortran is still quite common in HPC, both C and C++ are seeing increased use. This doesn't need to be an issue, but frequently is in practice. Remember the quote from John Levesque presented in Chapter 1: "Ask not what your compiler can do for you, ask what you can do for your compiler." The core of the problem is that C and C++ are very "low level" languages compared to Fortran, and this results in less usable information being available to the compiler with C and C++. Here, "low level" implies being close to the hardware with little abstraction (*how* to do something) while "high level" implies having a greater degree of abstraction away from actual hardware (*what* to do).

```
 4.             void daxpy2z(long n, double a, double *x, double *y, double *z)
 5.             {
 6.               long i;
 7.
 8.  + r4-----<   for(i=0; i<n; i++) {
 9.    r4           y[i] = a*x[i] + y[i];
10.    r4           z[i] = a*x[i] + y[i];
11.    r4----->   }
12.             }

CC-6308 CC: VECTOR daxpy2z, File = daxpy2z.c, Line = 8
  A loop was not vectorized because the loop initialization would be too costly.
```

Excerpt 4.9.1 Assumed aliasing example in DAXPY-like C/C++ code.

As a simple example, compare a basic vector (array) in C to one in Fortran. In Fortran, this array is a high-level construct of the language itself. The Fortran compiler knows things about the array, such as its length and stride. Through the Fortran standard, the compiler can also know that the array is not aliased (referred to by a different variable name). In C, a dynamically allocated array is just a pointer, a simple address. It does not come with a length or stride the compiler knows anything about. Importantly, the address pointed to could be aliased, as many pointer variables can point to the same place in memory. This means the compiler does not get as much usable information from C code as from Fortran code.

Aliasing is a particularly important issue, as it causes very real problems for the compiler, especially with respect to vectorization. When vectorizing C/C++ codes, there are two common solutions to the aliasing problem. First, the compiler can be informed that specific variables are not aliased through the use of the `restrict` keyword. Second, the compiler can be told to ignore assumed vector dependencies through the use of the `ivdep` pragma. Note that `restrict` may be better in terms of helping the compiler. This is due to the fact that the `ivdep` pragma instructs the compiler to ignore assumed *vector* dependencies, but may say nothing about scalar dependencies for some compilers. On the other hand, the `restrict` keyword tells the compiler there is no aliasing at all, irrespective of scalar or vector status.

```
 4.              void daxpy2z(long n, double a,
 5                            double *restrict x, double *restrict y, double *restrict z)
 6.              {
 7.                long i;
 8.
 9.     Vr2--<     for(i=0; i<n; i++) {
10.     Vr2          y[i] = a*x[i] + y[i];
11.     Vr2          z[i] = a*x[i] + y[i];
12.     Vr2-->     }
13.              }
```

Excerpt 4.9.2 Example showing assumed aliasing in DAXPY-like C/C++ code resolved with restrict keyword.

If it is found to be cumbersome to add the `restrict` keyword to large sections of code, the `-restrict` command-line option available with CCE can help. This flag instructs the compiler that the source code being compiled does not contain any aliasing. Note that this is a very heavy hammer; it can easily break codes which do contain aliasing. However, if one writes time-critical kernel functions carefully and places the kernel functions in separate source files, one can compile the most important routines with the `-restrict` flag, potentially saving some coding effort.

Excerpt 4.9.1 shows an extended DAXPY example where aliasing is an issue, as the compiler has to assume it is possible for `x` and `y` to point to the same place in memory. This assumption puts limits on how the compiler is able to translate this C code into efficient assembly. Excerpts 4.9.2 and 4.9.3

show the two previously mentioned approaches for informing the compiler that the arrays do not alias one another. As can be seen in the excerpts, once the compiler knows there are no aliasing issues, it vectorizes the loop.

While assumed aliasing can easily prevent the compiler from vectorizing a loop, it can cause the compiler to generate inefficient scalar code as well. To see an example of this, consider that aliasing in the extended DAXPY loop means the compiler needs to emit assembly code that works with both of the two possibilities with respect to the aliasing of `x` and `y` at the same time:

1. `x != y`: Writes to `y[i]` are independent of `x[i]`, so the compiler can reuse the original value of `x[i]` in the following statement computing `z[i]`.

2. `x == y`: Writes to `y[i]` are also writes to `x[i]`, so the compiler needs to use the new value of `x[i]` in the following statement computing `z[i]`.

To satisfy both these aliasing possibilities with one set of assembly instructions, the compiler will reload the value of `x[i]` from memory when computing `z[i]`, just to be sure it is using the new value if there is one.

```
 4.                 void daxpy2z(long n, double a, double *x, double *y, double *z)
 5.                 {
 6.                   long i;
 7.
 8.                 #pragma ivdep
 9.    Vr2-----<      for(i=0; i<n; i++) {
10.    Vr2              y[i] = a*x[i] + y[i];
11.    Vr2              z[i] = a*x[i] + y[i];
12.    Vr2----->      }
13.                 }
```

Excerpt 4.9.3 Example showing assumed aliasing in DAXPY-like C/C++ code resolved with ivdep pragma.

Consider the code presented in Excerpt 4.9.4 showing a purely scalar version of the extended DAXPY loop. To keep the assembly code generated from this C code short, the `nounroll` pragma is used. Excerpt 4.9.5 shows the x86 assembly code generated by the compiler from the C code containing assumed aliasing. Now compare this to the code which does not suffer from aliasing issues in Excerpt 4.9.6 as well as the assembly code generated from it by the compiler in Excerpt 4.9.7. A comparison of the assembly shows that the version with assumed aliasing is compiled to a loop body with 10 instructions containing 3 loads and 2 stores, while the same loop with no assumed aliasing is compiled to 8 instructions containing only 2 loads and 2 stores. The extra load is a particular issue here, as the `daxpy2z` function is bandwidth-limited. This example shows how aliasing can cause the compiler to generate suboptimal scalar code.

```
 4.            void daxpy2z(long n, double a, double *x, double *y, double *z)
 5.            {
 6.              long i;
 7.
 8.            #pragma novector
 9.            #pragma nounroll
10.  + 1-----<   for(i=0; i<n; i++) {
11.    1           y[i] = a*x[i] + y[i];
12.    1           z[i] = a*x[i] + y[i];
13.    1----->   }
14.            }
```

Excerpt 4.9.4 Scalar example showing assumed aliasing in DAXPY-like C/C++ code.

```
daxpy2z:
        testq           %rdi, %rdi
        jle             .LBBdaxpy2z_3
        xorl            %eax, %eax
.LBBdaxpy2z_2:
        vmovsd          (%rsi,%rax,8), %xmm1          // r1   <- x[i]
        vmovaps         %xmm0, %xmm2                  // r2   <- a
        vfmadd213sd     (%rdx,%rax,8), %xmm1, %xmm2   // r2   <- r2*r1 + y[i]
        vmovsd          %xmm2, (%rdx,%rax,8)          // y[i] <- r2
        vmovsd          (%rsi,%rax,8), %xmm1          // r1   <- x[i]
        vfmadd213sd     %xmm2, %xmm0, %xmm1           // r1   <- a*r1 + r2
        vmovsd          %xmm1, (%rcx,%rax,8)          // z[i] <- r1
        incq            %rax
        cmpq            %rdi, %rax
        jl              .LBBdaxpy2z_2
.LBBdaxpy2z_3:
        retq
```

Excerpt 4.9.5 Scalar example showing assumed aliasing in DAXPY-like C/C++ code compiled to assembly.

```
 4.          void daxpy2z(long n, double a,
 5.                       double *restrict x, double *restrict y, double *restrict z)
 6.          {
 7.            long i;
 8.
 9.          #pragma novector
10.          #pragma nounroll
11.  + 1--<    for(i=0; i<n; i++) {
12.    1         y[i] = a*x[i] + y[i];
13.    1         z[i] = a*x[i] + y[i];
14.    1-->    }
15.          }
```

Excerpt 4.9.6 Scalar example showing assumed aliasing in DAXPY-like C/C++ code resolved with restrict keyword.

Presenting the performance impact of the aliasing issue on the extended DAXPY example, Figure 4.9.1 shows the relative performance of scalar and vector versions of the extended DAXPY loop without aliasing (aliasing removed via the **restrict** keyword, automatic compiler unrolling allowed) as compared to a baseline version containing aliasing. While the vector performance results might be less surprising, it is informative to see that ignoring aliasing in scalar C/C++ code can leave a factor of 1.2 in performance on the table.

```
daxpy2z:
        testq           %rdi, %rdi
        jle             .LBBdaxpy2z_3
        xorl            %eax, %eax
.LBBdaxpy2z_2:
        vmulsd          (%rsi,%rax,8), %xmm0, %xmm1  // r1   <- a*x[i]
        vaddsd          (%rdx,%rax,8), %xmm1, %xmm2  // r2   <- r1 + y[i]
        vaddsd          %xmm2, %xmm1, %xmm1          // r1   <- r1 + r2
        vmovsd          %xmm2, (%rdx,%rax,8)         // y[i] <- r2
        vmovsd          %xmm1, (%rcx,%rax,8)         // z[i] <- r1
        incq            %rax
        cmpq            %rdi, %rax
        jl              .LBBdaxpy2z_2
.LBBdaxpy2z_3:
        retq
```

Excerpt 4.9.7 Scalar example showing assumed aliasing in DAXPY-like C/C++ code resolved with restrict keyword compiled to assembly.

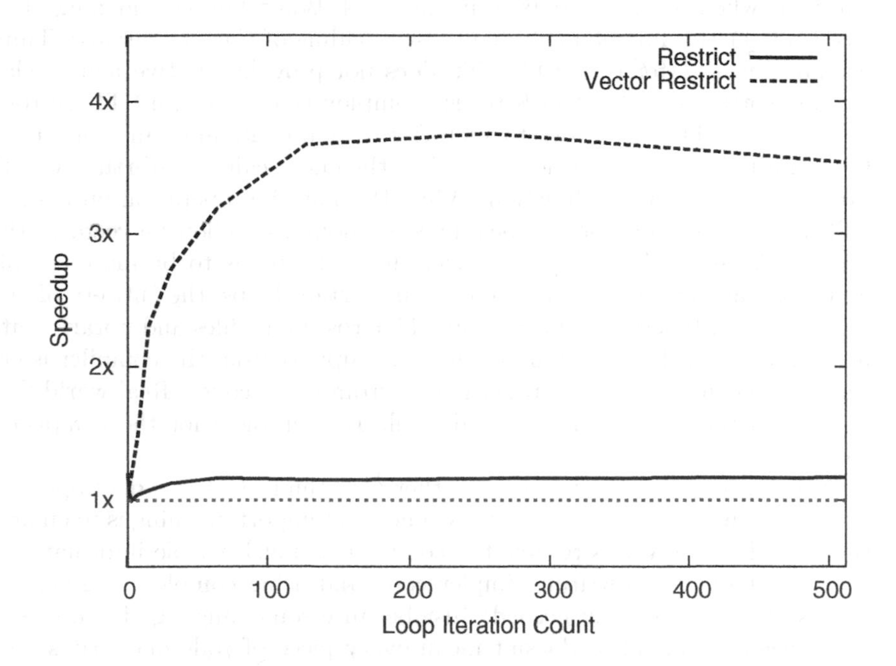

FIGURE 4.9.1 Speedup relative to extended DAXPY code with aliasing.

In terms of providing as much information as possible to the compiler, using C++ doesn't help things here. In particular, the C++ standard does not support the `restrict` keyword, though a number of compilers do have their own non-standard equivalents. This lack of standard support can make it harder to write performance portable C++ code. Also, there are times when having access to a C++ version of the `restrict` keyword still doesn't help much. Specifically, consider the case where objects are passed to a kernel function, and the object pointers/references are marked as restricted. This means the object pointers/references don't alias one another, but it doesn't mean the member variables holding the actual data inside the objects don't have aliasing issues. So, adding a keyword like `restrict` to the arguments of your C++ kernel function doesn't really help the compiler much. One would instead want to mark the declaration of the class's member variables themselves as being restricted. This may not be possible for an application developer using a library of classes created by others. Real-world C++ code often suffers from assumed aliasing issues with a much higher incidence than C code.

Additionally, the layers of function calls introduced when using C++ classes can also be a problem. For example, indexing into a matrix in C at position `[i][j]` (2D array) or `[i*width+j]` (1D array) of a 2D matrix is a simple operation, consisting of not much more than an offset calculation. In contrast, when using a matrix from the C++ Boost library, indexing at position `(i,j)` is a function call due to overloading of the `()` operator. Things are not much better with STL. STL does not provide a native matrix class, causing some C++ developers to use complex things like an STL vector of STL vectors. Thus, in order to vectorize a loop containing an index into a C++ matrix, the compiler needs to inline the entire call tree starting with the overloaded index/access function. While the indexing operation on a simple STL vector class can generally be inlined without issue allowing vectorization, real-world use of C++ classes in scientific codes tends to be more complex and more layered. In order to vectorize important loops, the entirety of large call trees with function definitions spread across many files and libraries often need to be inlined. Good inlining and IPA support from the compiler is critical to achieving reasonable performance from C++ codes. Real-world C++ code often presents a much more difficult IPA problem for the compiler as compared to C code.

When it comes to best practices, there are things C/C++ developers can do to avoid many of the worst pitfalls. The most important thing is to embrace the idea that one wants to give the compiler as much usable information as possible. This means writing simpler code and using complex language features only when actually needed. Just because some language features exist (e.g., classes, templates) doesn't mean every piece of code one writes needs to use them. If a simple 2D array (or better yet, a 1D array appropriately indexed) can do the job well, avoid adding complexity by using something like an STL vector of STL vectors. This may seem obvious to some readers, but judging by the way a number of real-world C++ HPC codes have been written, a significant number of developers appear to have embraced the opposite position, seemingly operating with the mind-set that *everything* in C++ *must* be an object. Keeping time-critical C/C++ kernels simple and free of aliasing to help the compiler will pay dividends in terms of performance.

4.10 COMPILER SCALAR OPTIMIZATIONS

4.10.1 Strength Reduction

Compilers can often replace an explicit use of an operation with a less expensive iterative operation in a loop. In particular, a multiply can be replaced with an addition, and an exponentiation can be replaced with a multiply. Excerpts 4.10.1 and 4.10.2 show an exponentiation replaced with a multiply.

```
DO I = 1,10
  A(I) = X**I
END DO
```

Excerpt 4.10.1 Strength reduction: original exponentiation.

```
XTEMP = X
DO I = 1,10
  A(I) = XTEMP
  XTEMP = XTEMP * X
END DO
```

Excerpt 4.10.2 Strength reduction: compiler optimized multiplication.

Similarly, the compiler can replace a multiply with an add as seen in Excerpts 4.10.3 and 4.10.4.

```
DO I = 1,10
  A(I) = X*I
END DO
```

Excerpt 4.10.3 Strength reduction: original multiplication.

```
XTEMP = X
DO I = 1,10
  A(I) = XTEMP
  XTEMP = XTEMP + X
END DO
```

Excerpt 4.10.4 Strength reduction: compiler optimized addition.

Strength reduction can also be used to optimize array indexing within a loop, as shown in Excerpts 4.10.5 through 4.10.8.

```
DIMENSION A(100,10)
DO I = 1,10
  A(3,I) = 0.0
END DO
```

Excerpt 4.10.5 Strength reduction: original indexing.

The compiler generates code to calculate the address of each element of the array A. The pseudo code in Excerpt 4.10.6 demonstrates the straightforward calculation of the memory address, taking into account that Fortran indexing begins with 1 and assuming each element of A is 8 bytes.

```
DO I = 1,10
  address = addr(A) + (3-1)*8 + (I-1)*100*8
  memory(address) = 0.0
END DO
```

Excerpt 4.10.6 Strength reduction: original indexing pseudo code.

The expression (I-1)*100*8 increases by 800 each time through the loop. The compiler can use this to eliminate the multiplies and optimize the address calculation as seen in Excerpt 4.10.7.

```
address = addr(A) + (3-1)*8 + 1*100*8
DO I = 1,10
  memory(address) = 0.0
  address = address + 800
END DO
```

Excerpt 4.10.7 Strength reduction: optimized indexing pseudo code v1.

The compiler will do the math at compile time and generate a simple set of the starting address as in Excerpt 4.10.8.

```
address = addr(A) + 816
```

Excerpt 4.10.8 Strength reduction: optimized indexing pseudo code v2.

This optimization is very important since the majority of time spent in most programs is in loops containing repeated accesses of multidimensional arrays. Array dimensions are often variables rather than constants. The examples in Excerpts 4.10.9 through 4.10.11 are more typical.

```
DIMENSION A(N1,N2,N3)
DO I = 1,N2
  A(M1,I,M3) = 0.0
END DO
```

Excerpt 4.10.9 Strength reduction: original indexing code.

```
DIMENSION A(N1,N2,N3)
DO I = 1,N2
  address = addr(A)     + (M1-1)*8       +
            (I-1)*8*N1 + (M3-1)*8*N1*N2
  memory(address) = 0.0
END DO
```

Excerpt 4.10.10 Strength reduction: brute force index translation.

```
DIMENSION A(N1,N2,N3)
address = addr(A) + 8*((M1 + N1*(I + N2*M3)) - (1 + N1*1+N2))
DO I = 1,N2
  memory(address) = 0.0
  address = address + N1
END DO
```

Excerpt 4.10.11 Strength reduction: optimized indexing.

The code generated within the loop, where most of the time is spent, is much simpler and faster than the brute force calculation.

4.10.2 Avoiding Floating Point Exponents

The two lines of code in Excerpt 4.10.12 are nearly identical, but the compiler will translate the first into Excerpt 4.10.13 and the second into Excerpt 4.10.14. Clearly, the second form will execute much slower than the first, so a simple rule to follow is to use integer exponents whenever possible.

```
X = A**3
Y = A**3.0
```

Excerpt 4.10.12 Integer and floating point exponentiation.

```
X = A*A*A
```

Excerpt 4.10.13 Compiler optimized integer exponentiation.

```
Y = EXP(LOG(A)*3.0)
```

Excerpt 4.10.14 Compiler produced floating-point exponentiation.

4.10.3 Common Subexpression Elimination

Compilers try to avoid calculating the same expression unnecessarily over and over when it can generate code to evaluate it just once, save it, and reuse it as needed. It can do this for expressions where none of the input values change from one use to another. Often the single result can be left in a register, saving additional access time. The compiler has to assume that the order and grouping (commutatively and associativity) of the subexpressions can be changed without altering the result (this is not always true with finite precision floating-point arithmetic). For example, Excerpt 4.10.15 might be compiled to Excerpt 4.10.16.

```
X = A*B*C*D
Y = E*B*F*D
```

Excerpt 4.10.15 Optimizable code containing redundant calculation.

```
T = B*D
X = A*C*T
Y = E*F*T
```

Excerpt 4.10.16 Compiler optimized code removing redundancies.

If T is saved in a register this optimization saves one multiply and two memory accesses. Of course, if B or D are modified between the set of X and Y, this optimization is not valid and will not be performed. The compiler can apply the distributive, associative, and commutative laws in a complex fashion to discover common subexpressions.

Note that array syntax may prevent the compiler from saving results in registers. The explicit loop code in Excerpt 4.10.17 can be transformed using a scalar T that is stored in a register after common subexpression elimination as seen in Excerpt 4.10.18.

If the temporary scalar T is stored in a register rather than memory, this saves one multiply and two memory accesses per loop iteration. If the user writes the code using array syntax rather than an explicit loop, the optimization looks similar to the scalar code above. But an array of length 100 must be allocated to hold the temporary T, which is now an array instead of a scalar or register. There is also a memory access for the store of the array T and the subsequent load, so 100 multiplies are saved but no memory accesses.

```
DIMENSION A(100), B(100), C(100), D(100), X(100), Y(100)
DO I = 1,100
  X(I) = A(I) * B(I) * C(I) * D(I)
  Y(I) = E(I) * B(I) * F(I) * D(I)
END DO
```

Excerpt 4.10.17 Optimizable code containing redundant calculation.

```
DO I = 1,100
  T = B(I) * D(I)
  X(I) = A(I) * C(I) * T
  Y(I) = E(I) * F(I) * T
END DO
```

Excerpt 4.10.18 Optimized code removing redundancies.

4.11 EXERCISES

4.1 Try the following example with your compiler:

```
DO I = 1,100
  DO J = 1,100
    DO K = 1,100
      A(I,J,K) = B(I,J,K) + C(I,J,K)
    ENDDO
  ENDDO
ENDDO
```

And then:

```
DO K = 1,100
  DO J = 1,100
    DO I = 1,100
      A(I,J,K) = B(I,J,K) + C(I,J,K)
    ENDDO
  ENDDO
ENDDO
```

Does your compiler perform this optimization automatically?

4.2 Given the Fortran 90 Array syntax example in this chapter, if the decomposition of parallel chunks were on the first and second dimension of the grid, at what sizes would the data being accessed in the code fit in level-2 cache? Assume that the level-2 cache is 512 KB and that the operands are 8-byte reals.

4.3 Why might the use of derived types degrade program performance?

4.4 Derived types do not always introduce inefficiencies, how would you rewrite the derived type in the example given in this chapter to allow for contiguous accessing of the arrays?

4.5 Would `X=Y**.5` run as fast as `X=SQRT(Y)`? What would be a better way of writing `X=Y**Z`? Try these on your compiler/system.

4.6 Why might the use of array syntax degrade program performance?

4.7 What constructs restrict the ability of the compiler to optimize the alignment of arrays in memory? What constructs gives the compiler the most flexibility in aligning arrays in memory?

4.8 What speedup factor is available from SSE instructions? How about GPGPUs?

4.9 Why is dependency analysis required to vectorize a loop?

4.10 Why might a `DO` loop get better performance than the equivalent array syntax?

4.11 When using array sections as arguments to a subroutine, what may degrade performance?

4.12 What is strength reduction? How can it be used to speed up array index calculations?

CHAPTER 5

Gathering Runtime Statistics for Optimizing

CONTENTS

5.1 FOREWORD BY JOHN LEVESQUE

While at Pacific Sierra Research we started developing an interactive code vectorizer/parallelizer called FORGE. What a wonderful piece of software. We would use runtime statistics gathered from running the application on an important problem, and then use that information for directing FORGE on identifying the most important parts of the application to examine for vectorizing/parallelizing. In addition, FORGE built a database representation of the program, so it could perform inter-procedural analysis. Even when subroutines and functions were found within a `DO` loop, it could perform parallel analysis. When it found something it did not understand, it would display that to the user and the user could over-ride FORGE's concerns and parallelize the loop. The output was OpenMP. This was the first instance of using runtime statistics to drive the optimization of an application.

This was also when I feel I made two of the biggest mistakes in my career. First, with respect to FORGE, I didn't understand how important it was to follow the Fortran standard. Second, we started our own company. FORGE only worked with Fortran 77 codes, and Fortran 90 killed us. Also we did not handle C, which started appearing in scientific applications. We moved into the 1990s with a new company without a thought that by the end of the decade the company would no longer exist.

5.2 INTRODUCTION

The first issue the reader must comprehend is that they are not measuring the performance of the application, they are measuring the operation of the application on an input problem. The problem being measured should be a problem of interest to be solved on a target system. So many are mislead by measuring a toy problem, and then working on optimizing the operation of the application on that problem – only to find out that the toy problem really was not typical of the important problems to be solved. The second issue the reader must comprehend is that premature estimation of what part of the application should be optimized can be a great waste of time, and gathering statistics on the operation of the application on a particular problem is very easy. Lastly, the reader should expect that numerous problem sets need to be instrumented to understand how the characteristics of the applications change given different inputs.

We will be using the performance toolkit from Cray Inc. While some of the capabilities of this toolkit are quite unique, most of the information we will be examining is available from numerous hardware and software vendors.

There are two major methods of instrumenting an application: sampling and tracing. In sampling runs, the profiler samples the program counter every so often. The default is typically around 1/100 of a second. This sampling interval can be modified, but 1/100 of a second seems to give good results. There are two major advantages to sampling. First, it has much lower overhead than tracing. Second, it can identify time used on a line within a routine. The second approach is tracing, where the profiler actually modifies the executable and keeps track of all of the routines called, who they are called from, and how much time is used within each call. An extension of tracing is to have the profiler also consider looping structures. While profiling looping structures introduces a significant amount of overhead, the information that can be obtained is extremely helpful and even necessary when looking to optimize for threading and vectorization.

More detailed information can be obtained with tracing. Sampling is preferred when the program being analyzed is very large with thousands of subroutines and functions. The examples we will use in this chapter are small enough that tracing will be fine.

If we are looking for parallelism within the MPI task, the best place is within looping structures. If an application is solving a complex science problem there will be plenty of looping structures within the application. We need to determine not only which are used most frequently but we also need to understand the loop iteration count. If a loop contains 90% of the computational time but has only one or two iterations depending upon the input problem, then it is not of interest.

5.3 WHAT'S IMPORTANT TO PROFILE

5.3.1 Profiling NAS BT

The following instrumentation is obtained from running one of the NAS Parallel Benchmarks: BT [1]. First we will examine the typical profile that lists the most important routines used in the execution of the sample problem.

TABLE 5.3.1 NAS BT basic tracing profile information.

Time%	Time	Imb. Time	Imb. Time%	Calls	Group Function PE=HIDE
100.0%	115.938333	--	--	167,965.0	Total
81.4%	94.387812	--	--	20,082.0	USER
28.6%	33.198440	0.905996	2.7%	6,275.0	z_solve_cell_
19.7%	22.798828	0.905880	3.8%	6,275.0	y_solve_cell_
18.1%	20.965508	0.398575	1.9%	6,275.0	x_solve_cell_
4.6%	5.367909	0.169083	3.1%	252.0	compute_rhs_
4.2%	4.910020	1.215859	19.9%	251.0	x_solve_
2.9%	3.359146	0.293244	8.0%	251.0	y_solve_
2.2%	2.603765	0.221997	7.9%	251.0	z_solve_
0.9%	1.036598	0.099347	8.8%	251.0	adi_
0.1%	0.147598	0.007259	4.7%	1.0	mpbt_
13.3%	15.367833	--	--	147,871.0	MPI
10.5%	12.116390	4.187081	25.7%	72,288.0	mpi_wait
1.5%	1.777612	0.469862	20.9%	37,656.0	mpi_isend
0.9%	1.073549	2.322925	68.5%	252.0	mpi_waitall
0.3%	0.396644	0.048068	10.8%	37,656.0	MPI_IRECV
5.3%	6.182627	--	--	11.0	MPI_SYNC
5.3%	6.159665	6.159561	100.0%	1.0	mpi_init_(sync)

Table 5.3.1 shows the top routines; the first column is the percentage of the total compute time. If this is an MPI job, this is averaged over all the MPI tasks. The second column is the number of samples, and the third column is imbalanced samples. That is, how much difference there is between the average and the min/max samples recorded across the MPI tasks. The fourth column is the percent of imbalance, and the last column is the name of the routine. The time is exclusive time. That is, it is the time excluding the routines called from within the routine that are traced. `Z_SOLVE`, `X_SOLVE`, and `Y_SOLVE` call low level routines that are not traced. We will see a breakout of these in later displays. So `Z_SOLVE_CELL` uses 28.6% of the total execution time of the run and has little load imbalance. There is also a summary of what MPI functions are using most of the time. In this case 10.5% of the time is spent in `MPI_WAIT`. Given the amount of load imbalance in the computational routines, much of this is probably due to some processors waiting for those processors that are using more time to execute the computational heavy routines.

TABLE 5.3.2 NAS BT profiling table containing loop statistics.

```
Table:  Inclusive and Exclusive Time in Loops (from -hprofile_generate)

  Loop|Loop Incl|     Time|Loop    | Loop| Loop| Loop|Function=/.LOOP[.]
  Incl|     Time|    (Loop| Hit    |Trips|Trips|Trips| PE=HIDE
 Time%|         |    Adj.)|        |  Avg|  Min|  Max|
|-----------------------------------------------------------
|98.2%|90.723414|0.000200|       1|200.0|  200|  200|mpbt_.LOOP.2.li.179
|30.3%|28.012002|0.002273|     201|  8.0|    8|    8|z_solve_.LOOP.1.li.32
|28.9%|26.710242|0.002115|     201|  8.0|    8|    8|y_solve_.LOOP.1.li.33
|28.4%|26.248891|0.002285|     201|  8.0|    8|    8|x_solve_.LOOP.1.li.34
|28.2%|26.046256|0.001245|   1,608| 20.0|   19|   21|z_solve_cell_.LOOP.1.li.413
|28.2%|26.045012|0.169854|  32,160| 20.0|   19|   21|z_solve_cell_.LOOP.2.li.414
|26.5%|24.487227|0.001034|   1,608| 20.0|   19|   21|y_solve_cell_.LOOP.1.li.413
|26.5%|24.486192|0.118459|  32,160| 20.0|   19|   21|y_solve_cell_.LOOP.2.li.414
|26.4%|24.385793|0.001132|   1,608| 20.0|   19|   21|x_solve_cell_.LOOP.1.li.418
|26.4%|24.384661|0.071892|  32,160| 20.0|   19|   21|x_solve_cell_.LOOP.2.li.419
|25.1%|23.236529|5.660015| 643,200| 20.0|   19|   21|z_solve_cell_.LOOP.6.li.709
|24.8%|22.878879|5.590501| 643,200| 20.0|   19|   21|x_solve_cell_.LOOP.5.li.703
|24.6%|22.782365|5.573173| 643,200| 20.0|   19|   21|y_solve_cell_.LOOP.6.li.708
| 3.1%| 2.844785|0.001925|     201|  8.0|    8|    8|x_solve_.LOOP.2.li.85
| 2.4%| 2.233616|0.000719|   1,608| 20.0|   19|   21|x_backsubstitute_.LOOP.5.li.371
| 2.4%| 2.232897|0.005968|  32,160| 20.0|   19|   21|x_backsubstitute_.LOOP.6.li.372
| 2.4%| 2.226929|0.111390| 643,200| 19.2|   19|   20|x_backsubstitute_.LOOP.7.li.373
```

Now consider Table 5.3.2 which has loop statistics. In this table the loops are once again ordered with the most time-consuming at the top. Now we have percentage of total compute time, inclusive and exclusive times, number of times the loop is encountered, and the average, minimum, and maximum loop iteration count. The loop notes the file and line number within the file. This is a little difficult to decipher since we would like to understand the loop nesting to determine which loops might be amenable to threading and what loops to consider for vectorization. Table 5.3.3 shows this – this is the loop structure within a call tree format.

From this display, we see how the computation proceeds. Within `Z_SOLVE_CELL`, three important loops with good iteration counts and using considerable time: the loops at lines 413, 414, and 709. Numerous routines are called from within these three loops. In particular the `MATVEC_SUB`, `MATMUL_SUB`, and `BINVCRHS` routines. From this table the first idea that comes to mind is to thread the loops at lines 413 and/or 414 and try to vectorize the loop at line 709. To vectorize we would probably have to inline the aforementioned routines. This is not a simple application, and the statistics that we have gathered immediately give us some ideas of how to get started on optimizing this application for the systems we mentioned in Chapter 2.

There are other features that are available from various profilers; hardware counters can be examined to determine the reason why certain loops may not be running as efficiently as possible. Table 5.3.4 shows the hardware counters from the NAS BT run discussed earlier. Unfortunately, KNL does not have the low level counters required to obtain this information.

TABLE 5.3.3 NAS BT profiling table with calltree.

```
Table:  Function Calltree View

  Time% |        Time |        Calls |Calltree
        |             |              | PE=HIDE

 100.0% | 66.820769 |           -- |Total
|-----------------------------------------------------------------
| 100.0% | 66.820750 |          2.0 |mpbt_
|  98.1% | 65.528378 |           -- | mpbt_.LOOP.2.li.179
3  97.7% | 65.282257 |        200.0 |  adi_
||||--------------------------------------------------------------
4|||  32.2% | 21.514638 |       400.0 |z_solve_
|||||-------------------------------------------------------------
5||||  29.1% | 19.473641 |            -- |z_solve_.LOOP.1.li.32
||||||------------------------------------------------------------
6|||||  26.2% | 17.529788 |     3,200.0 |z_solve_cell_
|||||||-----------------------------------------------------------
7||||||  13.8% |  9.233916 |            -- |z_solve_cell_.LOOP.1.li.413
8||||||        |           |               | z_solve_cell_.LOOP.2.li.414
9||||||  13.7% |  9.157534 |            -- |  z_solve_cell_.LOOP.6.li.709
||||||||||--------------------------------------------------------
10||||||||    5.9% |  3.941049 | 25,600,000.0 |matvec_sub_
10||||||||    5.8% |  3.860825 | 25,600,000.0 |matmul_sub_
10||||||||    2.0% |  1.355660 | 12,800,000.0 |binvcrhs_
||||||||||========================================================
7||||||  12.4% |  8.290755 |      3,200.0 |z_solve_cell_(exclusive)
|||||||===========================================================
6|||||   2.6% |  1.754893 |      2,800.0 |mpi_wait
||||||============================================================
5||||   3.0% |  2.036245 |            -- |z_solve_.LOOP.2.li.82
6||||   2.4% |  1.606913 |      3,200.0 | z_backsubstitute_
|||||=============================================================
4|||  30.8% | 20.553328 |       400.0 |x_solve_
|||||-------------------------------------------------------------
5||||  26.5% | 17.721693 |            -- |x_solve_.LOOP.1.li.34
||||||------------------------------------------------------------
6|||||  23.8% | 15.877105 |     3,200.0 |x_solve_cell_
|||||||-----------------------------------------------------------
7||||||  13.4% |  8.937288 |            -- |x_solve_cell_.LOOP.1.li.418
8||||||        |           |               | x_solve_cell_.LOOP.2.li.419
9||||||  13.3% |  8.870677 |            -- |  x_solve_cell_.LOOP.5.li.703
||||||||||--------------------------------------------------------
10||||||||    5.7% |  3.776963 | 25,600,000.0 |matmul_sub_
10||||||||    5.6% |  3.740623 | 25,600,000.0 |matvec_sub_
10||||||||    2.0% |  1.353092 | 12,800,000.0 |binvcrhs_
||||||||||========================================================
7||||||  10.4% |  6.934907 |      3,200.0 |x_solve_cell_(exclusive)
```

First we see that `Z_SOLVE_CELL` uses 28.6% of the time on KNL. The first five counters are obtained from hardware counters in the hardware, while the reminder of the statistics are derived from the available counters. In this case, the `LLC_MISSES` is level-2 cache, even if the run was using MCDRAM as cache. The LLC cache hit to miss ratio is okay, but unfortunately there is little reuse. Remember, if all the elements of a cache line are used once the line is fetched to cache, then the cache hit would be $7/8 = 87.5\%$. This is due to taking a miss for the first element of each cache line and then having the next 7 available in cache when required. Good reuse would give us a number > 98%.

TABLE 5.3.4 NAS Parallel Benchmark hardware counters on KNL.

```
USER / z_solve_cell_
----------------------------------------------------------------------------------
Time                                  28.6%     33.198440 secs
Imb. Time                              2.7%      0.905996 secs
Calls                          189.015 /sec      6,275.0 calls
UNHALTED_CORE_CYCLES                      49,321,883,033
UNHALTED_REFERENCE_CYCLES                 46,033,753,450
INSTRUCTION_RETIRED                       31,869,032,416
LLC_REFERENCES                             2,946,584,399
LLC_MISSES                                   149,491,215
LLC cache hit,miss ratio  94.9% hits              5.1% misses
Average Time per Call                          0.005291 secs
```

TABLE 5.3.5 NAS Parallel Benchmark hardware counters on Broadwell.

```
USER / z_solve_cell_
----------------------------------------------------------------------------------
Time                                                  25.1%      9.302125 secs
Imb. Time                                              2.3%      0.221623 secs
Calls                                674.577 /sec               6,275.0 calls
CPU_CLK_THREAD_UNHALTED:THREAD_P                         55,457,683,887
CPU_CLK_THREAD_UNHALTED:REF_XCLK                          2,000,820,106
DTLB_LOAD_MISSES:MISS_CAUSES_A_WALK                           6,243,105
DTLB_STORE_MISSES:MISS_CAUSES_A_WALK                          3,543,140
L1D:REPLACEMENT                                           4,204,226,248
L2_RQSTS:ALL_DEMAND_DATA_RD                               1,102,666,810
L2_RQSTS:DEMAND_DATA_RD_HIT                                 317,518,313
MEM_UOPS_RETIRED:ALL_LOADS                               22,979,057,828
FP_ARITH:SCALAR_DOUBLE                                    8,378,817,757
FP_ARITH:128B_PACKED_DOUBLE                                 158,841,150
FP_ARITH:256B_PACKED_DOUBLE                               8,871,772,296
CPU_CLK                                  2.77GHz
HW FP Ops / User time            4,749.838M/sec      44,183,589,241 ops   10.0%peak(DP)
Total DP ops                     4,749.838M/sec      44,183,589,241 ops
Computational intensity              0.80 ops/cycle            1.92 ops/ref
MFLOPS (aggregate)             2,968,648.81M/sec
TLB utilization                  2,348.10 refs/miss            4.59 avg uses
D1 cache hit,miss ratios            81.7% hits                18.3% misses
D1 cache utilization (misses)        5.47 refs/miss            0.68 avg hits
D2 cache hit,miss ratio             81.3% hits                18.7% misses
D1+D2 cache hit,miss ratio          96.6% hits                 3.4% misses
D1+D2 cache utilization             29.27 refs/miss            3.66 avg hits
D2 to D1 bandwidth             7,235.061MiB/sec   70,570,675,863 bytes
Average Time per Call                                      0.001482 secs
```

On the Xeon we can get much better data from the hardware counters. Table 5.3.5 presents the output from running on a Broadwell processor. On the Broadwell, `Z_SOLVE_CELL` uses 25.1% of the time. After the data concerning the load imbalance and number of calls, we have 11 outputs from the hardware counters. Following that data, we get the derived data that gives interesting data from the given hardware counters. In this case, we are running about 10%

of peak performance, which is pretty good. Computational intensity is showing that 1.9 operations are performed for each operand fetched from memory.

The important items in this report include the TLB utilization which is very good. The page size for this run was 4096 bytes or 512 8-byte words. The hardware counters indicate that each entry of the TLB is producing over 2300 references. That is, each entry is being used an average of 4 to 5 times. The level-1 (`D1`) cache utilization is poor and the combined level-1 and level-2 (`D1+D2`) at 96.6% is okay – though it could be better.

TABLE 5.3.6 NAS Parallel Benchmark MPI profile information.

```
Table:  MPI Message Stats by Caller (limited entries shown)

    MPI | MPI Msg Bytes | MPI Msg | MsgSz |  16<= |  4KiB<= | 64KiB<= |Function
    Msg |               |   Count |   <16 | MsgSz |   MsgSz |   MsgSz | Caller
 Bytes% |               |         | Count |  <256 |  <64KiB |   <1MiB |  PE=[mmm]
        |               |         |       | Count |   Count |   Count |

 100.0% | 799,526,816.0 | 9,661.0 |   5.0 |   2.0 | 4,221.0 | 5,433.0 |Total
|-----------------------------------------------------------------------------
| 100.0% | 799,526,700.0 | 9,654.0 |   0.0 |   0.0 | 4,221.0 | 5,433.0 |mpi_isend
||----------------------------------------------------------------------------
||  34.8% | 278,317,620.0 | 1,212.0 |   0.0 |   0.0 |     0.0 | 1,212.0 |copy_faces_
3|  34.6% | 276,939,810.0 | 1,206.0 |   0.0 |   0.0 |     0.0 | 1,206.0 | adi_
4|        |               |         |       |       |         |         |  mpbt_
|||||-------------------------------------------------------------------------
5||||  34.8% | 277,894,560.0 | 1,206.0 |   0.0 |   0.0 |     0.0 | 1,206.0 |pe.19
5||||  34.7% | 277,251,360.0 | 1,206.0 |   0.0 |   0.0 |     0.0 | 1,206.0 |pe.59
5||||  34.4% | 274,710,720.0 | 1,206.0 |   0.0 |   0.0 |     0.0 | 1,206.0 |pe.0
|||||=========================================================================
||  18.6% | 148,916,880.0 | 1,407.0 |   0.0 |   0.0 |     0.0 | 1,407.0 |x_send_solve_in
3|        |               |         |       |       |         |         | x_solve_
4|        |               |         |       |       |         |         |  adi_
5|        |               |         |       |       |         |         |   mpbt_
||||||------------------------------------------------------------------------
6|||||  18.6% | 148,916,880.0 | 1,407.0 |   0.0 |   0.0 |     0.0 | 1,407.0 |pe.0
6|||||  18.6% | 148,916,880.0 | 1,407.0 |   0.0 |   0.0 |     0.0 | 1,407.0 |pe.32
6|||||  18.6% | 148,916,880.0 | 1,407.0 |   0.0 |   0.0 |     0.0 | 1,407.0 |pe.63
```

Additionally, information about MPI message sizes are available as indicated in Table 5.3.6. In this table we see that `MPI_ISEND` is responsible for 100% of the messaging and 34.8% of the messages result from the call to `MPI_ISEND` from `COPY_FACES`. The first column gives the percentage of the total message traffic, the total bytes sent, the total number of messages, and then bins containing different size messages. In this case all of the messages were greater than 64 KB and less than 1 MB. Additionally the table gives us the MPI tasks that are responsible for the maximum, minimum, and average bytes transferred. An additional 18.6% of the calls to `MPI_ISEND` come from `X_SEND_SOLVE` which is called from `X_SOLVE` and its load imbalance information.

5.3.2 Profiling VH1

The next example, VH1, shows why we cannot rely on simple subroutine profiling. Table 5.3.7 is the profile of a popular astrophysics application VH1. Immediately we are drawn to **parabola_**, it uses 29.8% of the computation time, the only drawback is that it is called close to 32 million times, so the time per call is small. The other two important routines are only called 3.5 million times. There are loops somewhere; however, we cannot tell where. We need to profile the loops to get an idea of how we might optimize this application.

TABLE 5.3.7 VH1 simple subroutine profiling.

```
Table:  Profile by Function Group and Function

  Time%  |       Time  |     Imb.  |   Imb.  |       Calls  |Group
         |             |     Time  |  Time%  |              | Function
         |             |           |         |              |  PE=HIDE

 100.0%  | 194.481232  |        --  |      --  | 83189234.5  |Total
|-----------------------------------------------------------------------------
|  92.0%  | 179.010043  |        --  |      --  | 83156757.2  |USER
||----------------------------------------------------------------------------
||  29.8%  |  58.047136  | 0.809840  |    1.4%  | 31836672.0  |parabola_
||  16.5%  |  32.154693  | 1.814456  |    5.4%  |  3537408.0  |riemann_
||  10.6%  |  20.629389  | 0.194786  |    0.9%  |  3537408.0  |remap_
||   5.0%  |   9.645887  | 0.133508  |    1.4%  |  3537408.0  |evolve_
||   4.7%  |   9.218396  | 0.162341  |    1.7%  |  7074816.0  |paraset_
||   4.6%  |   8.857818  | 0.239105  |    2.6%  | 10612224.0  |forces_
||   4.4%  |   8.483707  | 0.308509  |    3.5%  | 10612224.0  |volume_
||   4.3%  |   8.417469  | 0.184731  |    2.2%  |  1768704.0  |ppmlr_
||   3.2%  |   6.185626  | 0.067999  |    1.1%  |  3537408.0  |states_
||   2.6%  |   5.015934  | 0.070624  |    1.4%  |  3537408.0  |flatten_
[...]
```

The following call tree in Table 5.3.8 has been obtained while profiling loops – this immediately shows us the important structure of the application. We see that 20.7% of the time is spent in a double nested loop that calls **parabola**, **riemann**, and **remap**. In this test these two loops are each 16 and they include a significant amount of computation – a good candidate for threading.

Table 5.3.9 shows the looping table which indicates some good lower level loops that are good candidates for vectorization. From this display we also see three other routines `SWEEPZ`, `SWEEPX1`, and `SWEEPX2` that have similar looping structures to `SWEEPY`. Further investigation shows all four routines call the important computational routines. Once again we needed the high level view of the loop structure in the application to determine how best to approach the optimization.

TABLE 5.3.8 VH1 profile with call tree.

```
Table:  Function Calltree View

  Time% |        Time |        Calls |Calltree
        |             |              | PE=HIDE

 100.0% | 194.481232 | 83189234.5 |Total
|-----------------------------------------------------
| 100.0% | 194.481188 | 83189034.5 |vhone_
||----------------------------------------------------
||  25.8% |  50.165489 | 13882484.0 |sweepy_
|||---------------------------------------------------
3||  20.7% |  40.195080 | 13854848.0 |sweepy_.LOOP.1.li.32
4||        |             |             | sweepy_.LOOP.2.li.33
5||        |             |             |  ppmlr_
|||||-------------------------------------------------
6||||    7.9% |  15.416847 |  5306112.0 |remap_
||||||------------------------------------------------
7|||||    4.8% |    9.274438 |  3537408.0 |parabola_
7|||||    2.3% |    4.454351 |   589568.0 |remap_(exclusive)
||||||================================================
6||||    5.1% |    9.878556 |   589568.0 |riemann_
6||||    2.4% |    4.706512 |  1768704.0 |parabola_
6||||    2.1% |    4.062440 |  2947840.0 |evolve_
6||||    1.0% |    1.980040 |  1179136.0 |states_
|||||=================================================
3||    2.3% |    4.550284 |      9212.0 |mpi_alltoall
3||    1.5% |    3.011031 |      9212.0 |sweepy_(exclusive)
3||    1.2% |    2.409094 |      9212.0 |mpi_alltoall_(sync)
```

TABLE 5.3.9 VH1 profile with loop statistics.

```
Table:  Inclusive and Exclusive Time in Loops (from -hprofile_generate)

 Loop Incl |   Time w/ |  Loop Hit |  Loop |  Loop |  Loop |Function=/.LOOP[.]
      Time |     Loops |           | Trips | Trips | Trips | PE=HIDE
     Total |           |           |   Avg |   Min |   Max |
|-------------------------------------------------------------------------------
| 56.388068 |  0.001549 |      2303 |  16.0 |    16 |    16 |sweepx2_.LOOP.1.li.28
| 56.386520 |  0.343100 |     36848 |  16.0 |    16 |    16 |sweepx2_.LOOP.2.li.29
| 55.689951 |  0.001792 |      2303 |  16.0 |    16 |    16 |sweepx1_.LOOP.1.li.28
| 55.688159 |  0.333424 |     36848 |  16.0 |    16 |    16 |sweepx1_.LOOP.2.li.29
| 48.171764 |  0.003925 |      4606 |  16.0 |    16 |    16 |sweepy_.LOOP.1.li.32
| 48.167840 |  0.194751 |     73696 |   4.0 |     4 |     4 |sweepy_.LOOP.2.li.33
| 47.462810 |  0.002631 |      2303 |  16.0 |    16 |    16 |sweepz_.LOOP.05.li.48
| 47.460178 |  0.186751 |     36848 |   4.0 |     4 |     4 |sweepz_.LOOP.06.li.49
| 26.596712 | 17.956960 |   1768704 | 135.0 |    71 |   263 |riemann_.LOOP.2.li.63
|  8.639752 |  8.639752 | 238775040 |  12.0 |    12 |    12 |riemann_.LOOP.3.li.64
|  7.715116 |  7.715116 |  15918336 | 132.0 |    66 |   264 |parabola_.LOOP.6.li.67
|  5.517043 |  5.517043 |  15918336 | 132.0 |    66 |   264 |parabola_.LOOP.7.li.75
|  4.691813 |  4.691813 |  15918336 | 133.0 |    67 |   265 |parabola_.LOOP.4.li.44
|  4.232493 |  4.232493 |   1768704 | 129.0 |    65 |   257 |remap_.LOOP.7.li.83
|  3.102253 |  3.102253 |  15918336 | 134.0 |    68 |   266 |parabola_.LOOP.2.li.30
|  2.995213 |  2.995213 |  15918336 | 132.0 |    66 |   264 |parabola_.LOOP.5.li.53
|  2.984931 |  2.984931 |  15918336 | 132.0 |    66 |   264 |parabola_.LOOP.8.li.84
|  2.231720 |  2.231720 |   1768704 | 135.0 |    71 |   263 |riemann_.LOOP.1.li.44
[...]
```

5.4 CONCLUSION

The sampling and tracing profiles give us a tremendous amount of useful information for identifying the bottlenecks in the program and potential looping structures we can investigate for vectorization and threading. Several items can be derived from the profiles generated. The most important elements to look for are

1. Load imbalance is without a doubt the most important feature to examine initially. If an application is significantly load imbalanced, then scaling to higher node counts will not be feasible. There can be several different causes of load imbalance. Most of the time it is due to poor decomposition of the application over the nodes in the system. Cray's Perftools will generate information about load imbalance and suggest a mapping of the MPI tasks to the nodes to maximize communication within a node and minimize communication between nodes. Additionally, load-imbalance can be caused by imbalance in the computation. A very good example is on Ocean model that retains all the grid blocks in the mesh regardless if there is ocean in the grid block or land. Those grid blocks that only have land do not have anything to compute and it is ideal to group those grid blocks that do not have any compute to perform with those that have a lot to compute. By doing that grouping, more memory bandwidth on the node can be utilized by the grid blocks that have a lot of compute.

2. Identification of looping structures in the application. When porting an application to KNL and/or a GPU system, it is important to understand the looping structure for identification of high-level loops that can be good candidates for parallelization and understanding the loop iteration counts to identify the best candidates for vectorization.

3. Identification of MPI message passing and where most of the communication is being performed. One example was the use of the MPI statistics to identify that numerous MPI all-to-alls were being called for obtaining global sums for a number of mesh quantities. By combining 6 MPI all-to-alls into a single MPI all-to-all that sums an array of 6 quantities, a significant amount of MPI time was eliminated.

4. Employing the sampling line profile, one can identify hot spots within a routine. This is an excellent way to identify loops that take a lot of time because they are not being vectorized.

Consider Table 5.4.1 for an example of element number 4 in the list above. Note that lines 77 and 78 used 5.2% of the total runtime. The compiler listing in Excerpt 5.4.1 indicates that this section of code contained a loop that was not vectorized.

TABLE 5.4.1 VH1 sampling profile with line information.

```
||  10.2% |   790.8 |     -- |     -- |riemann_
3|        |         |        |        | scratch/levesque/VH1_version1_orig/riemann.f90
||||-----------------------------------------------------------------------------
4|||   1.9% |   148.9 |  42.1 | 22.1% |line.77
4|||   3.3% |   258.8 |  67.2 | 20.6% |line.78
```

```
63.+1---< do l = lmin, lmax
64.+1 2-<   do n = 1, 12
65. 1 2       pmold(l) = pmid(l)
66. 1 2       wlft (l) = 1.0 + gamfac1*(pmid(l) - plft(l)) * plfti(l)
67. 1 2       wrgh (l) = 1.0 + gamfac1*(pmid(l) - prgh(l)) * prghi(l)
68. 1 2       wlft (l) = clft(l) * sqrt(wlft(l))
69. 1 2       wrgh (l) = crgh(l) * sqrt(wrgh(l))
70. 1 2       zlft (l) = 4.0 * vlft(l) * wlft(l) * wlft(l)
71. 1 2       zrgh (l) = 4.0 * vrgh(l) * wrgh(l) * wrgh(l)
72. 1 2       zlft (l) = -zlft(l) * wlft(l)/(zlft(l) - gamfac2*(pmid(l) - plft(l)))
73. 1 2       zrgh (l) =  zrgh(l) * wrgh(l)/(zrgh(l) - gamfac2*(pmid(l) - prgh(l)))
74. 1 2       umidl(l) = ulft(l) - (pmid(l) - plft(l)) / wlft(l)
75. 1 2       umidr(l) = urgh(l) + (pmid(l) - prgh(l)) / wrgh(l)
76. 1 2       pmid (l) = pmid(l)+(umidr(l)-umidl(l))*(zlft(l)*zrgh(l))/(zrgh(l)-zlft(l))
77. 1 2       pmid (l) = max(smallp,pmid(l))
78. 1 2       if (abs(pmid(l)-pmold(l))/pmid(l) < tol ) exit
79. 1 2->   enddo
80. 1---> enddo

ftn-6254 ftn: VECTOR RIEMANN, File = riemann.f90, Line = 64
  A loop on line 64 was not vectorized because of a recurrence on "pmid" line 77.
```

Excerpt 5.4.1 Compiler listing showing lack of vectorization of VH1 loop.

5.5 EXERCISES

5.1 Compare and contrast sampling and tracing profiles.

5.2 Why is it important to profile an application with a problem set that is representative of those used in production?

5.3 Why is it useful to obtain a profile which includes loop trip counts (consider `-hprofile_generate` or similar)?

5.4 What are hardware performance counters and what kinds of information do they provide?

5.5 *Construct a case study*: Select an application and problem set for further examination.

  a. Obtain a sampling profile of the application.

  b. Obtain a tracing profile of the application.

  c. Obtain a profile of the application including a call-tree.

d. Obtain a profile of the application including loop statistics.

e. Obtain a profile of the application with hardware counter data.

f. From the above profiles, identify:
   i. Degree of load-imbalance.
   ii. Important routines and looping structures.
   iii. Locations of any significant MPI communication.

CHAPTER 6

Utilization of Available Memory Bandwidth

CONTENTS

6.1 FOREWORD BY JOHN LEVESQUE

The name of the new company was Applied Parallel Research, the principals were myself, Gene Wagenbreth, and Richard Friedman. We also contracted work out to Einstein and Associates. The first big project we had in the new company was to develop a parallelizer for the Thinking Machine's Connection Machine. This product, called CMAX, would take as input Fortran and generate code that would run on a combination of the Connection Machine and the front-end machine. Scalar code was so slow on the Connection Machine, that it was better to run it on the front end system and transfer arrays back and forth between the front-end and the Connection Machine. Of course, the idea was to have most if not all of the major computation parallelized for the Connection Machine. We also got a contract with IBM to generate code for the initial Power 4, which was four Power systems sharing memory. However, the memory was not cache coherent. When we generated parallel code, the translated code had to handle cache coherency within the parallel loop in software. We started off pretty well, got to a high of 14 people, and then the lack of handling the new Fortran standard killed us. In 1998, I left Applied Parallel Research and went to work for Marc Snir at IBM Research.

6.2 INTRODUCTION

None of the existing architectures have enough bandwidth to main memory to satisfy most computational kernels. This has been the situation for the past 20 years and the problem is only getting worse. First we have seen cache hierarchies introduced to address this problem, and now we see the addition of 3D stacked memory technology (e.g., MCDRAM) that supplies higher bandwidth to the processors. Each of these hardware facilitators is non-trivial to employ for improving effective bandwidth. If an application does not utilize the cache hierarchy effectively, it will quickly become memory bandwidth limited and not realize its potential performance on the target system.

6.3 IMPORTANCE OF CACHE OPTIMIZATION

As chips built for high performance computing become denser, they become more complicated and more difficult to utilize effectively. The latest chips from all the vendors have a significant number of cores and more powerful instructions that can produce more results each clock cycle. To supply operands to this increased computational power, a very complicated memory hierarchy is used to mitigate a relatively slow bandwidth to main memory. For example, one of the options on the latest XC40 node from Cray Inc. has two Intel Haswell sockets at 2.3 GHZ, each with 16 cores, for a total of 32 cores on the node; each core is capable of producing 16 floating-point results (8 adds and 8 multiplys) each clock cycle, for a total FLOP (floating-point operations) rate of 1177.6 GFLOPS (billion floating-point operations per second). Each individual core can produce around 32.8 GFLOPS.

<table>
<tr><td colspan="2">$Core_0$</td><td colspan="2">$Core_1$</td><td colspan="2">$Core_i$</td><td colspan="2">$Core_n$</td></tr>
<tr><td>32KB L1D</td><td>32KB L1I</td><td>L1D</td><td>L1I</td><td>L1D</td><td>L1I</td><td>L1D</td><td>L1I</td></tr>
<tr><td colspan="2">256KB L2</td><td colspan="2">L2</td><td colspan="2">L2</td><td colspan="2">L2</td></tr>
<tr><td colspan="8">Shared L3 Cache</td></tr>
</table>

FIGURE 6.3.1 Haswell cache hierarchy.

A simple graphic of the Haswell memory hierarchy is given in Figure 6.3.1, it can supply different bandwidths depending upon where the operands reside, as summarized in Table 6.3.1.

TABLE 6.3.1 Haswell cache and memory bandwidth per core.

Cache/Memory	GB/Second/Core
L1	120
L2	50
L3	24
DDR3 – 2133	4

6.4 VARIABLE ANALYSIS IN MULTIPLE LOOPS

When multiple DO loops are present in an important kernel, benefit may be obtained by counting the amount of data used in the loops to see if strip mining can be used to reduce the memory footprint and obtain better cache performance. Consider the simple example presented in Excerpt 6.4.1 illustrating how strip mining can be used to achieve better performance through better cache utilization. In the first triple-nested DO loop, we used NZ*NY*NX elements of the A array. Then in the second triple-nested DO loop, the same array is used. Now this is a silly example, since it first multiples A by 2.0 and then multiples the result by 0.5. However, imagine multiple DO loops that perform an operation on an array or arrays in one loop and then other operations on the same arrays in other loops, similar to the previous example. How much data is loaded into cache in the first multiple-nested loop, and then what data is still around when we get to the second multiple-nested loop?

```
do iz = 1,nz
  do iy = 1,ny
    do ix = 1, nx
      a(ix,iy,iz) = a(ix,iy,iz)*2.0
    enddo
  enddo
enddo
do iz = 1,nz
  do iy = 1,ny
    do ix = 1, nx
      a(ix,iy,iz) = a(ix,iy,iz)*0.5
    enddo
  enddo
enddo
```

Excerpt 6.4.1 Example code before strip mining is applied.

In this case, NX=NY=300 and NZ=1024, so the total amount of data accessed in each loop is 703 MB, which is much larger than the caches available to a single processor. On KNL it would fit into a MCDRAM of 16 GB. However, consider that we are going to run this example on a number of cores on the node. Table 6.4.1 gives the amount of data referenced as the number of cores increase.

Since there is only 16 Gbytes of MCDRAM on KNL as soon as we use more than 32 cores, we will run out of MCDRAM. All the tests at 16 and below will run out of MCDRAM. Now consider the restructuring presented in Excerpt 6.4.2.

TABLE 6.4.1 Total amount of data accessed per loop.

Cores	NX	NY	NZ	MB
1	300	300	1024	703
2	300	300	1024	1406
4	300	300	1024	2812
8	300	300	1024	5625
16	300	300	1024	11250
32	300	300	1024	22500
64	300	300	1024	45000

```
do ic = 1,nc
  do iz = 1 + (ic-1)*nz/nc,ic*nz/nc
    do iy = 1,ny
      do ix = 1, nx
        a(ix,iy,iz) = a(ix,iy,iz)*2.0
      enddo
    enddo
  enddo
  do iz = 1 + (ic-1)*nz/nc,ic*nz/nc
    do iy = 1,ny
      do ix = 1, nx
        a(ix,iy,iz) = a(ix,iy,iz)*0.5
      enddo
    enddo
  enddo
enddo
```

Excerpt 6.4.2 Example code after strip mining is applied.

The outermost iz loop is being broken into nc chunks. As nc increases, we will see a point were we get cache reuse from the first loop to the second loop. For example, when nc is 256, each chunk of 300 × 300 × 4 = 2812.5 MB fits into the lower level caches on Xeon and KNL. Figure 6.4.1 shows the impact of chunking this loop on KNL running in Quadrant/Cache mode. Then we run the same test on KNL in SNC4/Cache mode, shown in Figure 6.4.2. Quadrant mode should do better since the bandwidth available to all cores is not limited by the quadrant containing the core. In SNC4 mode, all the cores within a single quadrant can only get at 1/4 of the bandwidth.

There is little difference from Quadrant to SNC4, though Quadrant is somewhat faster. We see a nice speedup of up to a factor of 2 as the blocking goes from 1 (no blocking) to 512, with the 64 to 128 getting most of the performance gain.

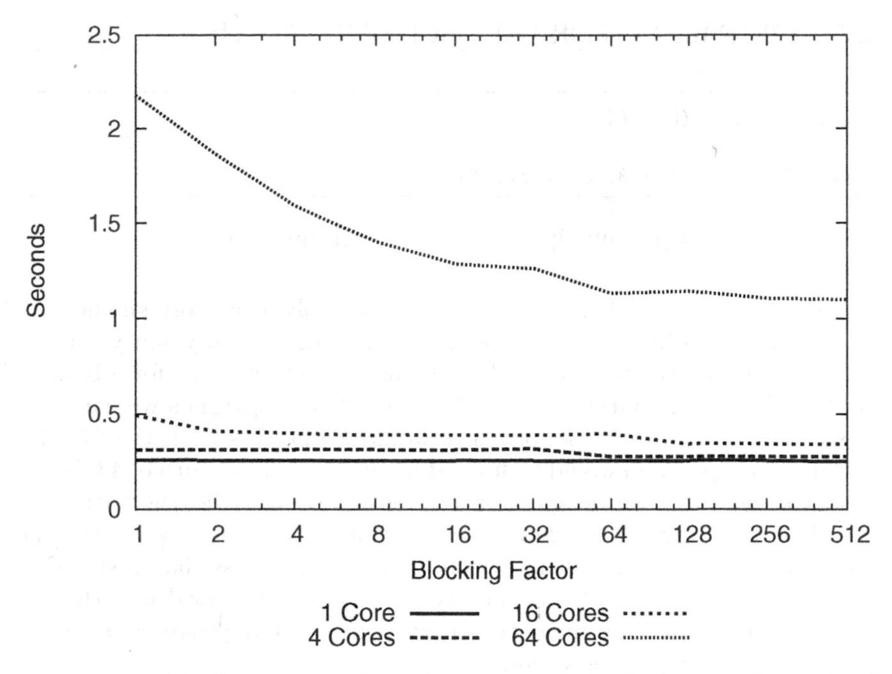

FIGURE 6.4.1 Performance of strip mine example in quadrant/cache mode.

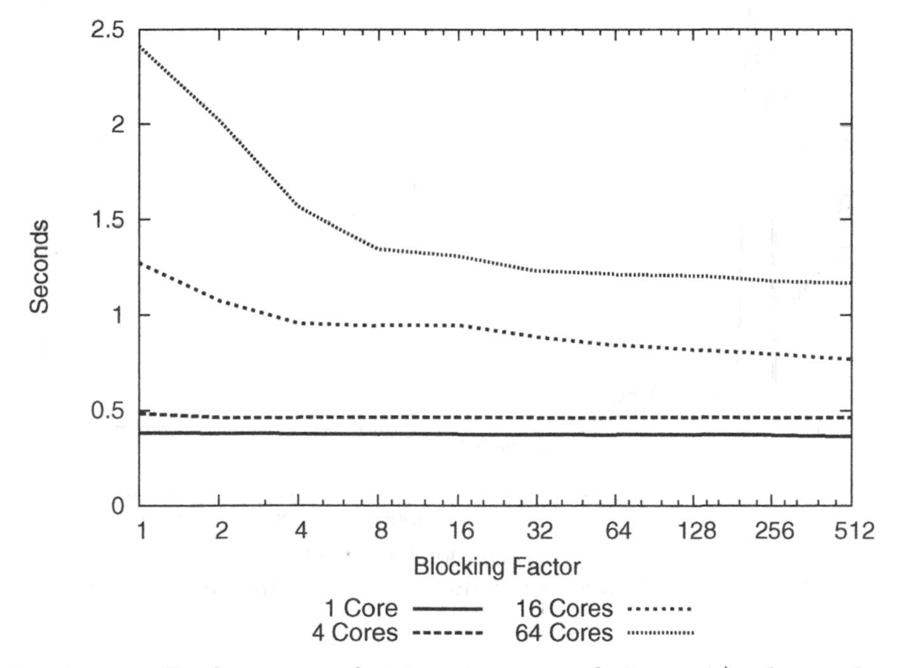

FIGURE 6.4.2 Performance of strip mine example in snc4/cache mode.

6.5 OPTIMIZING FOR THE CACHE HIERARCHY

```
REAL*8 X(N),Y(N),Z(N)

X(1:N) = Y(1:N) + 3.156 * Z(1:N)
```

Excerpt 6.5.1 Simple multiply and add example code.

Since Excerpt 6.5.1 has an add and a multiply, each core should be able to achieve 27 GFLOPS for this loop. However, the memory subsystem would have to supply two operands and store one each clock cycle, for a bandwidth of 3 * 8 bytes * 2.3 GHz = 81.5 GB/second. If the operands were all coming from main memory, the performance would be far less than the 27 GFlops, because the operation would be limited by memory bandwidth of 4 GB/sec. If, on the other hand, all the operands resided in level-1 cache, the operation has a sufficient bandwidth of 120 GB/sec to achieve the advertised performance. Interestingly, scalar processing which might produce less than a single result per clock cycle would also be limited by memory bandwidth if the results were coming from main memory and would achieve approximately the same performance as vector processing.

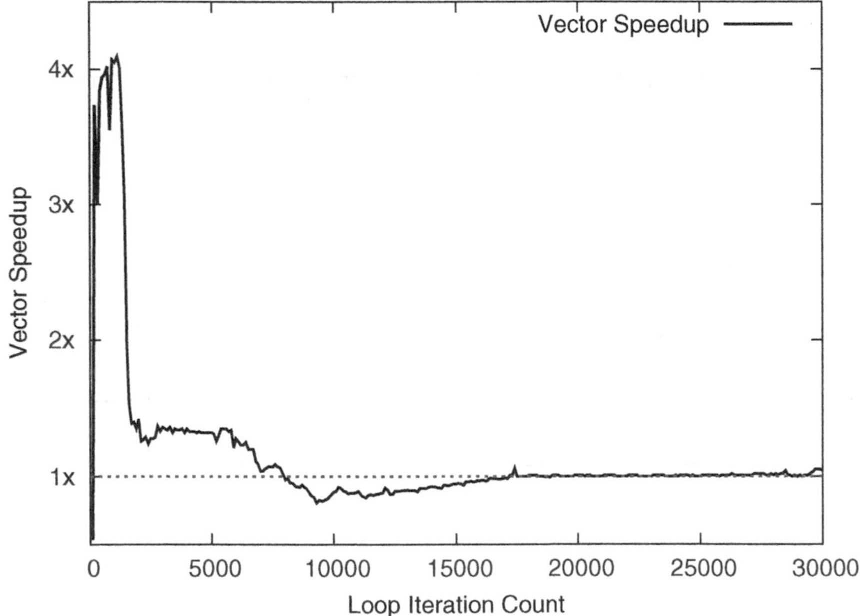

FIGURE 6.5.1 Vector speedup at increasing loop interation counts.

Figure 6.5.1 shows the ratio of vector performance to scalar performance for this operation. Each point represents the execution of the example for a particular loop iteration count. Each execution was performed 100 times so the time would measure the memory transfer time for operands already in cache for shorter loop iteration counts. Notice at around a loop iteration count of 1100, the vector performance drops significantly. At this point the total amount of memory required by the operation is 1100 * 8 bytes per operand * 3 operands = 26400 bytes. The level-1 cache on the system is 32768 bytes. We are at the area where hardware prefetching may spill us over into level-2 cache. Then at a loop iteration count around 9000, scalar performance actually starts beating vector performance. At this point we are spilling into level-3 cache. Then, as the loop iteration count increases, scalar and vector performance are comparable.

Now consider employing multiple threads on this operation. The example is actually the STREAM Triad kernel used for testing the bandwidth of various memory systems. Performance for this system using all the cores on the node has been shown to be 119 GB/second. This is the performance when all of the operands are coming from main memory and illustrates that only a few cores are required to saturate the bandwidth from main memory. If the operands reside in level-1 cache, the STREAM Triad would be an amazing 2608 GB/second.

Therefore, vectorization for employing the more powerful instructions and threading for utilizing more cores on the node is only successful if the algorithm is primarily using operands out of low-level caches. We conclude that the most important optimization for an application is to obtain the most efficient utilization of operands that are fetched into the low-level caches, and that threading and/or vectorization will not improve the performance if the code is memory bandwidth limited.

Before discussing examples in detail, it is important to understand how the cache works. All transfers from main memory to the processor are performed at the cache line length of 512 bits, 64 bytes, or 8 eight-byte operands. The cache is a very limited resource and can become abused in many different ways. First, there is a hardware prefetcher that determines the pattern of a series of memory accesses and strives to fetch ahead of the fetches generated by the compiler. In addition, the compiler itself may prefetch operands for the next pass through the inner loop with software prefetch instructions. While a user may be able to control the compiler, it is not possible to turn off hardware prefetching. The issue with prefetching is that the hardware and/or the compiler may fetch operands into the caches that are never used. Later we will talk about blocking, and if prefetching is not controlled, it can degrade any performance gain we achieve from blocking.

The ultimate goal of cache optimization is to achieve significant cache reuse. Cache reuse is a term used to describe how many operands are accessed in the computation. In the previous example, `Y` and `Z` are fetched, and if `N` is sufficiently long the first cache line of `Y(1:8)` and `Z(1:8)` brings up eight elements of each array and they are all used. However, there is no reuse. Each element is used once, and then there is no additional use of the element. When a loop has more array references than computations and there is no reuse, the code tends to be limited by memory bandwidth. There are examples of even worse cache utilization. Consider the operation in Excerpt 6.5.2.

```
REAL*8 X(N,N),Y(N,N),Z(N,N)

X(J,1:N) = Y(J,1:N) + 3.156 * Z(J,1:N)
```

Excerpt 6.5.2 Code showing poor cache utilization.

In this case we are accessing the arrays with a stride that is equal to `N`. If `N` is greater than or equal to 8, only one element of the cache line is used in the computation. The memory bandwidth required in this example is 8 times higher than the previous example.

Most applications today attempting to solve complex problems require a wide range of operations and memory access pattens. The ability to take advantage of the cache architecture is highly dependent upon the algorithm(s) being addressed. For example, many applications may utilize a solver, either iterative or direct. Solvers typically have to perform matrix-matrix products and matrix-vector products. Matrix-matrix products are referred to as level-3 BLAS and can be structured to have very high cache reuse. Matrix-vector products are level-2 BLAS operations and they have little cache reuse. The Triad example above is a level-2 BLAS operation and has no cache reuse. When using highly optimized libraries, the linear algebra operations tend to be finely tuned to take advantage of cache reuse whenever possible. The matrix product kernel in Excerpt 6.5.3 is an interesting example of reuse.

```
DO I = 1, N1
  DO J = 1, N2
    A(I,J) = 0.0
    DO K = 1, N3
      A(I,J) = A(I,J) + B(I,K) * C(K,J)
    END DO
  END DO
END DO
```

Excerpt 6.5.3 Simple matrix product implementation code.

While this is not the most optimal way to write a matrix product, it definitely shows the reuse that exists. For each `A(I,J)` point there are `N3` multiply/adds performed. Of course, the compiler will not fetch and store `A(I,J)` `N3` times, it will generate code that keeps it in a register and accumulate into that register. To illustrate the reuse, assume that `N2` and `N3` are both 4. If we unroll both inner loops the resulting loop would become that seen in Excerpt 6.5.4.

```
DO I = 1, N1
  A(I,1) = B(I,1) * C(1,1) + B(I,2) * C(2,1)
*        + B(I,3) * C(3,1) + B(I,4) * C(4,1)
  A(I,2) = B(I,1) * C(1,2) + B(I,2) * C(2,2)
*        + B(I,3) * C(3,2) + B(I,4) * C(4,2)
  A(I,3) = B(I,1) * C(1,3) + B(I,2) * C(2,3)
*        + B(I,3) * C(3,3) + B(I,4) * C(4,3)
  A(I,4) = B(I,1) * C(1,4) + B(I,2) * C(2,4)
*        + B(I,3) * C(3,4) + B(I,4) * C(4,4)
END DO
```

Excerpt 6.5.4 Unrolled matrix product implementation.

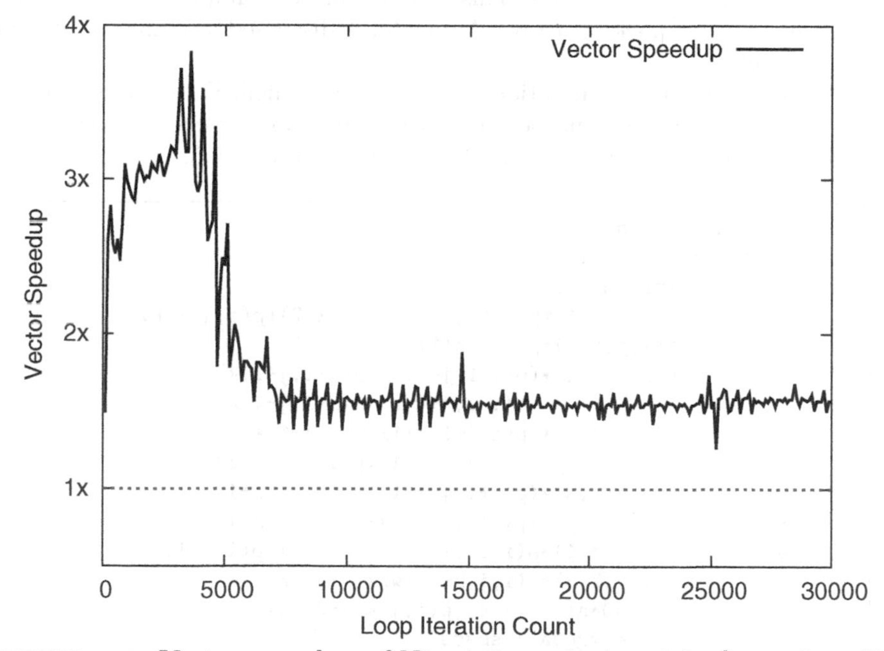

FIGURE 6.5.2 Vector speedup of N1×4 times 4×4 matrix for various N1 sizes.

Notice the reuse of B and C. For each fetch of a cache line of 8 elements of B and each store of a cache line of 8 elements of A, 96 (8*12) adds and 128 multiplies are performed. The optimized BLAS libraries utilize unrolling extensively when optimizing such kernels.

Also note that Figure 6.5.2 shows that vector performance is always faster than scalar performance. However, it is fastest when the operands all reside within level-1 cache. Actually the largest ratio between vector and scalar is at a loop iteration count of about 3000. At this point, the arrays reside within the level-1 and level-2 caches. An N of 3000 requires 3000 * 8 bytes per operand * 8 operands (A and B) + 8 bytes per operand * 16 operands (C) = 294912 bytes. As the loop iteration count increases, we see a degradation at 7000 where the operands are now coming from level-3 cache. At this point, the ratio stays constant with vector 1.5 times faster than scalar. While cache reuse improves the performance from vectorization, the example still degrades significantly when the operands are coming from the relativity low-bandwidth level-3 cache.

There is another potential issue using low-level caches effectively. The level-1 and level-2 caches on the Haswell chip are 8-way associative. This means that the cache is divided into eight associativity levels. Each level has 4096 bytes (32768/8) or 512 words of double precision. When two cache lines are separated by exactly 512 words, they cannot reside in the same associativity level. They must reside in different levels. This can become very important when trying to manage cache blocking arrays. Appendix A discussed associativity in much more detail.

Another important operation is a difference stencil. For example, the primary kernel in the Himeno benchmark is a nine-point stencil, which has the potential of excellent reuse as can be seen in Excerpt 6.5.5.

```
      DO 100 k=kst,kend
        DO 100 j=2,jmax-1
          DO 100 i=2,imax-1
            s0 = a(i,j,k,1)*p(i+1,j,k)+a(i,j,k,2)*p(i,j+1,k)
      1        + a(i,j,k,3)*p(i,j,k+1)
      3        + b(i,j,k,1)*(p(i+1,j+1,k)-p(i+1,j-1,k)
      *                    -p(i-1,j+1,k)+p(i-1,j-1,k))
      4        + b(i,j,k,2)*(p(i,j+1,k+1)-p(i,j-1,k+1)
      *                    -p(i,j+1,k-1)+p(i,j-1,k-1))
      5        + b(i,j,k,3)*(p(i+1,j,k+1)-p(i-1,j,k+1)
      *                    -p(i+1,j,k-1)+p(i-1,j,k-1))
      6        + c(i,j,k,1)*p(i-1,j,k)+c(i,j,k,2)*p(i,j-1,k)
      *        + c(i,j,k,3)*p(i,j,k-1)+wrk1(i,j,k)
               ss = (s0*a(i,j,k,4)-p(i,j,k))*bnd(i,j,k)
               wgosa = wgosa + ss*ss
               wrk2(i,j,k) = p(i,j,k) + omega * ss
  100       CONTINUE
```

Excerpt 6.5.5 Example nine-point stencil from Himeno.

Algorithmically, each point in the `p` array will be accessed nine times. Reusing the points `i-1`, `i` and `i+1` is trivial and most likely are in the same cache line. Reusing the points on different `i` lines is more challenging if the `i` line is long. Finally, reusing the points on each `k` plane is probably impossible unless one employs a blocking technique to perform the stencil in cubes that fit into low-level cache.

Additionally, the cache should be reserved for those arrays that do have reuse, and arrays that do not have reuse should be handled in such a way that they do not pollute the cache. For example, `a`, `b`, `c`, `bnd`, `wrk1` and `wrk2` have no reuse. To avoid having these arrays take up appreciable space in the lower-level caches, one would like to use non-temporal loads, which moves data directly from memory to the registers. Unfortunately, these systems only have non-temporal stores, which move an operand from registers to memory without polluting the cache. None of the systems discussed have non-temporal loads which would help a great deal in this case.

```
      DO kc=kst,kend,kchunk
         K1 = kc
         K2 = min(kc+kchunk-1,kend)
      DO jc=2,jmax-1,jchunk
         J1 = jc
         J2 = min(jc+jchunk-1,jmax-1)
      DO ic=2,imax-1,ichunk
         I1 = ic
         I2 = min(ic+ichunk-1,imax-1)
         DO k = K1, K2
         DO j = J1, J2
         DO I = I1, I2
         s0=a(i,j,k,1)*p(i+1,j,k)+a(i,j,k,2)*p(i,j+1,k)
     1          +a(i,j,k,3)*p(i,j,k+1)
     3          +b(i,j,k,1)*(p(i+1,j+1,k)-p(i+1,j-1,k)
     *                      -p(i-1,j+1,k)+p(i-1,j-1,k))
     4          +b(i,j,k,2)*(p(i,j+1,k+1)-p(i,j-1,k+1)
     *                      -p(i,j+1,k-1)+p(i,j-1,k-1))
     5          +b(i,j,k,3)*(p(i+1,j,k+1)-p(i-1,j,k+1)
     *                      -p(i+1,j,k-1)+p(i-1,j,k-1))
     6          +c(i,j,k,1)*p(i-1,j,k)+c(i,j,k,2)*p(i,j-1,k)
     *          +c(i,j,k,3)*p(i,j,k-1)+wrk1(i,j,k)
         ss=(s0*a(i,j,k,4)-p(i,j,k))*bnd(i,j,k)
         wgosa=wgosa+ss*ss
         wrk2(i,j,k)=p(i,j,k)+omega * ss
         END DO ; END DO ; END DO
     END DO ; END DO ; END DO
```

Excerpt 6.5.6 Cache-blocked nine-point stencil from Himeno.

In the next example, the extent of each of the three loops is 128. The total amount of data that is accessed in the execution of the loop is much larger than that contained in the low-level caches. We therefore investigate blocking the example to perform the computation in chunks that will fit into level-1 cache. To optimize the cache reuse of the Himeno kernel, one can introduce three additional loops as shown in Excerpt 6.5.6.

What should we make `kchunk`, `jchunk`, and `ichunk`? Level-1 cache has 32 KB. If this is double precision, then level-1 cache can hold 4096 operands. The minimum number of cache lines that we need for this computation is 9 (`I,J,K; I,J-1,K; I,J+1,K; I,J,K-1; I,J-1,K-1; I,J+1,K-1; I,J,K+1; I,J-1,K+1; I,J+1,K+1`) as well as 12 for `A,B,C,BND,WRK1`, and `WRK2`; each cache line contains 8 operands for a total of 864 operands. We will see that vectorization of the operations within the kernel will be more performant if the `I` length of the inner loop is larger. Consider having `ichunk` = 16, `jchunk` = 4, and `kchunk` =4. This will give us a total of 672 cache lines, or 5376 operands, required to perform the inner triple-nested loop and that will not fit into level-1 cache. Given the number of array references in the loop that do not have cache reuse blocking, this multinested loop will not obtain a performance increase.

Given the criticality of utilizing cache more efficiently, one must understand how much data can be held within the low level caches. Let's investigate a sample triple-nested loop for which we would like to maximize cache utilization. Excerpt 6.5.7 is a looping structure that might be encountered in a complex application.

What values for the loop bounds will fit best into a typical Xeon cache architecture? For example, on the Intel Haswell chip, the cache structures have the sizes shown in Table 6.5.1.

TABLE 6.5.1 Haswell cache sizes in different units.

Units	Level-1 Size	Level-2 Size	Level-3 Size
Bytes	32768	262144	8388608
64-bit operands	4096	32768	1048576
16-element vectors	256	2048	65536

Since vectorization of the innermost loop of a multinested loop is often desired, the size of the cache should be thought of as a multiple of vectors. As we see in the table, the number of vectors that fit into level-1 cache is only 256. Level-2 cache contains eight times that, and level-3 cache significantly more. Given the locality, we should strive to reside within the level-1 and level-2 cache. Level-3 is distributed all over the node and would cause cache coherency issues if the example is threaded.

In the proposed loop structure, the three inner `I` loops could achieve nice reuse in level-1 cache if and only if the number of iterations on the `I` loops were less than or equal to 16 and no more than 256 vectors were employed in the three loops. If the computation in the inner `I` loops contained references to `J`, `J-1`, and `J+1`, we would like to get some reuse from one iteration of the

J loop to the next. This may still be possible, or we may have to block up the I loop to achieve that reuse. Take, for example, one of the major kernels in the NPB suite, MG, presented in Excerpt 6.5.7.

```
      do i3=2,n3-1
         do i2=2,n2-1
            do i1=1,n1
               u1(i1) = u(i1,i2-1,i3) + u(i1,i2+1,i3)
     >                + u(i1,i2,i3-1) + u(i1,i2,i3+1)
               u2(i1) = u(i1,i2-1,i3-1) + u(i1,i2+1,i3-1)
     >                + u(i1,i2-1,i3+1) + u(i1,i2+1,i3+1)
            enddo
            do i1=2,n1-1
               r(i1,i2,i3) = v(i1,i2,i3)
     >                     - a(0) * u(i1,i2,i3)
     >                     - a(2) * ( u2(i1) + u1(i1-1) + u1(i1+1) )
     >                     - a(3) * ( u2(i1-1) + u2(i1+1) )
            enddo
         enddo
      enddo
```

Excerpt 6.5.7 MG kernel from the NPB suite before cache blocking.

In the innermost loop we see there are 13 vectors used, two of which are set and then used in the second innermost loop. Additionally, there is reuse on the second loop from indexing at i2-1, i2, and i2+1. Ideally we would like to take advantage of the reuse by ensuring that U(:,I2,-) is still in level-1 cache on the next two iterations of I2. How can we ensure that we not only get good vectorization, but also that I1 is short enough to get some reuse on subsequent iterations of the I2 loop by avoiding overflowing level-1 cache? Consider the rewrite presented in Excerpt 6.5.8.

So we would then have $13 * 4 = 52$ vectors in the inner I2, I1 loop. This is well under our limit of 256. Exercise caution when getting too close to the size of the level-1 cache. Hardware and software prefetching is performed by the hardware and compiler respectively. Cutting too close to the size of level-1 cache may result in cache overflow due to the prefetching. While software prefetching can be turned off, hardware prefetching is more of an issue.

We probably cannot do anything with the i3 loop in level-1 cache. However, we may be able to get some reuse by blocking i3 for level-2 cache. Since we have eight times the size in level-2 we could also block the i3 loop on 4 as we did the i2 loop. The results were not very encouraging on Haswell, as seen in Table 6.5.2.

Blocking did end up with a factor of two improvement when we employed OpenMP on the i3 loop. If we take a page from the WRF study mentioned in Chapter 2, we would want to collapse the outer two loops and create two-dimensional tiles. This restructuring is presented in Excerpt 6.5.9.

```
!$OMP PARALLEL DO PRIVATE(u1,u2,ii2s,ii2e,ii1s,ii1e)
      do i3=2,n3-1
        do ii2 = 2, n2-1, 4
         ii2s = ii2
         ii2e = min(n2-1,ii2s+4-1)
         do ii1 = 1, n1, 16
         ii1s = ii1
         ii1e = min(n1,ii1s+16-1)
         do i2=ii2s, ii2e
            do i1=ii1s,min(ii1e+1,n1)
               u1(i1) = u(i1,i2-1,i3) + u(i1,i2+1,i3)
     >                + u(i1,i2,i3-1) + u(i1,i2,i3+1)
               u2(i1) = u(i1,i2-1,i3-1) + u(i1,i2+1,i3-1)
     >                + u(i1,i2-1,i3+1) + u(i1,i2+1,i3+1)
            enddo
            do i1=max(ii1s,2),min(ii1e,n1-1)
               r(i1,i2,i3) = v(i1,i2,i3)
     >                     - a(0) * u(i1,i2,i3)
     >                     - a(2) * ( u2(i1) + u1(i1-1) + u1(i1+1) )
     >                     - a(3) * ( u2(i1-1) + u2(i1+1) )
            enddo
         enddo
      enddo
```

Excerpt 6.5.8 MG kernel from the NPB suite after cache blocking.

TABLE 6.5.2 Performance in seconds of MG kernel from NPB after cache blocking.

Threads	Original	Blocked
1	5.81E-03	6.56E-03
2	3.32E-03	3.61E-03
4	1.84E-03	1.96E-03
8	1.53E-03	1.12E-03
16	1.62E-03	6.85E-04

The IDX array contains the starting location of each of the tiles. In a 128-cubed grid using tiles of 8 by 8, we have a total of 256 tiles. Now the timings shown in Table 6.5.3 are a little better for lower thread counts. For this complex a computation, it is difficult to block for level-1 cache. What we are seeing is improvement in level-2 cache, especially as we get some contention with 8 to 16 threads trying to access operands in the level-3 cache at the same time.

```
!$OMP PARALLEL DO PRIVATE(u1,u2,i3s,i3e,i2s,i2e,ii1s,ii1e)
      do iii = 1, n2/ic*n3/ic
       i3s = max(2,((iii-1)/(n3/ic))*ic+1)
       i2s = max(2,mod(iii-1,n2/ic)*ic+1)
       i3e = min(i3s+ic-1,n3-1)
       i2e = min(i2s+ic-1,n2-1)
      do i3=i3s,i3e
       do i2 = i2s,i2e
         do ii1 = 1, n1, iic
         ii1s = ii1
         ii1e = min(n1,ii1s+iic-1)
            do i1=ii1s+1,min(ii1e+1,n1)
               u1(i1) = u(i1,i2-1,i3) + u(i1,i2+1,i3)
     >                + u(i1,i2,i3-1) + u(i1,i2,i3+1)
               u2(i1) = u(i1,i2-1,i3-1) + u(i1,i2+1,i3-1)
     >                + u(i1,i2-1,i3+1) + u(i1,i2+1,i3+1)
            enddo
            do i1=ii1s+1,min(ii1e,n1-1)
               r(i1,i2,i3) = v(i1,i2,i3)
     >                     - a(0) * u(i1,i2,i3)
     >                     - a(2) * ( u2(i1) + u1(i1-1) + u1(i1+1) )
     >                     - a(3) * ( u2(i1-1) + u2(i1+1) )
            enddo
         enddo
         enddo
         enddo
      enddo
```

Excerpt 6.5.9 MG kernel from the NPB suite after cache blocking and loop collapsing.

TABLE 6.5.3 Performance in seconds of MG kernel from NPB after cache blocking and loop collapsing.

Threads	Original	Blocked
1	5.81E-03	5.10E-03
2	3.32E-03	3.05E-03
4	1.84E-03	1.73E-03
8	1.53E-03	9.85E-04
16	1.62E-03	6.28E-04

6.6 COMBINING MULTIPLE LOOPS

Another valuable cache block strategy is to restructure multiple multinested DO loops into outer-loops with multiple inner loops. For example, consider the code in Excerpt 6.6.1 from the Leslie3d code, a computational fluid dynamics application. There are multiple restructurings that will improve this loop significantly. Earlier in this chapter, small loops were unrolled inside larger `DO` loops. Our first step will be to unroll `NN` inside of the `K`, `J`, and `I` looping structure, as shown in Excerpt 6.6.2, the rewrite of the first quadruply nested loop.

```
      DO NN = 1,5
        DO K = K1,K2
          KK  =  K + KADD
          KBD = KK - KBDD
          KCD = KK + KBDD
          DO J = J1,J2
            DO I = I1,I2
              QAV(I,J,K,NN) = R6I * (2.0D0  * Q(I,J,KBD,NN,N) +
>                                   5.0D0 * Q(I,J, KK,NN,N) -
>                                           Q(I,J,KCD,NN,N))
      END DO ; END DO ; END DO ; END DO
      DO K = K1,K2
        DO J = J1,J2
          DO I = I1,I2
            UAV(I,J,K) = QAV(I,J,K,2) / QAV(I,J,K,1)
      END DO ; END DO ; END DO
      DO K = K1,K2
        DO J = J1,J2
          DO I = I1,I2
            VAV(I,J,K) = QAV(I,J,K,3) / QAV(I,J,K,1)
      END DO ; END DO ; END DO
      DO K = K1,K2
        DO J = J1,J2
          DO I = I1,I2
            WAV(I,J,K) = QAV(I,J,K,4) / QAV(I,J,K,1)
      END DO ; END DO ; END DO
      DO K = K1,K2
        DO J = J1,J2
          DO I = I1,I2
            RKE  = 0.5D0 * (UAV(I,J,K) * UAV(I,J,K) +
>                          VAV(I,J,K) * VAV(I,J,K) +
>                          WAV(I,J,K) * WAV(I,J,K))
            EI = QAV(I,J,K,5) / QAV(I,J,K,1) - RKE
            TAV(I,J,K) = (EI - HFAV(I,J,K,1)) / HFAV(I,J,K,3)
            PAV(I,J,K) = QAV(I,J,K,1) * HFAV(I,J,K,4) * TAV(I,J,K)
      END DO ; END DO ; END DO
```

Excerpt 6.6.1 Example loops from original Leslie3d code.

In the next refactoring step, we will move the `K` and `J` loops to the outside, leaving the `I` loop on the inside. The result of this transformation is presented in Excerpt 6.6.3.

Several beneficial optimizations have been performed in this loop. One major optimization is that we have been able to eliminate several divides, by merging all the `I` loops together. This results in a savings in having to perform the `1/QAV(I,J,K,1)` only once instead of three times as in the original. This is due to the compiler replacing the divide with a reciprocal approximation, `1/QAV(I,J,K,1)` and then performing the divides by multiplying this by the numerator. This particular optimization may lose some precision and can be inhibited with appropriate compiler flags if needed.

```
      DO K = K1,K2
        KK  =  K + KADD
        KBD = KK - KBDD
        KCD = KK + KBDD
        DO J = J1,J2
          DO I = I1,I2
            QAV(I,J,K,1) = R6I * (2.0D0  * Q(I,J,KBD,1,N) +
     >                            5.0D0 * Q(I,J, KK,1,N) -
     >                                    Q(I,J,KCD,1,N))
            QAV(I,J,K,2) = R6I * (2.0D0  * Q(I,J,KBD,2,N) +
     >                            5.0D0 * Q(I,J, KK,2,N) -
     >                                    Q(I,J,KCD,2,N))
            QAV(I,J,K,3) = R6I * (2.0D0  * Q(I,J,KBD,3,N) +
     >                            5.0D0 * Q(I,J, KK,3,N) -
     >                                    Q(I,J,KCD,3,N))
            QAV(I,J,K,4) = R6I * (2.0D0  * Q(I,J,KBD,4,N) +
     >                            5.0D0 * Q(I,J, KK,4,N) -
     >                                    Q(I,J,KCD,4,N))
            QAV(I,J,K,5) = R6I * (2.0D0  * Q(I,J,KBD,5,N) +
     >                            5.0D0 * Q(I,J, KK,5,N) -
     >                                    Q(I,J,KCD,5,N))
      END DO ; END DO ; END DO
```

Excerpt 6.6.2 Optimized example loops from Leslie3d code.

```
      DO K = K1,K2
        KK  =  K + KADD
        KBD = KK - KBDD
        KCD = KK + KBDD
        DO J = J1,J2
          DO I = I1,I2
            QAV(I,J,K,1) = R6I * (2.0D0  * Q(I,J,KBD,1,N) +
     >                            5.0D0 * Q(I,J, KK,1,N) -
     >                                    Q(I,J,KCD,1,N))
            QAV(I,J,K,2) = R6I * (2.0D0  * Q(I,J,KBD,2,N) +
     >                            5.0D0 * Q(I,J, KK,2,N) -
     >                                    Q(I,J,KCD,2,N))
            QAV(I,J,K,3) = R6I * (2.0D0  * Q(I,J,KBD,3,N) +
     >                            5.0D0 * Q(I,J, KK,3,N) -
     >                                    Q(I,J,KCD,3,N))
            QAV(I,J,K,4) = R6I * (2.0D0  * Q(I,J,KBD,4,N) +
     >                            5.0D0 * Q(I,J, KK,4,N) -
     >                                    Q(I,J,KCD,4,N))
            QAV(I,J,K,5) = R6I * (2.0D0  * Q(I,J,KBD,5,N) +
     >                            5.0D0 * Q(I,J, KK,5,N) -
     >                                    Q(I,J,KCD,5,N))
            UAV(I,J,K) = QAV(I,J,K,2) / QAV(I,J,K,1)
            VAV(I,J,K) = QAV(I,J,K,3) / QAV(I,J,K,1)
            WAV(I,J,K) = QAV(I,J,K,4) / QAV(I,J,K,1)
            RKE   = 0.5D0 * (UAV(I,J,K) * UAV(I,J,K) +
     >                       VAV(I,J,K) * VAV(I,J,K) +
     >                       WAV(I,J,K) * WAV(I,J,K))
            EI = QAV(I,J,K,5) / QAV(I,J,K,1) - RKE
            TAV(I,J,K) = (EI - HFAV(I,J,K,1)) / HFAV(I,J,K,3)
            PAV(I,J,K) = QAV(I,J,K,1) * HFAV(I,J,K,4) * TAV(I,J,K)
      END DO ; END DO ; END DO
```

Excerpt 6.6.3 Further optimized example loops from Leslie3d code.

6.7 CONCLUSION

This chapter has dealt with the most important aspect of achieving a performance gain on hybrid multi/manycore systems. Basically, they are all limited by memory bandwidth, and whatever one can do to optimize the use of available bandwidth will pay off in terms of better utilization of memory, vectorization and threading. Optimizations for attaining better utilization of memory bandwidth should be the first step in an overall optimization plan. As was pointed out in this chapter, vectorization and threading will not achieve significant performance increases if the application is memory bandwidth bound. Once the application is optimized for memory utilization, then vectorization and threading have a better chance of delivering reasonable performance.

6.8 EXERCISES

6.1 We used 8-byte floating point operands in this chapter; how many operands could be placed in level-1, 2, and 3 cache if one used 4-byte floating point operands instead? How about 16-byte operands?

6.2 When optimizing an application for a vector system like KNL, what is the first issue you should address?

6.3 What is hardware prefetching? What is software prefetching? When a multi-loop is blocked for enhancing cache reuse, how can prefetching destroy performance?

6.4 Can one obtain a performance increase by blocking the following multi-nested loop? Why?

```
DO I = 1, N1
  DO J = 1, N2
    DO K = 1, N3
      A(I,J) = A(I,J) + B(I,J) * C(I,J)
END DO ; END DO ; END DO
```

6.5 Can the following loop-nests be strip-mined? Why?

```
do ic = 1,nc
  do iz = 1 + (ic-1)*nz/nc,ic*nz/nc
    do iy = 1,ny
      do ix = 1, nx
        a(ix,iy,iz) = a(ix,iy,iz)*2.0
  enddo ; enddo ; enddo
  do iz = 1 + (ic-1)*nz/nc,ic*nz/nc
    do iy = 1,ny
      do ix = 1, nx
        a(ix,iy,iz) = a(ix+1,iy,iz)*0.5
  enddo ; enddo ; enddo
enddo
```

6.6 When is blocking with tiles better than blocking with planes?

CHAPTER 7

Vectorization

CONTENTS

7.1 FOREWORD BY JOHN LEVESQUE

The next chapter in my life was forming groups to help application developers port and optimize their applications for the leading supercomputers of the day. At IBM Research I started the Advanced Computer Technology Center (ACTC). Our task was to assist potential customers in moving from the Cray T3E to the IBM SP. Since SGI had bought Cray Research and killed further development on the T3E line, we were extremely successful. This was when Luiz DeRose and I started working together. Luiz is now the Director of Cray Inc's Programming Environment group. Luiz developed some excellent tools that our customers could use for porting and optimizing their applications. At this time we also hired Gene Wagenbreth as a consultant.

Gene was running Applied Parallel Research. Our principal support was from Pete Ungaro, who was a VP of supercomputing sales at IBM. We were so successful that IBM Research gave the group of about 15 people a reward. This was the end of 2000, and IBM started de-emphasizing HPC and wanted to break up the group and have the IBM Research part of the team play a reduced role.

Well, I could not believe that IBM would kill a project that was so successful, so I quit and moved to Austin, Texas to join a start-up that didn't quite start up.

In September 2001, I joined Cray Inc., once again a separate company targeting HPC. The first good HPC system that Cray Inc. delivered was the Cray X1, a massively parallel processor with very powerful nodes comprised of 16 powerful vector units. The disadvantage of the system was its scalar performance. I remember a meeting at Sandia National Laboratory (SNL) with Bill Camp and Jim Tompkins, where we were discussing the merits of the X1. Bill made the comment that they would buy an X1 if I would guarantee that the compiler would vectorize their major codes, many of which were large C++ frameworks. Well, I couldn't do that.

Subsequent negotiations resulted in Cray Inc. getting a contract to build Red Storm – the most successful MPP system since the Cray T3E. Red Storm was designed by Bill Camp and Jim Tompkins, and I believe it saved the company. A couple years later, Pete Ungaro called me up and we talked for several hours about things at Cray Inc. Six months later Pete Ungaro joined Cray Inc, and I witnessed the rebirth of HPC with the follow-on to Red Storm, the Cray XT line. I moved from Austin to Knoxville, Tennessee and started Cray's Supercomputing Center of Excellence at Oak Ridge National Laboratory, to assist users in porting and optimizing their applications for the Cray X1 and the Cray XTs, the line that lead first to Jaguar and then to Titan, the fastest computers in the world for a time. The first person I hired for the ORNL Center of Excellence, Jeff Larkin, was a great hire. First, I have always thought that Eagle Scouts were smart, hard workers, and Jeff was also a standup comedian – what a combination. Jeff was a hard worker and gave great presentations.

7.2 INTRODUCTION

This chapter will look at several small examples that were run on Knight's Landing (KNL) and Intel Haswell (HSW) systems. As discussed in earlier chapters, the KNL is somewhat more complicated because of its memory hierarchy. Therefore, timings will vary depending upon where the operands are stored prior to computation. The test program that runs the examples always flushes the cache prior to running the example – this gives the most realistic performance as the operands have to come from main memory and the only reuse will result from the natural reuse of the operands within the loop. While KNL's memory hierarchy can deliver high bandwidth, its latencies are longer than Haswell's. Therefore, our approach in flushing cache will hurt KNL performance more than Haswell's. To flush the cache, we simply access a very large array that is not involved in the computation to wipe out any previous execution of the kernel. All examples in this section use the Cray Compiling Environemnt (CCE) 8.5 and Intel compiler 17.0.1

An important note – KNL will never beat Haswell on any of these examples. We are only comparing a single core of KNL to a single core of Haswell. A KNL node has twice the number of cores as a Haswell node. When KNL is within a factor of two in a core to core comparison, the performance per node should be similar.

In order to generate vector instructions, the compiler must identify a construct in the application that performs operations on a series of operands. This is typically a looping structure, a `DO` loop in Fortran or a `for` loop in C and C++. Some compilers even try to convert a backward `goto` into a loop. Once a looping structure is found, the compiler performs data dependency analysis on the loop. To convert a series of operations into a vector operation, the operations must work on a set of operands that have been computed prior to the loop and/or are computed within the loop. The common data dependency issue is a loop-carried dependency, such as that seen in Excerpt 7.2.1.

```
DO I = 2, N
  A(I) = A(I-1) + B(I) * C(I)
ENDDO
```

Excerpt 7.2.1 Small example loop with loop-carried dependency.

In this case, the product of `B(I) * C(I)` can be computed with a vector operation for all values of `I`; however, the sum of `A(I-1) + B(I) * C(I)` cannot be computed since the only value of `A(I-1)` that is available is `A(1)`. `A(2)` is computed the first pass through the loop and then is carried over to the second pass, and so on. The following section presents a set of rules that can be applied by a programmer to determine ahead of time if the loop can be vectorized by the compiler.

7.3 VECTORIZATION INHIBITORS

The following items will prevent vectorization by the compiler, so avoid them if at all possible.

1. Loop-carried dependencies.

   (a) Partial sums.

   ```
   DO I = 1, N
      SUM = SUM + B(I)
      A(I) = SUM
   ENDDO
   ```

Notice that the following is a special case of summation that is vectorized since only the last value of SUM is required.

```
DO I = 1, N
  SUM = SUM + B(I)
ENDDO
```

(b) Wrap around scalar.

```
XL = 0.0
DO I = 1, N
  XR = XL
  XL = SQRT(C(I)**2)*(D(I)**2)
  A(I) = (XR-XL)*B(I)
ENDDO
```

(c) Repeating store index.

```
DO I = 1, N
  A(MOD(I,10)) = A(MOD(I,10)) + SQRT(C(I)**2)*(D(I)**2)
ENDDO
```

2. Ambiguous subscripts that may lead to loop carried dependencies. For example, in Excerpt 7.2.1, replace the `A(I-1)` with `A(I+K)`. If `K` is less than 0, we have a loop carried dependency, if it is greater than or equal to 0, we do not have a loop carried dependency. Once again several cases of this may occur.

(a) Potential repeating store index.

```
DO I = 1, N
  A(IA(I)) = A(IA(I)) + SQRT(C(I)**2)*(D(I)**2)
ENDDO
```

(b) Potential overlap of index range on left and right side of =.

```
DO I = 1, N
  A(IA(I)) = A(I) + SQRT(C(I)**2)*(D(I)**2)
ENDDO
```

or

```
DO I = 1, N
  A(I) = A(IA(I)) + SQRT(C(I)**2)*(D(I)**2)
ENDDO
```

(c) Premature exit from the loop – or `EXIT` within a loop.

```
    DO I = 2, N
      DO INDEX = 2, TABLE_LENGTH
        IF(A(I).LT.TABLE(INDEX) THEN
          ALEFT = TABLE(INDEX-1)
          ARIGHT = TABLE(INDEX)
          GO TO 100
        ENDIF
      ENDDO
100   B(I) = ALEFT + (ARIGHT- A(I))/(ARIGHT-ALEFT)
    ENDDO
```

(d) Subroutine or function call within a loop.

```
DO I = 1, N
  A(I) = A(I) + SQRT(C(I)**2)*(D(I)**2)
  CALL CRUNCH(A(I))
ENDDO
```

7.4 VECTORIZATION REJECTION FROM INEFFICIENCIES

Things that may cause the compiler to reject vectorization due to inefficiencies:

1. Excessive gather/scatters within the loop.
2. Excessive striding within the loop.
3. Complex decision processes in the loop.
4. Loop that is too short.

Usually the compiler will alert the programmer of the potential inefficiency and suggest the use of a directive or pragma to over-ride the compiler's concern.

The compiler would like for the vector code that is generated after vectorizing a loop to run faster. Why would vector code run slower? Well, whenever there is indirect addressing and/or decision processes in a loop, there is overhead associated with the vectorization of the loop. Any indirect address on the right hand of the replacement sign would result in the overhead of fetching scalar elements into a contiguous array temporary. Consider the example loop shown in Excerpt 7.4.1.

```
DO I = 1, N
  A(IA(I)) = B(IB(I)) + C(IC(I))
ENDDO
```

Excerpt 7.4.1 Simple example loop with indirect addressing.

In this case a large amount of data must be moved to create the operands for a vector operation. That is overhead that would not be required if the loop was run in scalar mode. The benefit is a single vector add. Then the result must be scattered back out to memory. Ideally, when indirect addressing is encountered it only occurs in a few arrays and the amount of computation is much larger, so the overhead of data movement is amortized by the speedup in the computation. Striding is also an issue similar to indirect addressing. The next example in this chapter illustrates the poor performance when one tries vectorizing a loop which strides through or indirectly accesses an array. While strides and/or indirect addressing also slows down scalar code, there is additional overhead in the vectorized code for packing and unpacking the SIMD registers. This overhead is not incurred in the scalar code since scalar registers are employed.

7.4.1 Access Modes and Computational Intensity

Following is a simple set of loops that illustrate the performance difference from accessing arrays in different ways on KNL. Additionally, we introduce the concept of computational intensity. Computational intensity is simply the ratio of operations within the loop divided by the number of array references within the loop. As the computational intensity increases, the computation becomes less dependent upon memory bandwidth and performance will increase. Given different computational intensities, we measure the impact on performance of the different addressing modes for each loop. An example of this is presented in Figure 7.4.1. In the figure, the vector length is 461 for all runs, MFLOPS is the measurement, and higher is better.

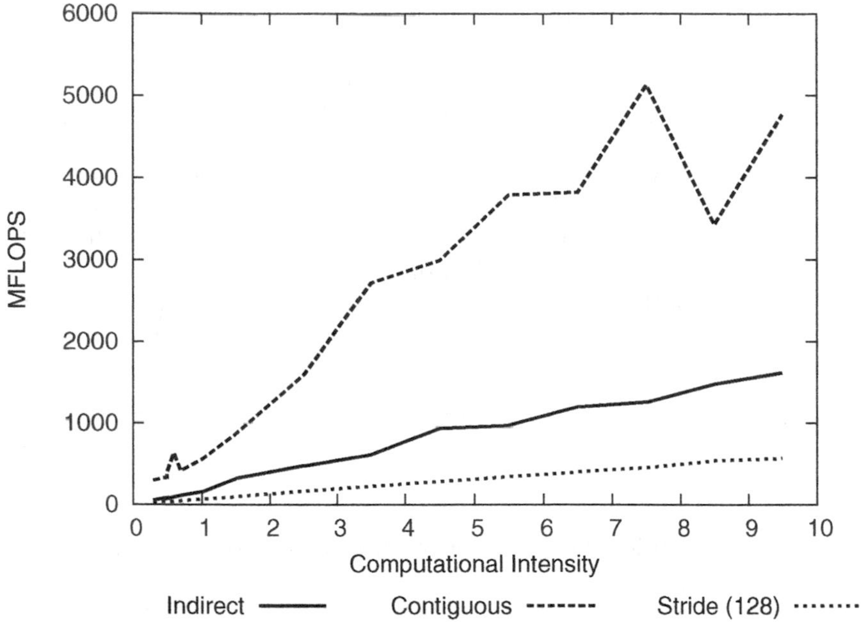

FIGURE 7.4.1 Performance of different memory access patterns with varying computational intensity.

In the figure, we see an increased performance as the computational intensity increases. The performance for the indirect addressing is less than the contiguous case, even when the computational intensity is large enough to amortize the memory bandwidth. The compiler does vectorize all three versions of each loop. Even though the code vectorizes, the overhead of dealing with the indirect addressing and striding degrades performance. The computational intensity in the indirect addressing examples does not consider the fetching of the address array, so technically there are more fetches than the contiguous case. For example, consider the loop presented earlier in Excerpt 7.4.1. While this computational intensity is plotted at 1/3, it is actually 1/6. In the strided case only one element is used in each cache because the stride is greater than 8. On the other hand, the indirect addressing examples may encounter several elements in a cache line.

As the computational intensity increases, the reliance on memory bandwidth decreases. A higher level of computational intensity comes from high-order polynomials and are representative of `log`, `exp`, `pow`; consider Table 7.4.1 for a comparison of different calculations with varying computational intensity.

TABLE 7.4.1 Example calculations with varying computational intensity.

Example Loop	Compute Intensity	Bottleneck
`A(:)= B(:)+C(:)`	0.333	Memory
`A(:)= C0*B(:)`	0.5	Memory
`A(:)= B(:)*C(:)+D(:)`	0.5	Memory
`A(:)= B(:)*C(:)+D(:)*E(:)`	0.6	Memory & Multiply
`A(:)= C0*B(:)+C(:)`	0.667	Memory
`A(:)= C0+B(:)*C1`	1.0	Still Memory
`A(:)= C0+B(:)*(C1+B(:)*C2)`	2.0	Add & Multiply
`A(:)= C0+B(:)*(C1+B(:)*(C2+B(:)*C3)`	3.0	Add & Multiply

Computational intensity is one of the important characteristics of an application. Today we see some applications that are able to achieve 60 to 70% of peak on a large MPP system while others are only achieving less than 5% of peak. The difference between these applications is computational intensity. The applications that are able to achieve a very high percentage of peak are typically dependent upon matrix multiply. For example, the HPL benchmark, used to determine the Top 500 list, is heavily dependent upon matrix $\times$ matrix multiply. On the Top 500 list, the percentages of peak that the large systems achieve is somewhere between 70 and 80% of peak. When the matrix multiply is performed by an optimal library, the computation intensity can approach 2. For example, later in this chapter we examine the traditional matrix multiply loop shown in Excerpt 7.4.2.

```
      DO 46030 J  = 1, N
       DO 46030 I = 1, N
        A(I,J) = 0.
46030 CONTINUE
      DO 46031   K = 1, N
       DO 46031  J = 1, N
        DO 46031 I = 1, N
         A(I,J) = A(I,J) + B(I,K) * C(K,J)
46031 CONTINUE
```

Excerpt 7.4.2 Code showing traditional matrix multiply loop.

What is the computational intensity for each invocation of the inner `DO 46031` loop? The `C(K,J)` variable is a scalar with respect to the loop, so we have two memory loads, `A(I,J)` and `B(I,J)`, one memory store `A(I,J)` and two floating point operations. The computational intensity is 2/3 (floating point operations divided by the number of memory operations). Is there any way the computational intensity of this loop can be improved? Consider the rewrite presented in Excerpt 7.4.3.

```
      DO 46032  J = 1, N
       DO 46032 I = 1, N
        A(I,J)=0.
46032 CONTINUE
C
      DO 46033   K = 1, N-5, 6
       DO 46033  J = 1, N
        DO 46033 I = 1, N
         A(I,J) = A(I,J) + B(I,K  ) * C(K  ,J)
     *                   + B(I,K+1) * C(K+1,J)
     *                   + B(I,K+2) * C(K+2,J)
     *                   + B(I,K+3) * C(K+3,J)
     *                   + B(I,K+4) * C(K+4,J)
     *                   + B(I,K+5) * C(K+5,J)
46033 CONTINUE
C
       DO 46034  KK = K, N
        DO 46034  J = 1, N
         DO 46034 I = 1, N
                 A(I,J) = A(I,J) + B(I,KK) * C(KK ,J)
46034 CONTINUE
```

Excerpt 7.4.3 Improved matrix multiply code.

Now look at the computational intensity of the inner `DO` loop. All of the `C`'s are scalars, we have 7 memory loads from all the `B`'s and `A(I,J)`, we have one store, and we have 12 floating point operations. The computational intensity of the rewritten loop is $12/8 = 1.5$. Optimized libraries take the unrolling of the `K` and `J` loop to extremes to reduce the dependency on memory bandwidth by raising the computational intensity.

In addition to unrolling to achieve higher performance for matrix multiply, cache blocking, to be discussed later, can also be employed to achieve very high computational performance.

7.4.2 Conditionals

Vectorization of loops that contain decision processes on today's systems may require so much overhead that the performance gain from vectorization is lost. The legacy vector computers had special hardware to handle vectorization of conditional blocks of code within loops controlled by `IF` statements. For example, the early Cray systems had vector mask registers. KNL and future generation Xeon chips have predicated execution which is a form of mask registers. However, Broadwell and Haswell do not have this hardware and the effect of vectorizing `DO` loops containing conditional code may be marginal on those systems. There are two ways to vectorize a loop with an `IF` statement. One is to generate code that computes all values of the loop index and then only store results when the `IF` condition is true. This is called a controlled store. For example, consider the code shown in Excerpt 7.4.4.

```
DO I=1,N
  IF(C(I).GE.0.0)B(I)=SQRT(A(I))
ENDDO
```

Excerpt 7.4.4 Simple example loop showing a conditional computation.

The controlled store approach would compute all the values for `I` $=$ $1, 2, 3, \cdots$,`N` and then only store the values where the condition `C(I).GE.0.0` is true. If `C(I)` is never true, this will give extremely poor performance.

If, on the other hand, a majority of the conditions are true, the benefit can be significant. Control stored treatment of `IF` statements has a problem in that the condition could be hiding a singularity. For example, consider the code presented in Excerpt 7.4.5.

```
DO I=1,N
  IF(A(I).GE.0.0)  B(I)=SQRT(A(I))
ENDDO
```

Excerpt 7.4.5 Example code showing a singularity being hidden by a conditional.

Here the `SQRT(A(I))` is not defined when `A(I)` is less than zero. Most smart compilers will handle this by artificially replacing `A(I)` with 1.0 whenever `A(I)` is less than zero, and take the `SQRT` of the resultant operand as shown in Excerpt 7.4.6.

```
DO I=1,N
  IF(A(I).LT.0.0)  TEMP(I)=1.0
  IF(A(I).GE.0.0)  TEMP(I)=A(I)
  IF(A(I).GE.0.0)  B(I)=SQRT(TEMP(I))
ENDDO
```

Excerpt 7.4.6 First method of fixing code with conditional singularity.

A second way is to compile code that gathers all of the operands for the cases when the condition is true, then perform the computation for the "true" path, and finally scatter the results out into the result arrays. Considering the previous `DO` loop (Excerpt 7.4.4), the compiler effectively performs the operations shown in Excerpt 7.4.7.

This is very ugly indeed. The first `DO` loop gathers the operands that are needed to perform the `SQRT`, this is called the gather loop. The second loop performs the operation on only those operands that are needed, having no problems with singularities. The third `DO` loop scatters the results back into the `B` array. The conditional store method has the overhead of the unnecessary operations performed when the condition is false. The gather/scatter approach avoids unnecessary computation, but it introduces data motion. Because of the overhead these methods introduce, most compilers will not try to vectorize loops containing conditional code. However, there are cases when it can be beneficial.

```
II = 1
DO I=1,N
  IF(A(I).GE.0.0) THEN
    TEMP(II) = A(II)
    II = II + 1
  ENDIF
ENDDO
DO I = 1,II-1
  TEMP(I) = SQRT(TEMP(I))
ENDDO
II = 1
DO I=1,N
  IF(A(I).GE.0.0) THEN
    B(I) = TEMP(II)
    II = II + 1
  ENDIF
ENDDO
```

Excerpt 7.4.7 Second method of fixing code with conditional singularity.

On older vector systems of the 1990s that had hardware gather/scatter support, many compilers would choose between the two approaches according to the number of operands that had to be accessed for the computation. Today, the controlled store approach is always used, employing the special predicated execution to avoid the singularities discussed above. Today's systems have limited gather/scatter approach which does not perform well enough to handle loops with conditional. The predicated execution approach will always be used.

```
DO I = 1, N
  IF(ENERGY(I).GT.0.0.AND.ENERGY(I).LT.2.5E-5) THEN
    [...]
  ENDIF
ENDDO
```

Excerpt 7.4.8 Example loop showing computation only performed for a specific energy range.

There is an additional way to handle conditionals that does not perform extra computation and does not move data. I call the approach "bit strings". It does take some setup; however, sometimes the setup is only needed once in a run and may be used many times. Say we have a computation that depends upon an energy range such as that seen in Excerpt 7.4.8. Once again this is a simple example to illustrate a point. Since the setup time for bit string may be time consuming, this would only be used when the computational loop(s) perform a lot of computation. It works even better when the bit string does not change, (e.g., material properties in a mesh). The setup for this loop is shown in Excerpt 7.4.9 and the computational loop is shown in Excerpt 7.4.10.

Years ago this optimization was performed for the material properties routines when vectorizing the CTH code, and it worked extremely well. This technique can be applied to computations where you have many different possible paths, especially if the setup only needs to be computed once.

```
ISTRINGS = 0
INSTRING = .FALSE.
ISS(:) = 0
ISE(:) = 0
DO I = 1,N
  LTEST = ENERGY(I).GT.0.0.AND.ENERGY(I).LT.2.5E-5
  IF(LTEST.AND..NOT.INSTRING) THEN
    ISTRINGS = ISTRINGS + 1
    ISS(ISTRINGS) = I
    INSTRING = .TRUE.
  ELSEIF(INSTRING.AND..NOT.LTEST) THEN
    ISE(ISTRINGS) = I - 1
    INSTRING = .FALSE.
  ENDIF
ENDDO
IF(ISE(ISTRINGS).EQ.0.AND.ISTRINGS.GE.1) ISE(ISTRINGS) = N
```

Excerpt 7.4.9 Example bit string setup loop.

```
DO II = 1, ISTRINGS
 DO I = ISS(ISTRINGS),ISE(ISTRINGS)
   [...]
 ENDDO
ENDDO
```

Excerpt 7.4.10 Example bit string computational loop.

7.5 STRIDING VERSUS CONTIGUOUS ACCESSING

In the next example, we show the difference between accessing the arrays with a stride versus accessing the arrays contiguously. The original code has some interesting twists, as seen in Excerpt 7.5.1.

```
61.  + 1 2-------------<        DO 41090 K = KA, KE, -1
62.    1 2 iVp---------<          DO 41090 J = JA, JE
63.  + 1 2 iVp i-------<            DO 41090 I = IA, IE
64.    1 2 iVp i                      A(K,L,I,J) = A(K,L,I,J) - B(J,1,I,k)*A(K+1,L,I,1)
65.    1 2 iVp i                 *    - B(J,2,I,k)*A(K+1,L,I,2) - B(J,3,I,k)*A(K+1,L,I,3)
66.    1 2 iVp i                 *    - B(J,4,I,k)*A(K+1,L,I,4) - B(J,5,I,k)*A(K+1,L,I,5)
67.    1 2 iVp i----->>> 41090 CONTINUE
```

Excerpt 7.5.1 Original strided code.

Notice that the compiler chose to vectorize the middle loop, even though the loop iteration count is less than or equal to 8 which the compiler could determine from the dimension statement in the test routine. The compiler is partially vectorizing the loop due to strides in the example. Several reasons for the compiler's loop selection follow:

1. There are some contiguous accesses on `J` while all the accesses on the inner loop are strided. The compiler is inverting the `I` and `J` loop (indicated by the `i`) and then partially vectorizing on `J` (indicated by `Vp`).

2. The outer loop has a loop-carried dependency on `K` (iterating with a step of -1) and is setting `A(K,L,I,J)` while using several elements of `A(K+1,L,I,:)`. Notice that the sign on the loop is negative 1. If J takes on the values 1-5 then the value of A(K,L,I,J) changes in each pass of the K loop and subsequent iterations of the K loop use the changed value, thus we have a loop carried dependency. Note that:
```
A(KA,  L,I,J) <== f(A(KA+1,L,I,J))
A(KA-1,L,I,J) <== f(A(KA,  L,I,J))
```

3. The SIMD instructions on KNL have a vector length of 8 DP floats, twice the width of the vector instructions on Haswell.

Intel's analysis is somewhat different than Cray's. Table 7.5.1 is the output from Intel's compilation, showing how it treated this loop-nest. It did not vectorize anything due to the excessive amount of striding in the loop.

TABLE 7.5.1 Intel compiler report from original strided code.

```
LOOP BEGIN at lp41090.f(61,10)
 remark #15344: loop was not vectorized: vector dependence prevents vectorization.
                First dependence is shown below. Use level 5 report for details
 remark #15346: vector dependence: assumed FLOW dependence between A(K,8,I,J) (64:13)
                and A(K+1,8,I,1) (64:13)
 LOOP BEGIN at lp41090.f(62,12)
  remark #15335: loop was not vectorized: vectorization possible but seems inefficient.
                 Use vector always directive or -vec-threshold0 to override
  LOOP BEGIN at lp41090.f(63,14)
   remark #15335: loop was not vectorized: vectorization possible but seems inefficient.
                  Use vector always directive or -vec-threshold0 to override
  LOOP END
 LOOP END
LOOP END
```

Since the longest loop is on `I`, it would be wise to have the longest loop as the innermost subscript. This would give contiguous accessing, and vectorization would be more profitable – of course, this is a change that would have to be performed throughout the application. The restructured version of this code is presented in Excerpt 7.5.2. In the restructured code, the compiler vectorizes the `I` loop – the `r2` indicates that it also unrolled the `I` loop by 2. Now we have good vectorization and should see good performance gain. And the Intel compiler also vectorizes the rewritten loop, as shown in Table 7.5.2.

```
 98.  + 1 2----------<         DO 41091 K = KA, KE, -1
 99.  + 1 2 3--------<           DO 41091 J = JA, JE
100.    1 2 3 Vr2----<             DO 41091 I = IA, IE
101.    1 2 3 Vr2                    AA(I,K,L,J) = AA(I,K,L,J) - BB(I,J,1,K)*AA(I,K+1,L,1)
102.    1 2 3 Vr2            *     - BB(I,J,2,K)*AA(I,K+1,L,2) - BB(I,J,3,K)*AA(I,K+1,L,3)
103.    1 2 3 Vr2            *     - BB(I,J,4,K)*AA(I,K+1,L,4) - BB(I,J,5,K)*AA(I,K+1,L,5)
104.    1 2 3 Vr2-->>> 41091 CONTINUE
```

Excerpt 7.5.2 Restructured contiguous code.

TABLE 7.5.2 Intel compiler report from restructured contiguous code.

```
LOOP BEGIN at lp41090.f(98,10)
   remark #15542: loop was not vectorized: inner loop was already vectorized
   LOOP BEGIN at lp41090.f(99,12)
      remark #15542: loop was not vectorized: inner loop was already vectorized
      LOOP BEGIN at lp41090.f(100,14)
         remark #15300: LOOP WAS VECTORIZED
      LOOP END
      LOOP BEGIN at lp41090.f(100,14)
      <Remainder loop for vectorization>
      LOOP END
   LOOP END
LOOP END
```

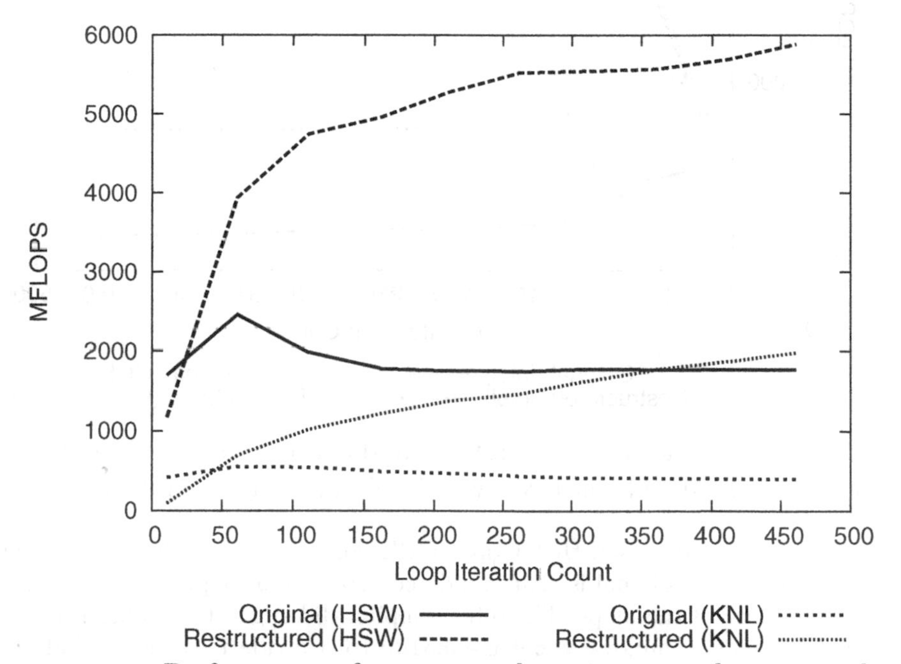

FIGURE 7.5.1 Performance of restructured contiguous code compared to original strided code.

Original and restructured performance is presented in Figure 7.5.1. Indeed we have a speedup of about 3 on Haswell and 4 to 5 on KNL. This is a good example which shows that while the vector length is four on Haswell and 8 on KNL, performance grows as the vector length increases and tends to flatten out around a length over 200 on Haswell and is still increasing at our last vector length of 461 on KNL. Longer vector lengths are beneficial even though the SIMD unit is small. This is primarily due to efficiencies from the hardware pre-fetcher and streaming the data through the cache/registers/functional units.

Figure 7.5.2 compares the Cray compiler and the Intel compiler running on a KNL system. The timings are reasonable given that both compilers vectorize the innermost loop in the restructured code and cannot do much with the original.

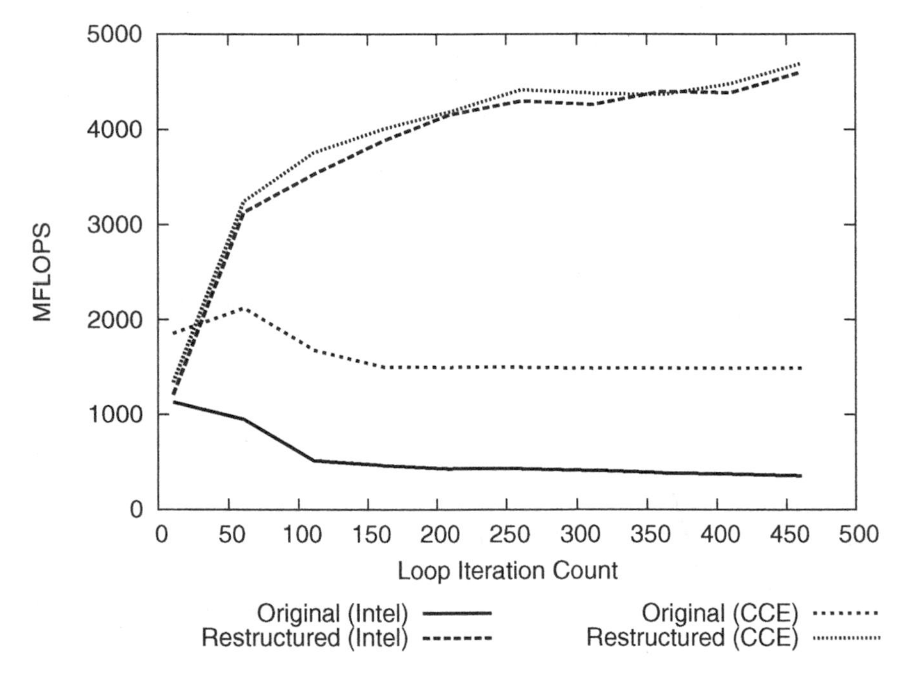

FIGURE 7.5.2 Performance of restructured contiguous code compared to original strided code on KNL with Intel and Cray compilers.

The principal reason that Cray's CCE compiler out-performed the Intel compiler on the original is that it did vectorize the `J` loop, and Intel did not vectorize any of the loops. The Intel compiler did suggest in Table 7.5.1 that the users could employ `-vec-threshold0` to override its concern and force vectorization.

The speedup on KNL is as much as a factor of 5. The vector version is only a factor of 3 slower than Haswell in vector mode and a factor of 4.5 in scalar mode. Why is KNL not faster and closer to Haswell? The principal reason for the difference is the clock cycle and the latencies in the memory hierarchy.

The Haswell system is 1.6 times faster in cycle time (2.3 GHz versus 1.4GHz). The latencies to memory are about another 1.5 difference, and finally the micro-architecture of the Xeon can handle more instructions per clock cycle than the KNL. As we vectorize the code, the factor of two advantage that KNL has over Haswell – vector length of 8 versus 4 – comes into play. As we utilize more cores we can approach and even improve upon the performance of the Haswell system. Are there more optimizations that can be applied to this loop? We will leave this as an assignment for the reader.

7.6 WRAP-AROUND SCALAR

The next example is illustrative of an old scalar optimization that inhibits vectorization. The original loop presented in Excerpt 7.6.1 computes and uses PF, and then the next pass through the loop uses the previous value. This is to avoid having to retest and set PF twice in each loop iteration.

```
 64.      1                PF = 0.0
 65.    + 1 2-----<        DO 44030 I = 2, N
 66.      1 2               AV    = B(I) * RV
 67.      1 2               PB    = PF
 68.      1 2               PF    = C(I)
 69.      1 2               IF ((D(I) + D(I+1)) .LT. 0.) PF = -C(I+1)
 70.      1 2               AA    = E(I) - E(I-1) + F(I) - F(I-1)
 71.      1 2             1       + G(I) + G(I-1) - H(I) - H(I-1)
 72.      1 2               BB    = R(I) + S(I-1) + T(I) + T(I-1)
 73.      1 2             1       - U(I) - U(I-1) + V(I) + V(I-1)
 74.      1 2             2       - W(I) + W(I-1) - X(I) + X(I-1)
 75.      1 2               A(I) = AV * (AA + BB + PF - PB + Y(I) - Z(I)) + A(I)
 76.      1 2-----> 44030 CONTINUE
```

Excerpt 7.6.1 Original code with scalar optimization using PF.

The compiler does not vectorize the loop due to a loop-carried dependency with PF. When the compiler analyzes the loop, it must be able to access all the elements of an array when that array is referenced. In this case, the values of PF, which represent a vector of values, are not available the way the loop is currently written. Intel also complains about the same dependency, as shown in Table 7.6.1.

TABLE 7.6.1 Intel compiler report from original loop-nest.

```
LOOP BEGIN at lp44030.f(65,10)
   remark #15344: loop was not vectorized: vector dependence prevents vectorization.
                  First dependence is shown below. Use level 5 report for details
   remark #15346: vector dependence: assumed FLOW dependence between PF (68:8)
                  and PF (67:8)
   remark #25456: Number of Array Refs Scalar Replaced In Loop: 10
LOOP END
```

In the rewrite presented in Excerpt 7.6.2, `PF` is promoted to an array and the compiler is given the information needed to compute all the elements of the array as a vector. Notice that `VPF(I-1)` is referenced in the setting of `A(I)`. However, that value has already been computed.

```
102.    1 Vr2---<          DO 44031 I = 2, N
103.    1 Vr2                AV      = B(I) * RV
104.    1 Vr2                VPF(I) = C(I)
105.    1 Vr2                IF ((D(I) + D(I+1)) .LT. 0.) VPF(I) = -C(I+1)
106.    1 Vr2                AA    = E(I) - E(I-1) + F(I) - F(I-1)
107.    1 Vr2              1       + G(I) + G(I-1) - H(I) - H(I-1)
108.    1 Vr2                BB    = R(I) + S(I-1) + T(I) + T(I-1)
109.    1 Vr2              1       - U(I) - U(I-1) + V(I) + V(I-1)
110.    1 Vr2              2       - W(I) + W(I-1) - X(I) + X(I-1)
111.    1 Vr2                A(I) = AV * (AA + BB + VPF(I) - VPF(I-1) + Y(I) - Z(I)) + A(I)
112.    1 Vr2---> 44031 CONTINUE
```

Excerpt 7.6.2 Restructured code with PF promoted to an array.

Figure 7.6.1 shows that this rewrite gives a small speedup on the Haswell system and a slightly better improvement on KNL. One of the issues here is the alignment of `VPF` is a bit tricky. Two views of `VPF` are accessed, one shifted one place to the left compared to the other. Unfortunately, the Intel compiler still does not vectorize this example, as can be seen from Table 7.6.2. If we then place the `!$OMP SIMD` directive in front of the restructured loop, we get it to vectorize the loops, and the answers are correct. Table 7.6.3 shows the Intel compiler output indicating vectorization.

TABLE 7.6.2 Intel compiler report from restructured loop-nest.

```
LOOP BEGIN at lp44030.f(101,10)
   remark #15344: loop was not vectorized: vector dependence prevents vectorization.
                  First dependence is shown below. Use level 5 report for details
   remark #15346: vector dependence: assumed FLOW dependence between VPF(I) (104:37)
                  and VPF(I-1) (110:8)
   remark #25456: Number of Array Refs Scalar Replaced In Loop: 12
LOOP END
```

TABLE 7.6.3 Intel compiler report from restructured code with OMP SIMD directive.

```
LOOP BEGIN at lp44030.f(102,10)
   remark #15301: OpenMP SIMD LOOP WAS VECTORIZED
LOOP END
```

Figure 7.6.2 compares the Cray compiler and Intel compiler on this example. Both compilers generate pretty much identical code. The only difference is that Intel needed the `!$OMP SIMD` directive.

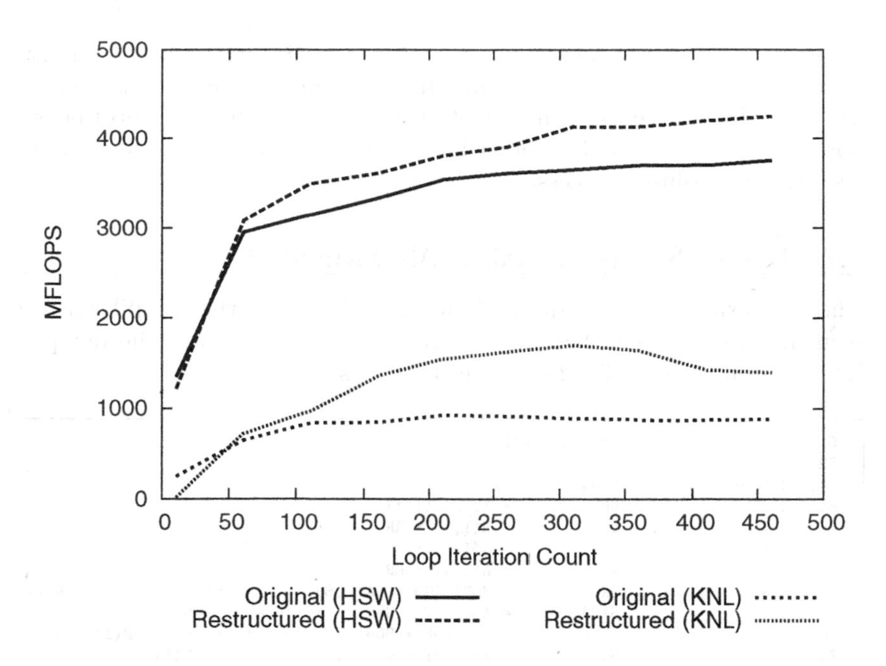

FIGURE 7.6.1 Performance of restructured code compared to original code with scalar optimization.

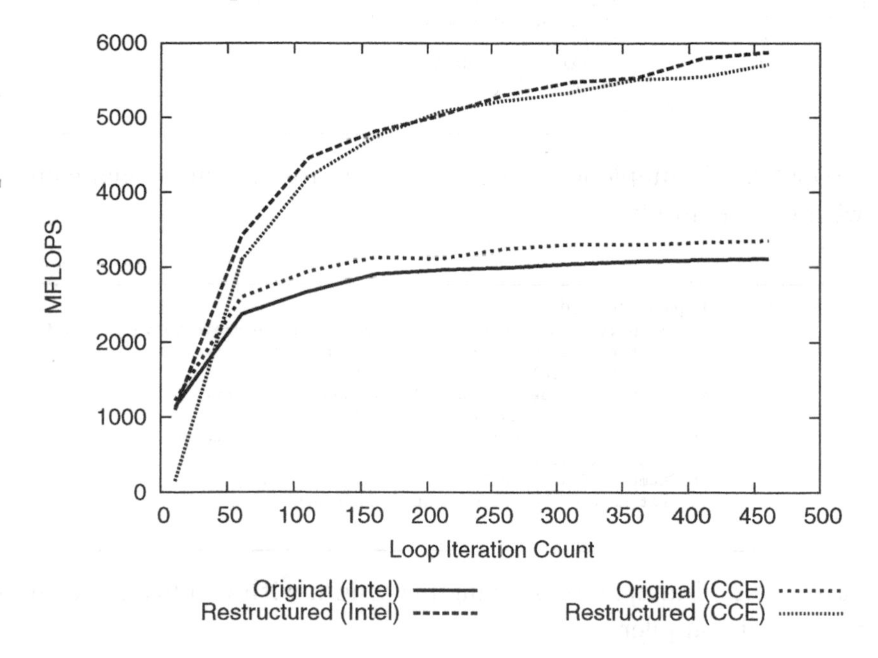

FIGURE 7.6.2 Performance of restructured code compared to original scalar code with Intel and Cray compilers.

Remember when using the `!$OMP SIMD` directive with Intel you must also have `-qopenmp` on the compile line. It is also important to understand that use of SIMD can generate incorrect answers if applied to a loop that should not be vectorized. SIMD is somewhat of a heavy hammer and should only be used when absolutely necessary.

7.7 LOOPS SAVING MAXIMA AND MINIMA

The next example occurs quite often. As indicated in the compiler listing file seen in Excerpt 7.7.1 the Cray compiler vectorizes it, and the optrpt from Intel seen in Excerpt 7.7.2 indicates that it is not vectorized.

```
 61.  1          C       THE ORIGINAL
 62.  1
 63.  1 V-----<         DO 44040 I = 2, N
 64.  1 V                RR          = 1. / A(I,1)
 65.  1 V                U           = A(I,2) * RR
 66.  1 V                V           = A(I,3) * RR
 67.  1 V                W           = A(I,4) * RR
 68.  1 V                SNDSP       = SQRT (GD * (A(I,5) * RR + .5* (U*U + V*V + W*W)))
 69.  1 V                SIGA        = ABS (XT + U*B(I) + V*C(I) + W*D(I))
 70.  1 V               *               + SNDSP * SQRT (B(I)**2 + C(I)**2 + D(I)**2)
 71.  1 V                SIGB        = ABS (YT + U*E(I) + V*F(I) + W*G(I))
 72.  1 V               *               + SNDSP * SQRT (E(I)**2 + F(I)**2 + G(I)**2)
 73.  1 V                SIGC        = ABS (ZT + U*H(I) + V*R(I) + W*S(I))
 74.  1 V               *               + SNDSP * SQRT (H(I)**2 + R(I)**2 + S(I)**2)
 75.  1 V                SIGABC      = AMAX1 (SIGA, SIGB, SIGC)
 76.  1 V                IF (SIGABC.GT.SIGMAX) THEN
 77.  1 V                IMAX        = I
 78.  1 V                SIGMAX      = SIGABC
 79.  1 V                ENDIF
 80.  1 V-----> 44040 CONTINUE
```

Excerpt 7.7.1 Example loop saving maximum with conditional compiled with Cray compiler.

```
LOOP BEGIN at lp44040.f(63,10)
   remark #15344: loop was not vectorized: vector dependence prevents vectorization
   remark #15346: vector dependence: assumed ANTI dependence between
                  sigmax (76:18) and sigmax (78:8)
   remark #15346: vector dependence: assumed FLOW dependence between
                  sigmax (78:8) and sigmax (76:18)
   remark #15346: vector dependence: assumed ANTI dependence between
                  sigmax (76:18) and sigmax (78:8)
   remark #25456: Number of Array Refs Scalar Replaced In Loop: 9
   remark #25015: Estimate of max trip count of loop=460
LOOP END
```

Excerpt 7.7.2 Example loop saving maximum with conditional compiled with Intel compiler.

```
 99.  1        C       THE RESTRUCTURED
100.  1
101.  1 V-----<       DO 44041 I = 2, N
102.  1 V              RR          = 1. / A(I,1)
103.  1 V              U           = A(I,2) * RR
104.  1 V              V           = A(I,3) * RR
105.  1 V              W           = A(I,4) * RR
106.  1 V              SNDSP       = SQRT (GD * (A(I,5) * RR + .5* (U*U + V*V + W*W)))
107.  1 V              SIGA        = ABS (XT + U*B(I) + V*C(I) + W*D(I))
108.  1 V            *               + SNDSP * SQRT (B(I)**2 + C(I)**2 + D(I)**2)
109.  1 V              SIGB        = ABS (YT + U*E(I) + V*F(I) + W*G(I))
110.  1 V            *               + SNDSP * SQRT (E(I)**2 + F(I)**2 + G(I)**2)
111.  1 V              SIGC        = ABS (ZT + U*H(I) + V*R(I) + W*S(I))
112.  1 V            *               + SNDSP * SQRT (H(I)**2 + R(I)**2 + S(I)**2)
113.  1 V              VSIGABC(I) = AMAX1 (SIGA, SIGB, SIGC)
114.  1 V-----> 44041 CONTINUE
115.  1
116.  1 A-----<       DO 44042 I = 2, N
117.  1 A              IF (VSIGABC(I) .GT. SIGMAX) THEN
118.  1 A               IMAX        = I
119.  1 A               SIGMAX      = VSIGABC(I)
120.  1 A              ENDIF
121.  1 A-----> 44042 CONTINUE
```

Excerpt 7.7.3 Example loop saving maximum with conditional split out into seperate loop compiled with Cray compiler.

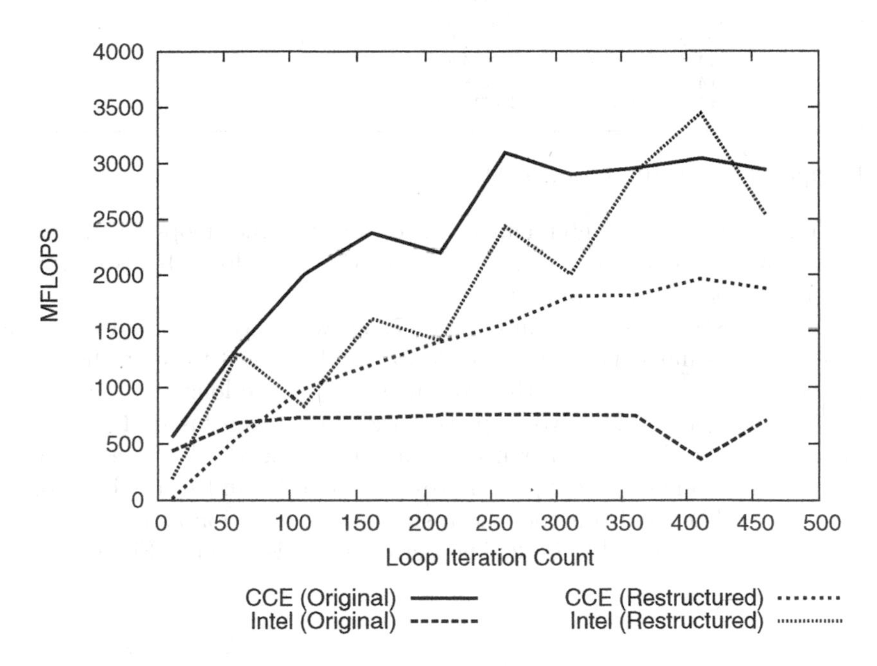

FIGURE 7.7.1 Performance of original and restructured max-save loop.

The restructuring seen in Excerpt 7.7.3 simply splits out the code finding the element that contains the maximum. Now both compilers vectorize the restructured version. However, we have forced some data motion by splitting the computation of the maximum out into its own loop with the temporary VSIGABC. Figure 7.7.1 shows the performance results from the Cray compiler and the Intel compiler on KNL. In this example, the original loop is faster on the Cray than the restructured. This is because the restructured has introduced more data motion from splitting the loop. Since the original was vectorized with the Cray compiler, it gave better performance. The A in the loop listing indicates that the compiler replaced that loop with a call to an optimized library routine.

7.8 MULTINESTED LOOP STRUCTURES

The following example is a contrived example that illustrates the best way to organize multinested loops. The original code in Excerpt 7.8.1 shows a triple-nested loop in which the longest loop is on the outside. The Cray compiler actually vectorizes the innermost K loop which is only five iterations and interchanges K with both the J and I loop. Intel chose to vectorize on the J loop after it unrolls the innermost K loop, as shown in Table 7.8.1.

```
49.      1 ir4----------<        DO 45020 I = 1, N
50.      1 ir4                    F(I) = A(I) + .5
51.    + 1 ir4 i--------<         DO 45020 J = 1, 10
52.      1 ir4 i                   D(I,J) = B(J) * F(I)
53.      1 ir4 i V------<          DO 45020 K = 1, 5
54.      1 ir4 i V                   C(K,I,J) = D(I,J) * E(K)
55.      1 ir4 i V---->>> 45020 CONTINUE
```

Excerpt 7.8.1 Original loop-nest code.

The Cray compiler interchanges the outer and inner loops on this example to increase the computational intensity of the inner loop. It also unrolls the outermost loop for the same reason.

In the restructured code in Excerpt 7.8.2, we try to break out any computation that is independent of other loops, and proceed to force the compiler to vectorize on I – the loop that we know will be the longest.

Now we get the first two loops vectorized on I and fused (f – fused with previous loop). The compiler insists on vectorizing the last statement on K due to the stride on C(K,I,J). However, it unrolls and fuses this loop with the others. The Intel compiler did a similar optimization as indicated in the diagnostics seen in Table 7.8.2. The performance is given in Figure 7.8.1.

TABLE 7.8.1 Intel compiler report from original loop-nest.

```
LOOP BEGIN at lp45020.f(49,10)
<Distributed chunk1>
   remark #25426: Loop Distributed (2 way)
   remark #15301: PARTIAL LOOP WAS VECTORIZED
LOOP END
LOOP BEGIN at lp45020.f(49,10)
<Remainder loop for vectorization, Distributed chunk1>
   remark #15301: REMAINDER LOOP WAS VECTORIZED
LOOP END
LOOP BEGIN at lp45020.f(49,10)
<Remainder loop for vectorization, Distributed chunk1>
LOOP END
LOOP BEGIN at lp45020.f(49,10)
<Distributed chunk2>
   remark #15542: loop was not vectorized: inner loop was already vectorized
   LOOP BEGIN at lp45020.f(51,11)
      remark #15300: LOOP WAS VECTORIZED
      remark #25456: Number of Array Refs Scalar Replaced In Loop: 5

      LOOP BEGIN at lp45020.f(53,12)
         remark #25436: completely unrolled by 5   (pre-vector)
      LOOP END
   LOOP END
   LOOP BEGIN at lp45020.f(51,11)
   <Remainder loop for vectorization>
      remark #25436: completely unrolled by 2
LOOP END ; LOOP END
```

TABLE 7.8.2 Intel compiler report from restructured loop-nest.

```
LOOP BEGIN at lp45020.f(74,10)
   remark #15300: LOOP WAS VECTORIZED
LOOP END
LOOP BEGIN at lp45020.f(74,10)
<Remainder loop for vectorization>
   remark #15301: REMAINDER LOOP WAS VECTORIZED
LOOP END
LOOP BEGIN at lp45020.f(74,10)
<Remainder loop for vectorization>
LOOP END
LOOP BEGIN at lp45020.f(78,10)
   remark #15542: loop was not vectorized: inner loop was already vectorized
   LOOP BEGIN at lp45020.f(79,11)
      remark #15300: LOOP WAS VECTORIZED
   LOOP END
   LOOP BEGIN at lp45020.f(79,11)
   <Remainder loop for vectorization>
      remark #15301: REMAINDER LOOP WAS VECTORIZED
   LOOP END
   LOOP BEGIN at lp45020.f(79,11)
   <Remainder loop for vectorization>
   LOOP END
   LOOP BEGIN at lp45020.f(79,11)
   <Peeled loop for vectorization>
      remark #25436: completely unrolled by 2
LOOP END ; LOOP END
```

```
 74.      1 Vr2----------<       DO 45021 I = 1,N
 75.      1 Vr2                   F(I) = A(I) + .5
 76.      1 Vr2----------> 45021 CONTINUE
 77.      1
 78.    + 1 f------------<       DO 45022 J = 1, 10
 79.      1 f Vr2--------<        DO 45022 I = 1, N
 80.      1 f Vr2                  D(I,J) = B(J) * F(I)
 81.      1 f Vr2------->> 45022 CONTINUE
 82.      1
 83.      1 iV-----------<       DO 45023 K = 1, 5
 84.    + 1 iV fi--------<        DO 45023 J = 1, 10
 85.    + 1 iV fi ir4----<         DO 45023 I = 1, N
 86.      1 iV fi ir4               C(K,I,J) = D(I,J) * E(K)
 87.      1 iV fi ir4-->>> 45023 CONTINUE
```

Excerpt 7.8.2 Restructured loop-nest code.

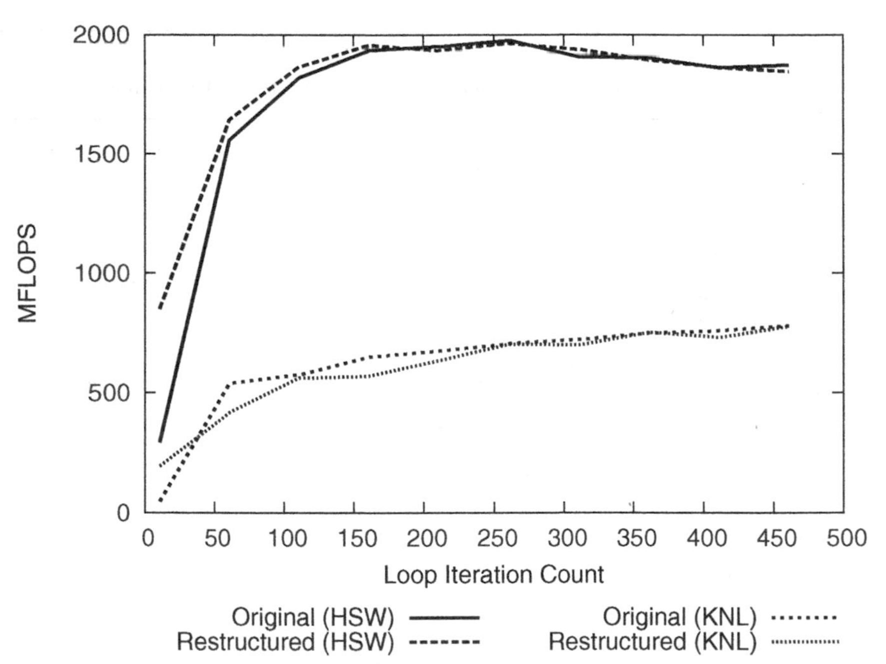

FIGURE 7.8.1 Performance of restructured loop-nest compared to original loop-nest using Cray compiler on Haswell and KNL.

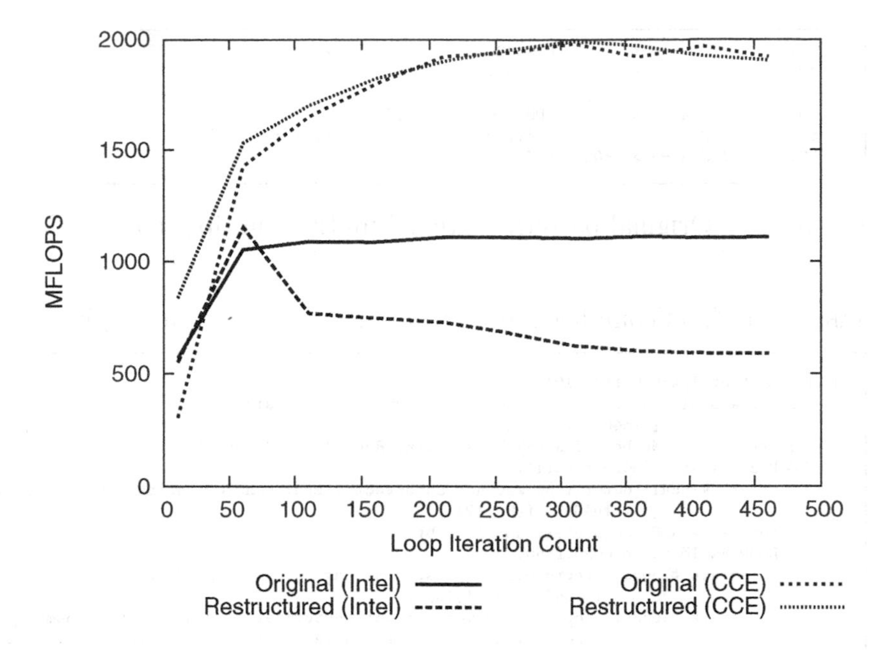

FIGURE 7.8.2 Performance of restructured loop-nest compared to original with Intel and Cray compilers.

This is another example that can probably be optimized further – another example for the reader. Perhaps a little better rewrite could help KNL a little more. Figure 7.8.2 shows reasonable performance for both compilers, since the rewrite helped both.

7.9 THERE'S MATMUL AND THEN THERE'S MATMUL

Matrix multiply is the best example to run on any supercomputer because it can be optimized so that it is not memory bandwidth limited. In fact, the Linpack benchmark is based on matrix multiply. Thus, compilers and library writers have fine-tuned matrix multiply to the maximum – as long as the matrices are square-ish. What about matrices that are not square? Well, most libraries do not do well on weird-shaped matrix multiply; that comes up a lot in the real world. Consider the original example code in Excerpt 7.9.1. Here we have a `Nx4` matrix times a 4x4 matrix. The compiler tried to be ultra-smart and replace the triple-nested loop with a call to DGEMM in the library. (`A` – pattern match). The Intel compiler did a lot of work on the example, as can be seen in Table 7.9.1. However, it did not vectorize anything, due to the short loops.

```
47.     1 A--------<        DO 46020 I = 1,N
48.     1 A 3------<         DO 46020 J = 1,4
49.     1 A 3                 A(I,J) = 0.
50.     1 A 3 4----<          DO 46020 K = 1,4
51.     1 A 3 4                A(I,J) = A(I,J) + B(I,K) * C(K,J)
52.     1 A 3 4-->>> 46020 CONTINUE
```

Excerpt 7.9.1 Original pattern-matched matrix multiply code.

TABLE 7.9.1 Intel compiler report from original matrix multiply code.

```
LOOP BEGIN at lp46020.f(47,10)
   remark #15541: outer loop was not auto-vectorized: consider using SIMD directive
                  [ lp46020.f(50,12) ]
   remark #25456: Number of Array Refs Scalar Replaced In Loop: 16
   LOOP BEGIN at lp46020.f(48,11)
      remark #15541: outer loop was not auto-vectorized: consider using SIMD directive
                     [ lp46020.f(50,12) ]
      remark #25436: completely unrolled by 4
      LOOP BEGIN at lp46020.f(50,12)
         remark #25085: Preprocess Loopnests: Moving Out Load and Store
                        [ lp46020.f(51,10) ]
         remark #15335: loop was not vectorized: vectorization possible but seems
                        inefficient. Use vector always directive or -vec-threshold0
                        to override
         remark #25436: completely unrolled by 4
      LOOP END
      LOOP BEGIN at lp46020.f(50,12)
      LOOP END
      LOOP BEGIN at lp46020.f(50,12)
      LOOP END
      LOOP BEGIN at lp46020.f(50,12)
      LOOP END
   LOOP END
LOOP END
```

```
71.     1 V--------<        DO 46021 I = 1, N
72.     1 V                  A(I,1) = B(I,1) * C(1,1) + B(I,2) * C(2,1)
73.     1 V               *         + B(I,3) * C(3,1) + B(I,4) * C(4,1)
74.     1 V                  A(I,2) = B(I,1) * C(1,2) + B(I,2) * C(2,2)
75.     1 V               *         + B(I,3) * C(3,2) + B(I,4) * C(4,2)
76.     1 V                  A(I,3) = B(I,1) * C(1,3) + B(I,2) * C(2,3)
77.     1 V               *         + B(I,3) * C(3,3) + B(I,4) * C(4,3)
78.     1 V                  A(I,4) = B(I,1) * C(1,4) + B(I,2) * C(2,4)
79.     1 V               *         + B(I,3) * C(3,4) + B(I,4) * C(4,4)
80.     1 V--------> 46021 CONTINUE
```

Excerpt 7.9.2 Restructured vectorized matrix multiply code.

In the rewrite presented in Excerpt 7.9.2, the two loops are manually unrolled by 4. Now there is a very nice compute-intensive vector loop, and the compiler simply vectorizes it. The Intel compiler handles the restructured very well, as can be seen in Table 7.9.2.

TABLE 7.9.2 Intel compiler report from restructured, vectorized matrix multiply code.

```
LOOP BEGIN at lp46020.f(71,10)
   remark #15300: LOOP WAS VECTORIZED
   remark #25456: Number of Array Refs Scalar Replaced In Loop: 12
LOOP END
```

As can be seen in Figure 7.9.1, we achieve a factor of 1.5 to 2 on the Haswell system and a factor of 4 on KNL. The saw-toothed curve on Haswell is due to the library trying to unroll the matrix multiply without knowing the bounds on the loops. The library probably unrolls with a non-optimal amount on the lower performing points.

Whenever you have short loops around longer loops, you should always unroll the small loops inside the longer loops. This particular optimization came up in a reservoir model I helped optimize in the 1980s for the early Cray vector systems. The problem was that the bounds on the `J` and `K` loops were either 1, 2, 3, or 4. The performance increase the application developer was getting was so good that he replicated the code four times and explicitly performed this optimization.

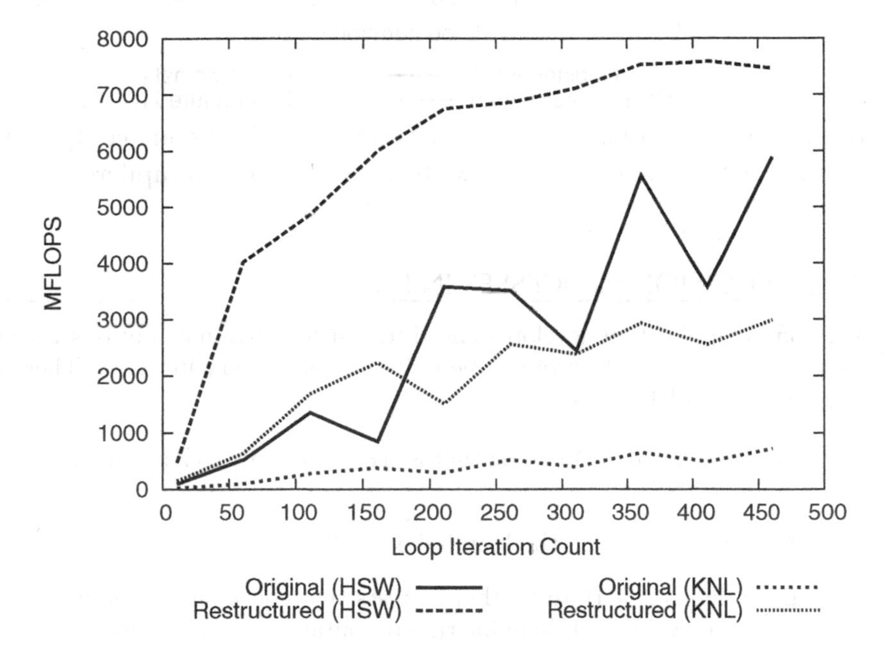

FIGURE 7.9.1 Performance of restructured vectorized code compared to original pattern-matched code.

Comparing the Cray compiler to Intel in Figure 7.9.2 shows that the substitution of the matrix multiply routine by the Cray compiler really performs poorly, while Intel's approach of running the example in scalar mode wins. In the restructured, Intel also beats the Cray compiler for shorter vector lengths.

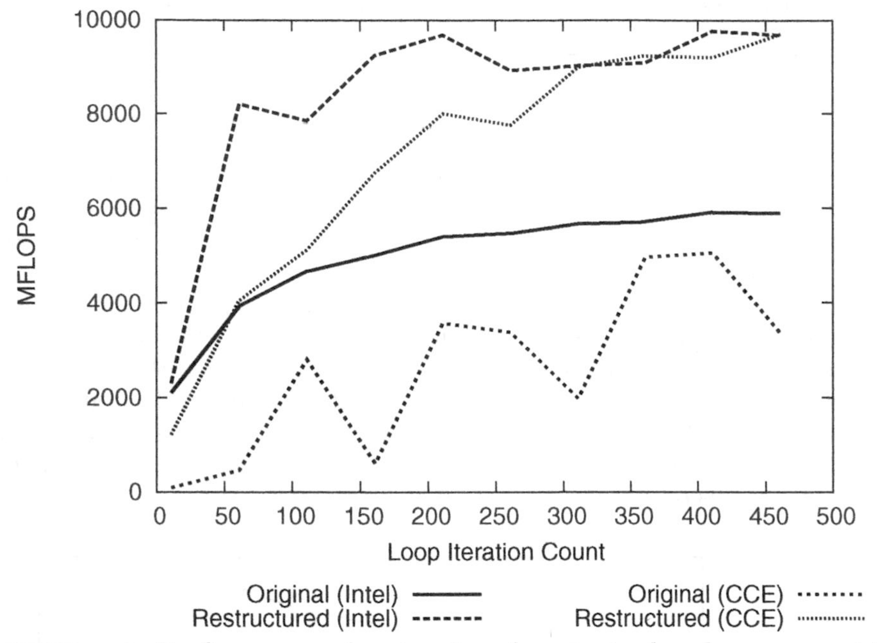

FIGURE 7.9.2 Performance of restructured vectorized code compared to original pattern-matched code with Intel and Cray compilers.

7.10 DECISION PROCESSES IN LOOPS

Very often, application developers need to test for certain conditions within a computation. These `IF` tests can be quite simple or very involved. There are different types of `IF` tests:

1. Loop-independent `IF` tests; that is, `IF` tests that check a constant within the loop.

2. `IF` tests checking on the loop index itself.

3. Loop-dependent `IF` tests; that is, tests that check on an existing and/or computed value and changing the computation according to the outcome of the test.

4. `IF` tests that cause premature exiting from the loop.

7.10.1 Loop-Independent Conditionals

```
55.     1               C      THE ORIGINAL
56.   + 1 2----------<         DO 47012   K = 2, N
57.   + 1 2 w--------<          DO 47011  J = 2, 3
58.     1 2 w Vcr2---<           DO 47010 I = 2, N
59.     1 2 w Vcr2                A(I,J) = (1. - PX - PY - PZ) * B(I,J,K)
60.     1 2 w Vcr2          1           + .5 * PX * ( B(I+1,J,K) + B(I-1,J,K) )
61.     1 2 w Vcr2          2           + .5 * PY * ( B(I,J+1,K) + B(I,J-1,K) )
62.     1 2 w Vcr2          3           + .5 * PZ * ( B(I,J,K+1) + B(I,J,K-1) )
63.     1 2 w Vcr2                IF (K .LT. 3)  GO TO 11
64.     1 2 w Vcr2                 IF (K .LT. N) GO TO 10
65.     1 2 w Vcr2                  B(I,J,K )   = A(I,J)
66.     1 2 w Vcr2         10      B(I,J,K-1)  = C(I,J)
67.     1 2 w Vcr2         11     C(I,J)       = A(I,J)
68.     1 2 w Vcr2---> 47010   CONTINUE
79.     1 2 w--------> 47011  CONTINUE
80.     1 2----------> 47012 CONTINUE
```

Excerpt 7.10.1 Original code using IF inside a loop-nest.

The first example, presented in Excerpt 7.10.1, will investigate a loop-independent test. The compiler does very well with this example. It unwinds the J loop (w – unwind) and vectorizes and unrolls the I loop by two (**r2** – unroll by 2). The Vc indicates that the compiler conditionally vectorizes the loop; it's probably dependent upon the value of K – if K is equal to N this loop has a loop-carried dependency – albeit the last iteration. The compiler certainly uses its entire toolbox of techniques on this example code.

The original code is restructured to use a loop-independent IF test instead – it is checking on an outer loop index and within this loop the IF is independent of I. The restructured code is presented in Excerpt 7.10.2, where the loop-independent IFs are split out from the loop. Intel diagnostics for this loop are also shown in Table 7.10.1. The Intel compiler did the same optimization that the Cray compiler did.

TABLE 7.10.1 Intel compiler report from restructured code with IF pulled out of loop-nest.

```
LOOP BEGIN at lp47010.f(99,10)
   remark #15542: loop was not vectorized: inner loop was already vectorized
   LOOP BEGIN at lp47010.f(100,11)
      remark #15542: loop was not vectorized: inner loop was already vectorized
      LOOP BEGIN at lp47010.f(101,12)
         remark #15300: LOOP WAS VECTORIZED
      LOOP END
      LOOP BEGIN at lp47010.f(101,12)
      <Remainder loop for vectorization>
      LOOP END
      LOOP BEGIN at lp47010.f(101,12)
      <Peeled loop for vectorization>
         remark #25436: completely unrolled by 3
      LOOP END
   LOOP END
```

```
 97.    1               C      THE RESTRUCTURED
 98.  + 1 2----------<         DO 47016   K = 2, N - 1
 99.  + 1 2 f--------<          DO 47013  J = 2, 3
100.    1 2 f fVr2---<           DO 47013 I = 2, N
101.    1 2 f fVr2                A(I,J) = (1. - PX - PY - PZ) * B(I,J,K)
102.    1 2 f fVr2            1          + .5 * PX * ( B(I+1,J,K) + B(I-1,J,K) )
103.    1 2 f fVr2            2          + .5 * PY * ( B(I,J+1,K) + B(I,J-1,K) )
104.    1 2 f fVr2            3          + .5 * PZ * ( B(I,J,K+1) + B(I,J,K-1) )
105.    1 2 f fVr2-->> 47013 CONTINUE
106.    1 2
107.    1 2                     IF (K .EQ. 2) THEN
108.    1 2
109.  + 1 2 f--------<           DO 47014  J =2, 3
110.    1 2 f f------<            DO 47014 I =2, N
111.    1 2 f f                    C(I,J)     = A(I,J)
112.    1 2 f f----->> 47014    CONTINUE
113.    1 2
114.    1 2                     ELSE
115.    1 2
116.  + 1 2 f--------<           DO 47015  J = 2, 3
117.    1 2 f f------<            DO 47015 I = 2, N
118.    1 2 f f                    B(I,J,K-1)  = C(I,J)
119.    1 2 f f                    C(I,J)      = A(I,J)
120.    1 2 f f----->> 47015    CONTINUE
121.    1 2
122.    1 2                     ENDIF
123.    1 2
124.    1 2----------> 47016 CONTINUE
125.    1                         K = N
126.  + 1 ir2----------<         DO 47017 I = 2, N
127.  + 1 ir2 iw-------<          DO 47017  J = 2, 3
128.    1 ir2 iw                   A(I,J) = (1. - PX - PY - PZ) * B(I,J,K)
129.    1 ir2 iw                1          + .5 * PX * ( B(I+1,J,K) + B(I-1,J,K) )
130.    1 ir2 iw                2          + .5 * PY * ( B(I,J+1,K) + B(I,J-1,K) )
131.    1 ir2 iw                3          + .5 * PZ * ( B(I,J,K+1) + B(I,J,K-1) )
132.    1 ir2 iw                   B(I,J,K)    = A(I,J)
133.    1 ir2 iw                   B(I,J,K-1)  = C(I,J)
134.    1 ir2 iw                   C(I,J)      = A(I,J)
135.    1 ir2 iw------>> 47017 CONTINUE
```

Excerpt 7.10.2 Restructured code with IF pulled out of loop-nest.

Restructuring has split out the conditionals from the loop. The real issue with this restructuring is that it can cause significant data motion. Whereas the original loop was getting good reuse of the A and B arrays, the splits result in cycling through the data twice for K.LT.N and three times for K.EQ.N. This will cause a performance degradation if the kernel does not fit into cache.

As can be seen in Figure 7.10.1, both versions of the loop run pretty much the same on KNL as well as Haswell. The original may not perform so well when built with a less intelligent compiler. This example is more a study of handling multinested loops, since the original is most optimal from a looping structure perspective, while the restructured uses a less optimal approach to get better vectorization. Comparing the Cray to Intel compiler indicates that the restructured does run slower on both compilers with the edge for each version going to Cray, as seen in Figure 7.10.2.

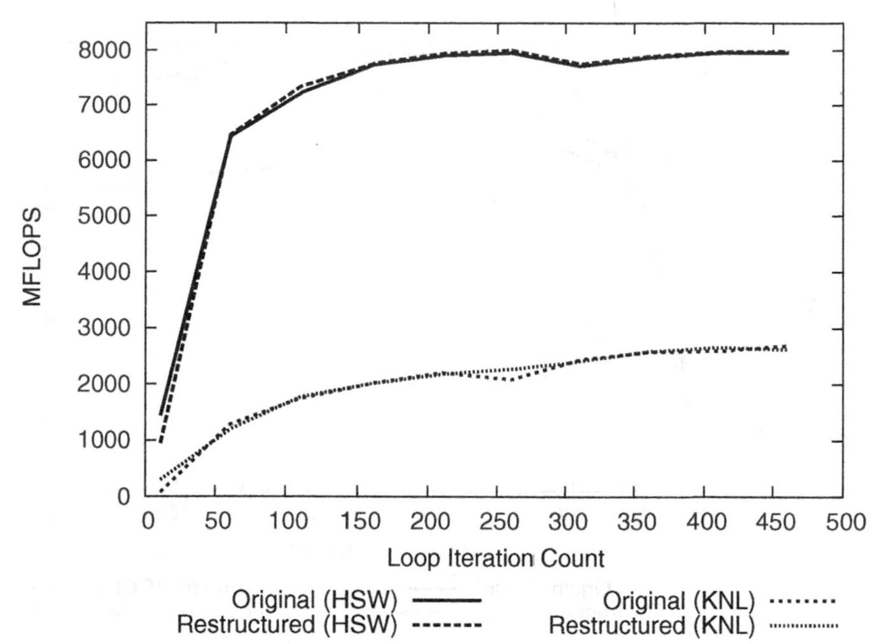

FIGURE 7.10.1 Performance of restructured code compared to original with IF statements within the loop-nest.

7.10.2 Conditionals Directly Testing Indicies

The next example, presented in Excerpt 7.10.3, is another where outer-loop indices are being tested for setting boundary conditions. While this example is rather long, it really illustrates some of the coding being used in the community. The compiler does not vectorize the `I` loop. The reason is given in the compiler notes. Intel diagnostics are not shown here. They are too verbose for this example code. It should suffice to say that Intel's compiler did pretty much the same as the Cray compiler.

In the restructured version shown in Excerpt 7.10.4, we will be splitting the `I` loop and separating the `IF` outside the loop. In doing so, we will have to generate some temporary arrays. It is extremely important that these temporaries are small enough to fit into level-1 cache. Ideally, it would be nice to have a cache variable, since we do not have to store the results back into memory.

Once again the Intel compiler optimized the restructured similar to the Cray compiler. Notice that we have nine independent arrays dimensioned on `I`. These get computed and used but are not required outside the loop. The compiler should recognize this and reuse cache nicely. Performance is shown in Figure 7.10.3.

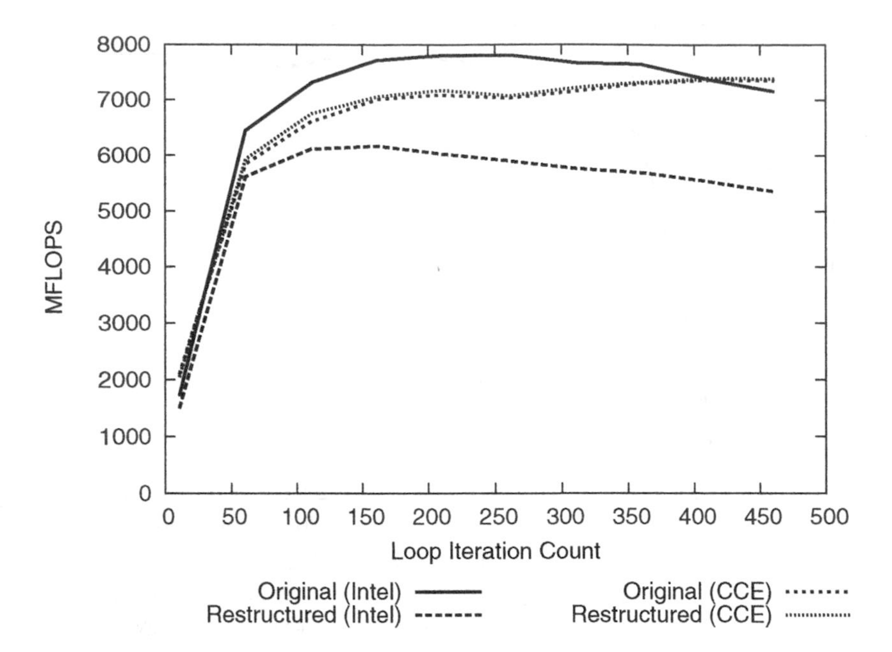

FIGURE 7.10.2 Performance of restructured code compared to original with IF statements within the loop-nest with Intel and Cray compilers.

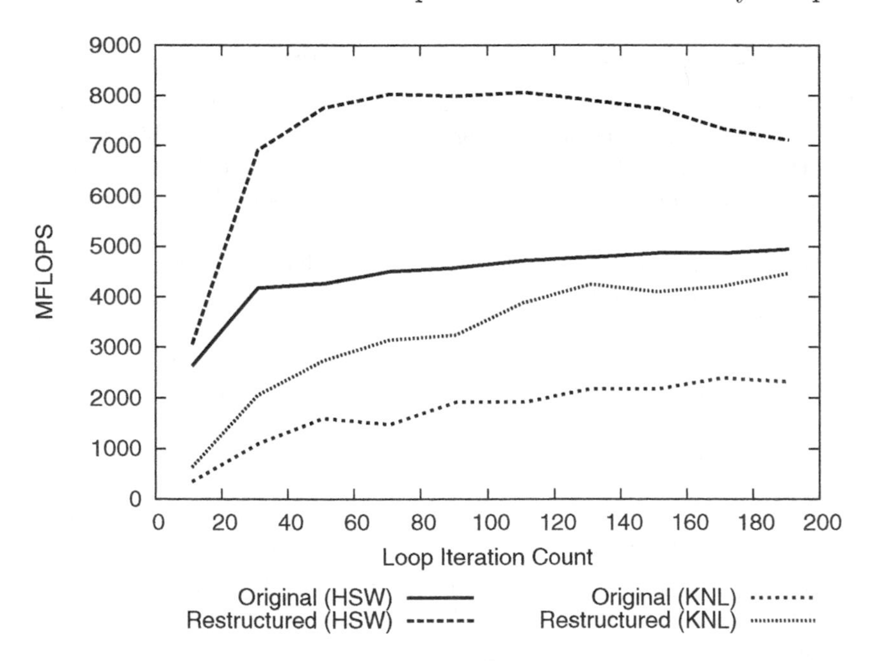

FIGURE 7.10.3 Performance of restructured code compared to original code with IF statements within the loop-nest.

```
 53.   1                 C     THE ORIGINAL
 54.   + 1 i-----------<       DO 47020   J = 1, JMAX
 55.   + 1 i i---------<        DO 47020  K = 1, KMAX
 56.   + 1 i i 4-------<         DO 47020 I = 1, IMAX
 57.     1 i i 4                  JP        = J + 1
 58.     1 i i 4                  JR        = J - 1
 59.     1 i i 4                  KP        = K + 1
 60.     1 i i 4                  KR        = K - 1
 61.     1 i i 4                  IP        = I + 1
 62.     1 i i 4                  IR        = I - 1
 63.     1 i i 4                  IF (J .EQ. 1)      GO TO 50
 64.     1 i i 4                   IF( J .EQ. JMAX) GO TO 51
 65.     1 i i 4                    XJ = ( A(I,JP,K) - A(I,JR,K) ) * DA2
 66.     1 i i 4                    YJ = ( B(I,JP,K) - B(I,JR,K) ) * DA2
 67.     1 i i 4                    ZJ = ( C(I,JP,K) - C(I,JR,K) ) * DA2
 68.     1 i i 4                    GO TO 70
 69.     1 i i 4             50   J1 = J + 1
 70.     1 i i 4                  J2 = J + 2
 71.     1 i i 4                  XJ = (-3. * A(I,J,K) + 4. * A(I,J1,K) - A(I,J2,K) ) * DA2
 72.     1 i i 4                  YJ = (-3. * B(I,J,K) + 4. * B(I,J1,K) - B(I,J2,K) ) * DA2
 73.     1 i i 4                  ZJ = (-3. * C(I,J,K) + 4. * C(I,J1,K) - C(I,J2,K) ) * DA2
 74.     1 i i 4                  GO TO 70
 75.     1 i i 4             51   J1 = J - 1
 76.     1 i i 4                  J2 = J - 2
 77.     1 i i 4                  XJ = ( 3. * A(I,J,K) - 4. * A(I,J1,K) + A(I,J2,K) ) * DA2
 78.     1 i i 4                  YJ = ( 3. * B(I,J,K) - 4. * B(I,J1,K) + B(I,J2,K) ) * DA2
 79.     1 i i 4                  ZJ = ( 3. * C(I,J,K) - 4. * C(I,J1,K) + C(I,J2,K) ) * DA2
 80.     1 i i 4             70   CONTINUE
 81.     1 i i 4                  IF (K .EQ. 1)      GO TO 52
 82.     1 i i 4                   IF (K .EQ. KMAX) GO TO 53
 83.     1 i i 4                    XK = ( A(I,J,KP) - A(I,J,KR) ) * DB2
 84.     1 i i 4                    YK = ( B(I,J,KP) - B(I,J,KR) ) * DB2
 85.     1 i i 4                    ZK = ( C(I,J,KP) - C(I,J,KR) ) * DB2
 86.     1 i i 4                    GO TO 71
 87.     1 i i 4             52   K1 = K + 1
 88.     1 i i 4                  K2 = K + 2
 89.     1 i i 4                  XK = (-3. * A(I,J,K) + 4. * A(I,J,K1) - A(I,J,K2) ) * DB2
 90.     1 i i 4                  YK = (-3. * B(I,J,K) + 4. * B(I,J,K1) - B(I,J,K2) ) * DB2
 91.     1 i i 4                  ZK = (-3. * C(I,J,K) + 4. * C(I,J,K1) - C(I,J,K2) ) * DB2
 92.     1 i i 4                  GO TO 71
 93.     1 i i 4             53   K1 = K - 1
 94.     1 i i 4                  K2 = K - 2
 95.     1 i i 4                  XK = ( 3. * A(I,J,K) - 4. * A(I,J,K1) + A(I,J,K2) ) * DB2
 96.     1 i i 4                  YK = ( 3. * B(I,J,K) - 4. * B(I,J,K1) + B(I,J,K2) ) * DB2
 97.     1 i i 4                  ZK = ( 3. * C(I,J,K) - 4. * C(I,J,K1) + C(I,J,K2) ) * DB2
 98.     1 i i 4             71   CONTINUE
 99.     1 i i 4                  IF (I .EQ. 1)      GO TO 54
100.     1 i i 4                   IF (I .EQ. IMAX) GO TO 55
101.     1 i i 4                    XI = ( A(IP,J,K) - A(IR,J,K) ) * DC2
102.     1 i i 4                    YI = ( B(IP,J,K) - B(IR,J,K) ) * DC2
103.     1 i i 4                    ZI = ( C(IP,J,K) - C(IR,J,K) ) * DC2
104.     1 i i 4                    GO TO 60
105.     1 i i 4             54   I1 = I + 1
106.     1 i i 4                  I2 = I + 2
107.     1 i i 4                  XI = (-3. * A(I,J,K) + 4. * A(I1,J,K) - A(I2,J,K) ) * DC2
108.     1 i i 4                  YI = (-3. * B(I,J,K) + 4. * B(I1,J,K) - B(I2,J,K) ) * DC2
109.     1 i i 4                  ZI = (-3. * C(I,J,K) + 4. * C(I1,J,K) - C(I2,J,K) ) * DC2
110.     1 i i 4                  GO TO 60
111.     1 i i 4             55   I1 = I - 1
112.     1 i i 4                  I2 = I - 2
113.     1 i i 4                  XI = ( 3. * A(I,J,K) - 4. * A(I1,J,K) + A(I2,J,K) ) * DC2
114.     1 i i 4                  YI = ( 3. * B(I,J,K) - 4. * B(I1,J,K) + B(I2,J,K) ) * DC2
115.     1 i i 4                  ZI = ( 3. * C(I,J,K) - 4. * C(I1,J,K) + C(I2,J,K) ) * DC2
116.     1 i i 4             60   CONTINUE
117.     1 i i 4                  DINV     = XJ * YK * ZI  +  YJ * ZK * XI  +  ZJ * XK * YI
118.     1 i i 4                *          - XJ * ZK * YI  -  YJ * XK * ZI  -  ZJ * YK * XI
119.     1 i i 4                  D(I,J,K) = 1. / (DINV + 1.E-20)
120      1 i i 4----->>> 47020 CONTINUE

ftn-6221 ftn: VECTOR File = lp47020.f, Line = 55
  A loop starting at line 56 uses a checked speculative load of "c".

ftn-6375 ftn: VECTOR File = lp47020.f, Line = 55
  A loop starting at line 56 would benefit from "!dir$ safe_address".
```

Excerpt 7.10.3 Original code with IF statements inside the loop-nest.

```
142.    1                  C      THE RESTRUCTURED
143.  + 1 2-----------<           DO 47029 J = 1, JMAX
144.  + 1 2 3---------<            DO 47029 K = 1, KMAX
145.    1 2 3                        IF(J.EQ.1)THEN
146.    1 2 3                        J1         = 2
147.    1 2 3                        J2         = 3
148.    1 2 3 Vr2-----<              DO 47021 I = 1, IMAX
149.    1 2 3 Vr2                     VAJ(I) = (-3. * A(I,J,K) + 4. * A(I,J1,K) - A(I,J2,K) ) * DA2
150.    1 2 3 Vr2                     VBJ(I) = (-3. * B(I,J,K) + 4. * B(I,J1,K) - B(I,J2,K) ) * DA2
151.    1 2 3 Vr2                     VCJ(I) = (-3. * C(I,J,K) + 4. * C(I,J1,K) - C(I,J2,K) ) * DA2
152.    1 2 3 Vr2-----> 47021        CONTINUE
153.    1 2 3                        ELSE IF(J.NE.JMAX) THEN
154.    1 2 3                        JP         = J+1
155.    1 2 3                        JR         = J-1
156.    1 2 3 Vr2-----<              DO 47022 I = 1, IMAX
157.    1 2 3 Vr2                     VAJ(I) = ( A(I,JP,K) - A(I,JR,K) ) * DA2
158.    1 2 3 Vr2                     VBJ(I) = ( B(I,JP,K) - B(I,JR,K) ) * DA2
159.    1 2 3 Vr2                     VCJ(I) = ( C(I,JP,K) - C(I,JR,K) ) * DA2
160.    1 2 3 Vr2-----> 47022        CONTINUE
161.    1 2 3                        ELSE
162.    1 2 3                        J1         = JMAX-1
163.    1 2 3                        J2         = JMAX-2
164.    1 2 3 Vr2-----<              DO 47023 I = 1, IMAX
165.    1 2 3 Vr2                     VAJ(I) = ( 3. * A(I,J,K) - 4. * A(I,J1,K) + A(I,J2,K) ) * DA2
166.    1 2 3 Vr2                     VBJ(I) = ( 3. * B(I,J,K) - 4. * B(I,J1,K) + B(I,J2,K) ) * DA2
167.    1 2 3 Vr2                     VCJ(I) = ( 3. * C(I,J,K) - 4. * C(I,J1,K) + C(I,J2,K) ) * DA2
168.    1 2 3 Vr2-----> 47023        CONTINUE
169.    1 2 3                        ENDIF
170.    1 2 3                        IF(K.EQ.1) THEN
171.    1 2 3                        K1         = 2
172.    1 2 3                        K2         = 3
173.    1 2 3 Vr2-----<              DO 47024 I = 1, IMAX
174.    1 2 3 Vr2                     VAK(I) = (-3. * A(I,J,K) + 4. * A(I,J,K1) - A(I,J,K2) ) * DB2
175.    1 2 3 Vr2                     VBK(I) = (-3. * B(I,J,K) + 4. * B(I,J,K1) - B(I,J,K2) ) * DB2
176.    1 2 3 Vr2                     VCK(I) = (-3. * C(I,J,K) + 4. * C(I,J,K1) - C(I,J,K2) ) * DB2
177.    1 2 3 Vr2-----> 47024        CONTINUE
178.    1 2 3                        ELSE IF(K.NE.KMAX)THEN
179.    1 2 3                        KP         = K + 1
180.    1 2 3                        KR         = K - 1
181.    1 2 3 Vr2-----<              DO 47025 I = 1, IMAX
182.    1 2 3 Vr2                     VAK(I) = ( A(I,J,KP) - A(I,J,KR) ) * DB2
183.    1 2 3 Vr2                     VBK(I) = ( B(I,J,KP) - B(I,J,KR) ) * DB2
184.    1 2 3 Vr2                     VCK(I) = ( C(I,J,KP) - C(I,J,KR) ) * DB2
185.    1 2 3 Vr2-----> 47025        CONTINUE
186.    1 2 3                        ELSE
187.    1 2 3                        K1         = KMAX - 1
188.    1 2 3                        K2         = KMAX - 2
189.    1 2 3 Vr2-----<              DO 47026 I = 1, IMAX
190.    1 2 3 Vr2                     VAK(I) = ( 3. * A(I,J,K) - 4. * A(I,J,K1) + A(I,J,K2) ) * DB2
191.    1 2 3 Vr2                     VBK(I) = ( 3. * B(I,J,K) - 4. * B(I,J,K1) + B(I,J,K2) ) * DB2
192.    1 2 3 Vr2                     VCK(I) = ( 3. * C(I,J,K) - 4. * C(I,J,K1) + C(I,J,K2) ) * DB2
193.    1 2 3 Vr2-----> 47026        CONTINUE
194.    1 2 3                        ENDIF
195.    1 2 3                        I = 1
196.    1 2 3                        I1         = 2
197.    1 2 3                        I2         = 3
198.    1 2 3                        VAI(I) = (-3. * A(I,J,K) + 4. * A(I1,J,K) - A(I2,J,K) ) * DC2
199.    1 2 3                        VBI(I) = (-3. * B(I,J,K) + 4. * B(I1,J,K) - B(I2,J,K) ) * DC2
200.    1 2 3                        VCI(I) = (-3. * C(I,J,K) + 4. * C(I1,J,K) - C(I2,J,K) ) * DC2
201.    1 2 3 Vr2-----<              DO 47027 I = 2, IMAX-1
202.    1 2 3 Vr2                     IP         = I + 1
203.    1 2 3 Vr2                     IR         = I - 1
204.    1 2 3 Vr2                     VAI(I) = ( A(IP,J,K) - A(IR,J,K) ) * DC2
205.    1 2 3 Vr2                     VBI(I) = ( B(IP,J,K) - B(IR,J,K) ) * DC2
206.    1 2 3 Vr2                     VCI(I) = ( C(IP,J,K) - C(IR,J,K) ) * DC2
207.    1 2 3 Vr2-----> 47027        CONTINUE
208.    1 2 3                        I = IMAX
209.    1 2 3                        I1         = IMAX - 1
210.    1 2 3                        I2         = IMAX - 2
211.    1 2 3                        VAI(I) = ( 3. * A(I,J,K) - 4. * A(I1,J,K) + A(I2,J,K) ) * DC2
212.    1 2 3                        VBI(I) = ( 3. * B(I,J,K) - 4. * B(I1,J,K) + B(I2,J,K) ) * DC2
213.    1 2 3                        VCI(I) = ( 3. * C(I,J,K) - 4. * C(I1,J,K) + C(I2,J,K) ) * DC2
214.    1 2 3 Vr2-----<              DO 47028 I = 1, IMAX
215.    1 2 3 Vr2                     DINV = VAJ(I) * VBK(I) * VCI(I) + VBJ(I) * VCK(I) * VAI(I)
216.    1 2 3 Vr2               1          + VCJ(I) * VAK(I) * VBI(I) - VAJ(I) * VCK(I) * VBI(I)
217.    1 2 3 Vr2               2          - VBJ(I) * VAK(I) * VCI(I) - VCJ(I) * VBK(I) * VAI(I)
218.    1 2 3 Vr2                     D(I,J,K) = 1. / (DINV + 1.E-20)
219.    1 2 3 Vr2-----> 47028         CONTINUE
220.    1 2 3-------->> 47029 CONTINUE
```

Excerpt 7.10.4 Restructured loop-nest with IFs pulled out of the I loop.

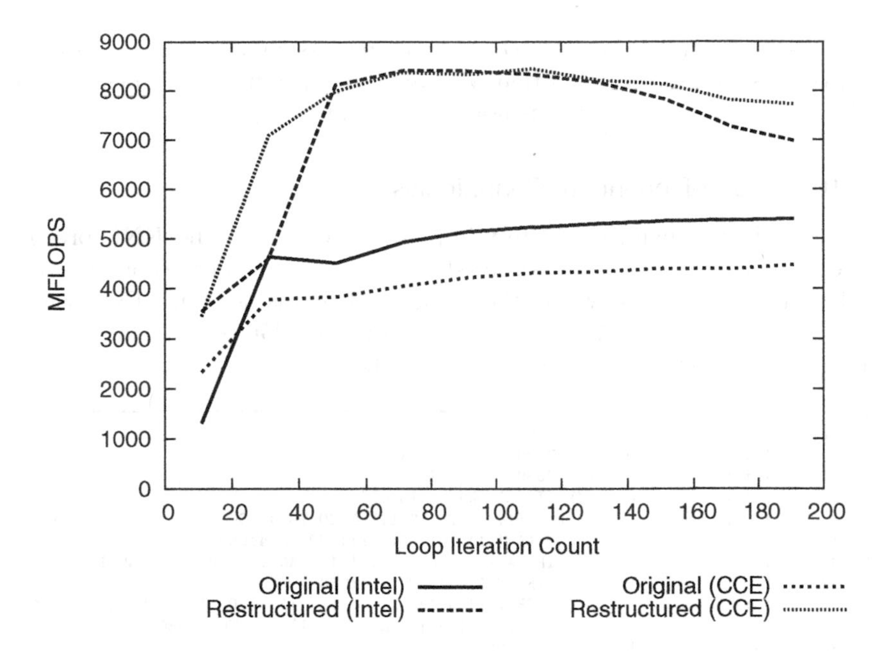

FIGURE 7.10.4 Performance of restructured code compared to original with IF statements within the loop-nest with Intel and Cray compilers.

The comparison between Intel and Cray is shown in Figure 7.10.4 with Cray doing better on the restructured and Intel doing better on the original (not by much).

We do get a factor of two performance gain from vectorizing the restructured code. Notice that performance maxes out at a loop iteration count of 71. If `I` is very large, we would want to strip-mine the `I` loop on some value around 60 to 70 to keep the temporary arrays small enough to fit into memory. Strip-mining is the technique of computing a chunk of the computation at a time. In this case we would introduce an outer loop on `II`; all the inner loops would run from `I = IS, IE`. This stip-mined version is shown in Excerpt 7.10.5.

```
  142.    1                 C      THE RESTRUCTURED
  143.  + 1 2-----------<          DO 47029 J = 1, JMAX
  144.  + 1 2 3---------<           DO 47029 K = 1, KMAX
                                     DO 47029 II = 1, IMAX, 64
                                      IS = II
                                      IE = min(IMAX, II+64-1)
 DO I = IS, II
```

Excerpt 7.10.5 Loop-nest with the addition of a strip-mining loop.

In this example, the KNL core is running less than a factor of two slower on the restructured version than Haswell. Once again, KNL does have twice the number of cores, so this is not really a bad thing.

7.10.3 Loop-Dependent Conditionals

When a loop-dependent `IF` test is present in a loop, the KNL predicated execution is used to set a mask register and then perform the computation for all values except those where the mask/condition is not true. Even with this feature, some complex `IF` tests are not vectorized. Consider the loop presented in Excerpt 7.10.6 containing numerous tests.

```
 82.     1                    SUM = 0.0
 83.   + 1 i-------<          DO 47080 J = 2, JMAX
 84.   + 1 i i-----<          DO 47080 I = 2, N
 85.     1 i i                 IF (I .EQ. N) GO TO 47080
 86.     1 i i                  IF (A(1,J) .LT. B(1,I)) GO TO 47080
 87.     1 i i                   IF (A(1,1) .GT. B(1,I)) GO TO 47080
 88.     1 i i                    IF (A(1,J) .GE. B(1,I+1) .AND. I .NE. N)  GO TO 500
 89.     1 i i                     IF (J.EQ.1) GO TO 47080
 90.     1 i i                      IF (A(1,J-1) .LT. B(1,I-1) .AND. I*J .NE. 1)  GO TO 500
 91.     1 i i                       IF (A(1,J-1) .LT. B(1,I)) GO TO 47080
 92.     1 i i          500      CONTINUE
 93.     1 i i                   P1    = C(1,I-1)
 94.     1 i i                   P2    = D(I-1)
 95.     1 i i                   DD    = B(1,I) - B(1,I-1)
 96.     1 i i                   P3    = (3.0 * E(I) - 2.0 * P2 - D(I)) / DD
 97.     1 i i                   P4    = ( P2 + D(I) - 2.0 * E(I)      ) / DD**2
 98.     1 i i                   SUMND = DD * (P1       + DD * (P2 / 2.
 99.     1 i i             *           + DD * (P3 / 3. + DD *  P4 / 4.) ) )
100.     1 i i                   SUM   = SUM + SUMND
101.     1 i i---->> 47080 CONTINUE
```

Excerpt 7.10.6 Example loop-nest with multiple conditionals.

In this case neither the Intel nor the Cray compiler has vectorized the loop. Given all of the conditions, the assumption is that the frequency of performing the computation is very low. The `i` in the listing is indicating that the loops are interchanged. Intel's diagnostics are interesting, in that it computes the benefit of vectorization and then decides not to vectorize the loop, as seen in Excerpt 7.10.7.

Whenever numerous complication decision processes are encountered, such as in this example, one can frequently benefit from vectorizing the conditional itself. Consider the rewrite of this particular loop presented in Excerpt 7.10.8.

Notice the code to guard against dividing by `DD` – if it is zero. While the result will not be correct, its value is never used. This is known as singularity handling. Once again the Cray compiler inverts the loops and vectorizes on `I`. Also, the Intel compiler has a completely different tradeoff analysis as shown in Excerpt 7.10.9. Of course, the performance will be dependent upon the percentage of truth for performing the computation. Figure 7.10.5 illustrates the performance for the two compilers on KNL. The figure shows very low

MFLOP numbers since the density of the decision is very low and so this example does little computation and a lot of memory accessing.

```
LOOP BEGIN at lp47080.f(83,10)
 remark #15305: vectorization support: vector length 2
 remark #15309: vectorization support: normalized vectorization overhead 0.229
 remark #15335: loop was not vectorized: vectorization possible but seems inefficient.
                Use vector always directive or -vec-threshold0 to override
 remark #15305: vectorization support: vector length 16
 remark #15309: vectorization support: normalized vectorization overhead 0.142
 remark #15456: masked unaligned unit stride loads: 4
 remark #15475: --- begin vector cost summary ---
 remark #15476: scalar cost: 12
 remark #15477: vector cost: 46.060
 remark #15478: estimated potential speedup: 0.120
 remark #15486: divides: 3
```

Excerpt 7.10.7 Optimization report from Intel compiler showing non-vectorized loop with conditionals.

```
  121.    1             C      THE RESTRUCTURED
  122.    1
  123.    1                   SUM = 0.0
  124.    1 i---------<       DO 47081 J = 2, JMAX
  125.    1 i iV------<        DO 47081 I = 2, N-1
  126.    1 i iV
  127.    1 i iV              LOG1 = A(1,J) .GE. B(1,I)
  128.    1 i iV              LOG2 = A(1,1) .LE. B(1,I)
  129.    1 i iV              LOG3 = A(1,J) .GE. B(1,I+1)
  130.    1 i iV              LOG4 = J .NE. 1
  131.    1 i iV              LOG5 = A(1,J-1) .LT. B(1,I-1)
  132.    1 i iV              LOG6 = A(1,J-1) .GE. B(1,I)
  133.    1 i iV              LOG7 = LOG1 .AND. LOG2 .AND. LOG3 .OR.
  134.    1 i iV             *       LOG1 .AND. LOG2 .AND. LOG4 .AND. LOG5 .OR.
  135.    1 i iV             *       LOG1 .AND. LOG2 .AND. LOG4 .AND. LOG6
  136.    1 i iV               P1     = C(1,I-1)
  137.    1 i iV               P2     = D(I-1)
  138.    1 i iV               DD     = B(1,I) - B(1,I-1)
  139.    1 i iV               IF(.NOT. LOG7) DD=1.0
  140.    1 i iV               P3     = (3.0 * E(I) - 2.0 * P2 - D(I)) / DD
  141.    1 i iV               P4     = ( P2 + D(I) - 2.0 * E(I)      ) / DD**2
  142.    1 i iV               SUMND = 0.0
  143.    1 i iV               IF(LOG7) SUMND = DD * (P1       + DD * (P2 / 2.
  144.    1 i iV             *                  + DD * (P3 / 3. + DD *  P4 / 4.) ) )
  145.    1 i iV               SUM    = SUM + SUMND
  146.    1 i iV----->> 47081 CONTINUE
```

Excerpt 7.10.8 Refactored example loop with multiple conditionals.

The Intel compiler does very well with the restructured and beats the Cray on the original as well. In this case, the frequency of truth is 1%, so the real advantage is that the decisions are vectorized in the restructured and not in the original. One thing about predicated execution is that if the mask is all zeros then the computation will not be performed. It is important to note that all of the loops with `IF` tests compute the number of floating point operations that are actually performed so the MFLOP numbers are real.

```
remark #15300: LOOP WAS VECTORIZED
remark #15450: unmasked unaligned unit stride loads: 7
remark #15475: --- begin vector cost summary ---
remark #15476: scalar cost: 190
remark #15477: vector cost: 36.750
remark #15478: estimated potential speedup: 4.830
remark #15486: divides: 3
remark #15488: --- end vector cost summary ---
remark #25456: Number of Array Refs Scalar Replaced In Loop: 5
remark #25015: Estimate of max trip count of loop=57
```

Excerpt 7.10.9 Refactored example with multiple conditionals with Intel compiler.

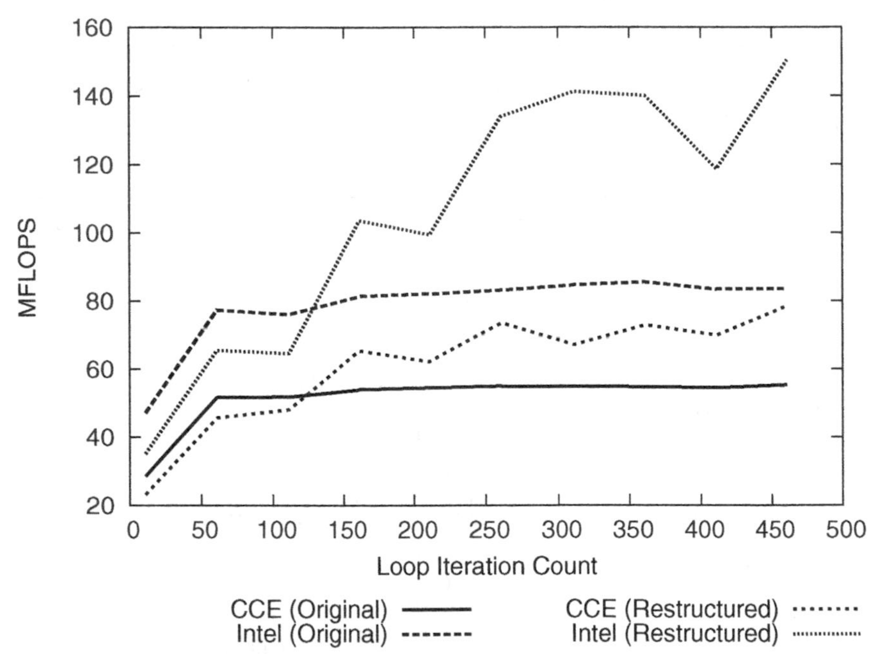

FIGURE 7.10.5 Performance of loop with multiple conditionals with Intel and CCE compilers.

7.10.4 Conditionals Causing Early Loop Exit

The next example presented in Excerpt 7.10.10 doesn't have a DO loop, it is written as an IF loop, testing on a value computed within the IF construct. Of course this could be rewritten as a DO loop with an EXIT from the loop. However, the compiler would still not vectorize such a loop. The restructured code shown in Excerpt 7.10.11 uses a very powerful technique called strip-mining for minimizing the amount of unnecessary computation.

Notice that chunks of 128 are computed rather than the entire length of N. Note that the chunk size can be reduced. Both compilers refuse to vectorize

```
51.    1 2          C      THE ORIGINAL
52.    1 2
53.    1 2                I = 0
54. +  1 2          47120 CONTINUE
55.    1 2                 I = I + 1
56.    1 2                 A(I) = B(I)**2 + .5 * C(I) * D(I) / E(I)
57.    1 2                IF (A(I) .GT. 0.) GO TO 47120
```

Excerpt 7.10.10 Example code showing loop with early exit condition.

```
76.    1 2          C      THE RESTRUCTURED
77.    1 2
78. +  1 2 3------<       DO 47123 II = 1, N, 128
79.    1 2 3                LENGTH = MINO (128, N-II+1)
80.    1 2 3 Vr2--<         DO 47121 I = 1, LENGTH
81.    1 2 3 Vr2             VA(I) = B(I+II-1)**2 + .5 * C(I+II-1) * D(I+II-1) / E(I+II-1)
82.    1 2 3 Vr2--> 47121   CONTINUE
83.    1 2 3
84. +  1 2 3 4----<         DO 47122 I = 1, LENGTH
85.    1 2 3 4               A(I+II-1) = VA(I)
86.    1 2 3 4               IF (A(I+II-1) .LE. 0.0) GO TO 47124
87.    1 2 3 4----> 47122   CONTINUE
88.    1 2 3------> 47123 CONTINUE
89.                 47124 CONTINUE
```

Excerpt 7.10.11 Example code showing refactored loop with strip-mining.

the IF loop in the original, and they both vectorize the restructured. Unfortunately, loop 47122 is not vectorized. The performance results are given in Figure 7.10.6, which plots the performance as a function of the percent of the computation that is performed prior to exiting the IF loop. The loop iteration count for this example is 461. The Intel compiler does extremely well with both of these loops, though it does not vectorize loop 47122. However, its handling of the scalar loop is very good.

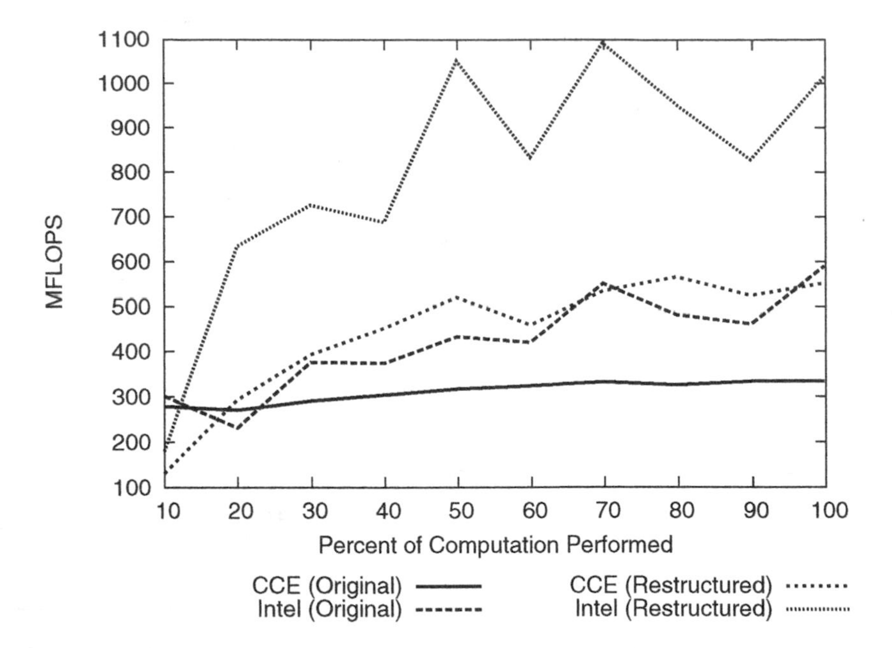

FIGURE 7.10.6 Performance on loop with early exit condition compared to refactored code with CCE and Intel.

7.11 HANDLING FUNCTION CALLS WITHIN LOOPS

```
      DO 48020 I = 1, N
        A(I) = B(I) * FUNC (D(I)) + C(I)
48020 CONTINUE

      FUNCTION FUNC (X)
        FUNC = X**2 + 2.0 / X
        RETURN
      END
```

Excerpt 7.11.1 Example loop containing subroutine call.

```
      FUNCX (X) = X**2 + 2.0 / X

      DO 48021 I = 1, N
       A(I) = B(I) * FUNCX (D(I)) + C(I)
48021 CONTINUE
```

Excerpt 7.11.2 Restructured loop with subroutine call in the form of a statement function.

Subroutine and function calls from within a DO will inhibit vectorization. Many cases where the called routines are inlined still fail to vectorize. To overcome the principal problem, programmers should think of calling a routine to update many elements of the array. In this way, the loop that should be vectorized is contained within the routine of interest. When an application consists of multinested loops, only one of the loops needs to be pulled down into the computational routines. Ideally, this loop will be updating at least 100 to 200 elements of the array. Calling a subroutine to perform an update on a single grid point is very inefficient. We will also see later how higher level loops will be good for OpenMP.

One can get the Cray compiler and the Intel compiler to vectorize these loops by invoking inter-procedural analysis. This is a very heavy hammer in that `-ipo` on Intel and `-hipa3` on CCE inlines all of the calls to all of the subroutines in the entire program. It increases the compile time significantly and is not recommended for very large programs.

To take an existing application and rewrite important `DO` loops that call subroutines, there are a number of options. First if the function/routine can be written as a single line update, consider using a statement function. Statement functions maintain modularity and the compiler will expand the statement function within the loop and then vectorize the loop. Consider the example loop presented in Excerpt 7.11.1. This function call can easily be turned into the statement function presented in Excerpt 7.11.2.

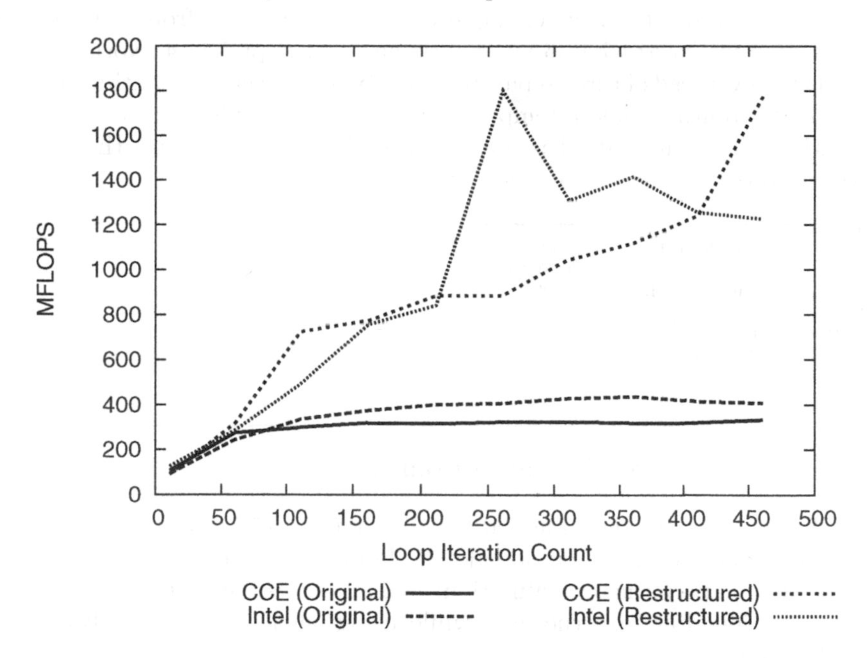

FIGURE 7.11.1 Performance of original loop with subroutine call and restructured loop.

The compiler generates excellent vector code for the restructured version. Figure 7.11.1 shows a comparison between the Cray compiler and the Intel compiler.

```
      DO 48060 I = 1, N
        AOLD = A(I)
        A(I) = UFUN (AOLD, B(I), SCA)
        C(I) = (A(I) + AOLD) * .5
48060 CONTINUE
```

Excerpt 7.11.3 Example loop containing subroutine call.

```
      DO 48061 I = 1, N
        VAOLD(I) = A(I)
48061 CONTINUE
      CALL VUFUN (N, VAOLD, B, SCA, A)
      DO 48062 I = 1, N
        C(I) = (A(I) + VAOLD(I)) * .5
48062 CONTINUE
```

Excerpt 7.11.4 Optimized loop with vector subroutine call.

Now consider the next example of a subroutine call from within a loop, presented in Excerpt 7.11.3 and rewritten in Excerpt 7.11.4. Notice that this adds the overhead of three separate loops, two in the calling routine and one in the called routine – and a temporary is used to pass information to the called routine. The routine itself is presented in Excerpt 7.11.5 with the rewritten, vector version presented in Excerpt 7.11.6.

```
      FUNCTION UFUN (X, Y, SCA)
        IF (SCA .GT. 1.0) GO TO 10
          UFUN = SQRT (X**2 + Y**2)
          GO TO 5
10      UFUN = 0.0
5       CONTINUE
        RETURN
      END
```

Excerpt 7.11.5 Subroutine called from target loop.

In this case, we get an added benefit that the `IF` becomes independent of the loop and can be taken outside the loop. Consequently, the optimization gives a good performance gain. Figure 7.11.2 shows the performance obtained on the original and restructured while using whole-program analysis for the Cray compiler

Finally, consider the next example loop presented in Excerpt 7.11.7. Here we have a more complicated example with two calls and the communication of a scalar variable through a `COMMON` block. The restructured version of the code is presented in Excerpt 7.11.8.

```
      SUBROUTINE VUFUN (N, X, Y, SCA, UFUN)
      DIMENSION X(*), Y(*), UFUN(*)
      IF (SCA .GT. 1.0) GO TO 15
       DO 10 I = 1, N
        UFUN(I) = SQRT (X(I)**2 + Y(I)**2)
   10  CONTINUE
      RETURN
   15 CONTINUE
       DO 20 I = 1, N
        UFUN(I) = 0.0
   20  CONTINUE
      RETURN
      END
```

Excerpt 7.11.6 Optimized subroutine called from restructured loop.

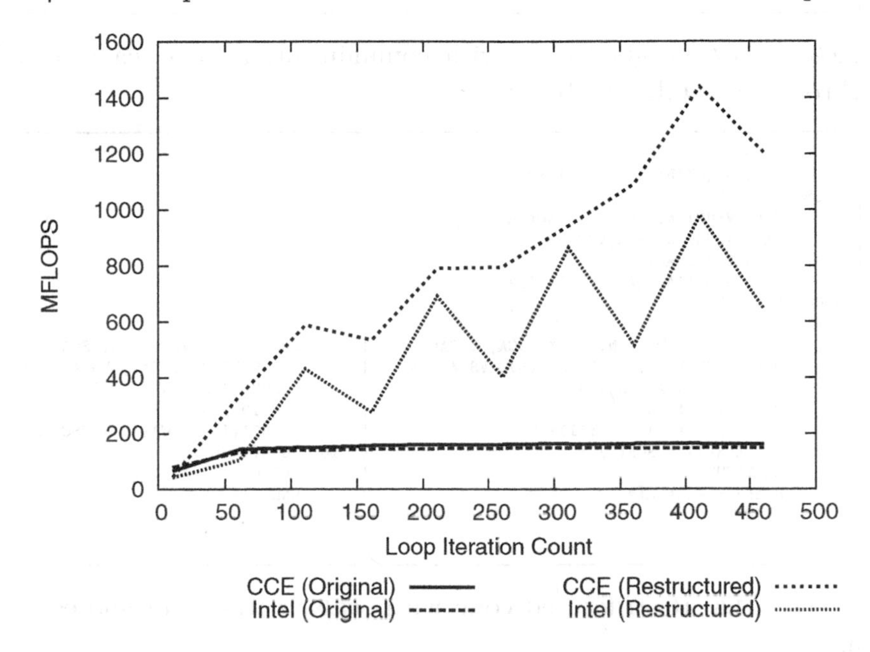

FIGURE 7.11.2 Performance original code with subroutine call and restructured code.

In this restructuring, care must be taken to ensure that any variable that carries information from one loop to another is promoted to an array. Also notice that the last value of SCALR was saved after the DO loop 10 in subroutine VSUB1. This is to take care of the case when this COMMON block is contained in another routine in the program. The restructured is now getting better performance on both of the compilers, as shown in Figure 7.11.3.

```
      COMMON /SCALAR/SCALR
      DO 48080 I = 1 , N
       A(I)=SQRT(B(I)**2+C(I)**2)
       SCA=A(I)**2+B(I)**2
       SCALR=SCA*2
       CALL SUB1(A(I),B(I),SCA)
       CALL SUB2(SCA)
       D(I)=SQRT(ABS(A(I)+SCA))
48080 CONTINUE

      SUBROUTINE SUB1 (X, Y, SCA)            |      SUBROUTINE SUB2 (SCA)
       COMMON /SCALAR/ SCALR                 |       COMMON /SCALAR/ SCALR
       SCA = X**2 + Y**2                     |       SCA = SCA + SCALR
       SCALR = SCA * 2                       |       RETURN
       RETURN                                |      END
      END
```

Excerpt 7.11.7 Original code with communication of a scalar variable through a COMMON block.

```
      DO 48081 I = 1 , N
       A(I)=SQRT(B(I)**2+C(I)**2)
48081 CONTINUE
      CALL VSUB1(N,A,B,VSCA,VSCALR)
      CALL VSUB2(N,VSCA,VSCALR)
      DO 48082 I = 1 , N
       D(I)=SQRT(ABS(A(I)+VSCA(I)))
48082 CONTINUE

     SUBROUTINE VSUB1 (N, X, Y, SCA, VSCALR)  |      SUBROUTINE VSUB2 (N, SCA, VSCALR)
      DIMENSION X(*), Y(*), SCA(*), VSCALR(*) |       DIMENSION SCA(*), VSCALR(*)
      COMMON /SCALAR/ SCALR                   |       COMMON /SCALAR/ SCALR
      DO 10 I = 1, N                          |       DO 10 I = 1, N
       SCA(I) = X(I)**2 +Y(I)**2              |         SCA(I) = SCA(I) + VSCALR(I)
       VSCALR(I) = SCA(I) * 2                 | 10    CONTINUE
  10  CONTINUE                                |       RETURN
      SCALR = VSCALR(N)                       |      END
      RETURN
     END
```

Excerpt 7.11.8 Restructured code with scalar variable promoted to an array.

In summary, when an application calls a routine from within a `DO` loop, the programmer has a number of available options. The most performant option will depend upon the specific code being optimized, so Table 7.11.1 can be used as a hint as to which should be applied to a number of loop types.

1. Turn it into a statement function.
2. Manually inline the called routine(s).
3. Split the do loop at the call(s) and call a vector version of the routine.

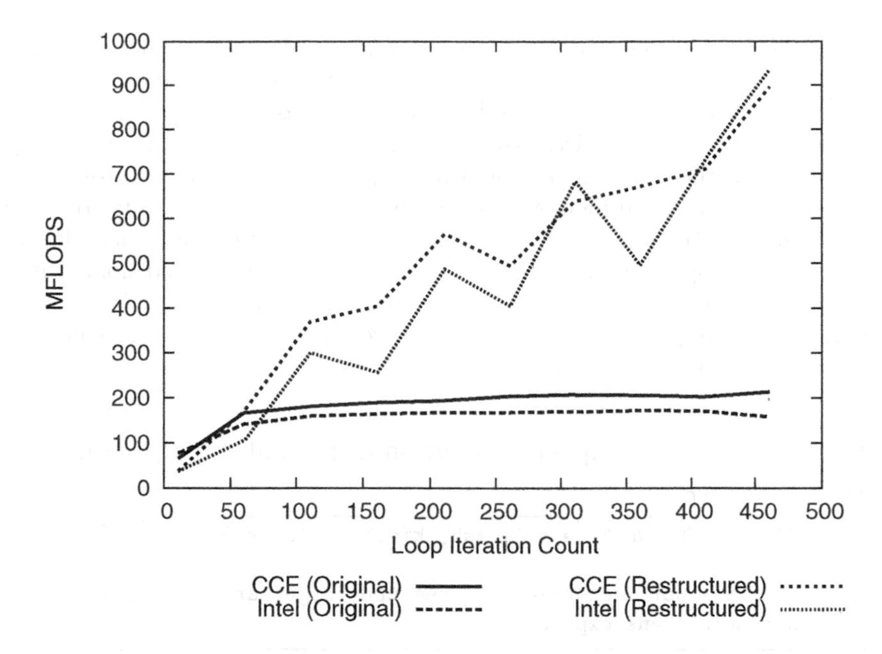

FIGURE 7.11.3 Performance original code with subroutine call and restructured code.

TABLE 7.11.1 Guidance on handling subroutine calls when vectorizing loops.

Called Subroutine Type	Statement Function	Inline	Vector Routine
Simple.	✓	✓	
Very large routine calling other routines.			✓
Very small.		✓	
When complications exist with `COMMON` and/or `MODULE`.			✓

7.12 RANK EXPANSION

When performing outer-loop vectorization (see Section 7.13), rank expansion can be an important issue. If one wanted to vectorize the outer loop of a loop-nest including a small call tree of functions, these functions would need to be inlined by the compiler. Most of the time, this isn't a problem for small call trees. However, if these functions (or the loops of the loop-nest themselves) contain non-scalar local variables allocated on the stack (i.e., arrays like `double B[3][8]`), the compiler can run into problems. This is not all that uncommon in real-world C/C++ code. The issue here is that each vector lane needs its own copy of `B`, and there isn't nearly enough register space to hold all of the copies efficiently. Thus, the compiler would need to allocate space for and expand these non-scalar variables, probably placing them on the stack. The extension of such variables is called "rank expansion". The problem is that rank expansion is fairly expensive, and it is very difficult for the compiler to know whether the cost is worth paying in order to allow vectorization of the loop. Stack-based memory allocation takes time and is likely to result in additional memory motion. Frequently, compilers will simply default to not vectorizing loops which contain variables requiring rank expansion. With the Cray compiler, one will see a message like that seen in Table 7.12.1 if rank expansion is required. In order to vectorize such loops, the compiler usually needs to be instructed to do so with pragmas.

TABLE 7.12.1 Cray compiler message showing rank expansion inhibiting vectorization.

```
CC-6315 CC: VECTOR File = lulesh_ckfe.cc, Line = 368

  A loop was not vectorized because the target array (xd_loc)
would require rank expansion.
```

For a concrete example of a loop nest with variables needing rank expansion, see Excerpt 7.12.1 from the LULESH (Livermore Unstructured Lagrangian Explicit Shock Hydrodynamics) code [17]. This loop over elements is a good candidate for outer-loop vectorization, as `numElem` can get quite large (many thousands of iterations here) and there is a lot of work done in each iteration. Once all the functions called inside the loop are inlined, the loop body will represent what was originally over 350 lines of C/C++ code. This is a great opportunity to vectorize a lot of code in a small amount of developer time. However, this is a complex loop and the compiler is not going to vectorize it 100% automatically. One needs a heavy hammer to make a reluctant compiler vectorize this loop. The best option is currently the `omp simd` pragma. This OpenMP pragma instructs the compiler that the loop is safe to vectorize and to do so if at all possible. Excerpt 7.12.2 shows the `omp simd` pragma applied to the original LULESH loop. In this case, the compiler can see that the `B`, `D`, and `*_loc` arrays are declared inside the loop and are thus local to each vector lane. It considers these variables to be scoped `private` in OpenMP, and performs rank expansion automatically. In cases where the compiler cannot automatically scope the variables, one needs to explicitly provide a `private` clause, as seen in Excerpt 7.12.3.

When compiled with CCE, the original version of the loop presented in Excerpt 7.12.1 ran in 0.845s, while the vector version seen in Excerpt 7.12.2 took only 0.472s, giving a factor of 1.8 speedup with the vector loop. The Intel compiler improved from 0.713s to 0.445s for a factor of 1.6 speedup. Is this the best that can be done for this loop-nest? The answer is "no," but keep in mind the amount of developer time involved. Adding the `omp simd` pragma for a factor of 1.8 in performance is very easy and quick to do compared to what needs to be done in order to get the most out of the loop-nest.

```
void CalcKinematicsForElems( Domain &domain, Real_t *vnew,
                             Real_t deltaTime, Index_t numElem )
{
  // loop over all elements
#pragma omp parallel for
  for( Index_t k=0 ; k<numElem ; ++k ) {
    Real_t D[6], B[3][8] ; /* shape function derivatives */
    Real_t x_loc[8],  y_loc[8],  z_loc[8] ;
    Real_t xd_loc[8], yd_loc[8], zd_loc[8] ;
    Real_t vol, relativeVol, detJ = Real_t(0.0) ;
    const Index_t* const elemToNode = domain.nodelist(k) ;

    // get nodal coordinates from global arrays and copy into local arrays.
    CollectDomainNodesToElemNodes(domain, elemToNode, x_loc, y_loc, z_loc);
    // volume calculations
    vol = CalcElemVolume(x_loc, y_loc, z_loc );
    relativeVol = vol / domain.volo(k) ;
    vnew[k] = relativeVol ;
    domain.delv(k) = relativeVol - domain.v(k) ;
    // set characteristic length
    domain.arealg(k) = CalcElemCharacteristicLength(x_loc, y_loc, z_loc, vol);
    // get nodal velocities from global array and copy into local arrays.
    for( Index_t lnode=0 ; lnode<8 ; ++lnode ) {
      Index_t gnode = elemToNode[lnode];
      xd_loc[lnode] = domain.xd(gnode);
      yd_loc[lnode] = domain.yd(gnode);
      zd_loc[lnode] = domain.zd(gnode);
    }
    Real_t dt2 = Real_t(0.5) * deltaTime;
    for ( Index_t j=0 ; j<8 ; ++j ) {
       x_loc[j] -= dt2 * xd_loc[j];
       y_loc[j] -= dt2 * yd_loc[j];
       z_loc[j] -= dt2 * zd_loc[j];
    }
    CalcElemShapeFunctionDerivatives( x_loc, y_loc, z_loc, B, &detJ );
    CalcElemVelocityGradient( xd_loc, yd_loc, zd_loc, B, detJ, D );
    // put velocity gradient quantities into their global arrays.
    domain.dxx(k) = D[0];
    domain.dyy(k) = D[1];
    domain.dzz(k) = D[2];
  }
}
```

Excerpt 7.12.1 Loop-nest from LULESH before vectorization.

```
  // loop over all elements
#pragma omp parallel for simd
  for( Index_t k=0 ; k<numElem ; ++k ) {
    Real_t D[6], B[3][8] ; /* shape function derivatives */
    Real_t x_loc[8],  y_loc[8],  z_loc[8] ;
    Real_t xd_loc[8], yd_loc[8], zd_loc[8] ;
[...]
  }
```

Excerpt 7.12.2 Loop-nest from LULESH with vectorization using omp simd pragma.

```
  Real_t D[6], B[3][8] ; /* shape function derivatives */
  Real_t x_loc[8],  y_loc[8],  z_loc[8] ;
  Real_t xd_loc[8], yd_loc[8], zd_loc[8] ;

  // loop over all elements
#pragma omp parallel for simd private(B,D,x_loc,y_loc,z_loc,xd_loc,yd_loc,zd_loc)
  for( Index_t k=0 ; k<numElem ; ++k ) {
[...]
  }
```

Excerpt 7.12.3 Loop-nest from LULESH with vectorization using omp simd pragma including private clause.

Return for a moment to the rank expansion issue. When the compiler rank expands a variable, how exactly is the additional memory allocated in terms of the variable's dimensions/stride? That is, is the rank-expanded variable extended by the SIMD width on the "left" (i.e., `D[VECL][6]`) or on the "right" (i.e., `D[6][VECL]`)? Figure 7.12.1 illustrates the differences in memory layout and striding seen when rank-expanding `D` on the left versus the right.

NewD[VL][6]:	D1[0:5]	D2[0:5]	D(...)[0:5]	D(VL)[0:5]
NewD[6][VL]:	D(1:VL)[0]	D(1:VL)[1]	D(1:VL)[...]	D(1:VL)[5]

FIGURE 7.12.1 Striding layout from rank expansion on left versus right.

In the loop shown in Excerpt 7.12.2, the compiler has to choose between the easy option and the difficult option. This is due to the fact that the rank-expanded variables are passed into existing functions as arguments. The code in the functions expects the arguments to be unstrided. Thus, it would be easier for the compiler to use rank expansion on the left for these variables, keeping the instances of the original array contiguous (merely replicating the original variable). While it's possible for the compiler to inline all the functions and then translate all the array indicies so as to work with rank-expanded variables with a stride, this may be asking too much of the compiler. The Cray compiler has some support for this transformation, but it may not be able to perform this optimization in all situations.

Importantly, rank expansion on the right makes the newly introduced vector index stride-1, which can improve performance depending on the way the arrays are accessed in the loop. The example loop-nest from `CalcKinematicsForElems` being optimized here will indeed perform better with arrays rank-expanded on the right. If your compiler does not support these more advanced rank-expansion features, one could always manually rank-expand as needed. However, a detailed description of the manual process of rank-expanding variables will not be given here, in favor of recommending the use of OpenMP SIMD support when possible.

7.13 OUTER LOOP VECTORIZATION

A powerful vectorization technique that effectively utilizes cache is to vectorize on outer loops of a multiloop case. Often times a seemingly bandwidth limited kernel can be restructured to achieve good cache reuse and improve something that vectorizes. The intent is to improve the memory access pattern to minimize data accesses in a vectorized loop. The following example shown in Excerpt 7.13.1 is for the NEK-CEM application from Argonne National Laboratory [23].

```
42.  + 1-----------<       do e = 1,EE
43.  + 1 2---------<          do k = 1,N+1
44.  + 1 2 3-------<          do j = 1,N+1
45.    1 2 3 iV----<          do i = 1,N+1
46.  + 1 2 3 iV i--<             do l=1,N+1
47.    1 2 3 iV i                   u1r(i,j,k,e) = u1r(i,j,k,e) + D(i,l) * u1(l,j,k,e)
48.    1 2 3 iV i                   u2r(i,j,k,e) = u2r(i,j,k,e) + D(i,l) * u2(l,j,k,e)
49.    1 2 3 iV i                   u3r(i,j,k,e) = u3r(i,j,k,e) + D(i,l) * u3(l,j,k,e)
50.    1 2 3 iV i                   u1s(i,j,k,e) = u1s(i,j,k,e) + D(j,l) * u1(i,l,k,e)
51.    1 2 3 iV i                   u2s(i,j,k,e) = u2s(i,j,k,e) + D(j,l) * u2(i,l,k,e)
52.    1 2 3 iV i                   u3s(i,j,k,e) = u3s(i,j,k,e) + D(j,l) * u3(i,l,k,e)
53.    1 2 3 iV i                   u1t(i,j,k,e) = u1t(i,j,k,e) + D(k,l) * u1(i,j,l,e)
54.    1 2 3 iV i                   u2t(i,j,k,e) = u2t(i,j,k,e) + D(k,l) * u2(i,j,l,e)
55.    1 2 3 iV i                   u3t(i,j,k,e) = u3t(i,j,k,e) + D(k,l) * u3(i,j,l,e)
56.    1 2 3 iV i-->             enddo
57.    1 2 3 iV-->>>          enddo ; enddo ; enddo
60.    1----------->       enddo
```

Excerpt 7.13.1 Original loop-nest from the NEK-CEM code with inner-loop vectorization.

Notice that the loop is summing into arrays within a loop that does not reference the array, thus representing numerous dot products being computed into the array. When the loop listing simply has a number for a given loop that indicates that no optimization was performed on the loop.

The rewrite presented in Excerpt 7.13.2 unwinds the dot product into a scalar, and then the compiler vectorizes on the loop outside of the dot product.

With the dimensions of the computation EE was 125 and N was 15. The performance on KNL and Broadwell is displayed in the Table 7.13.1. Notice the restructuring improves the kernel performance on broadwell as well as KNL.

TABLE 7.13.1 Performance of original and optimized NEK-CEM loop.

	Original	Restructured	Speedup
Broadwell	0.034s	0.006s	5.313
KNL	0.090s	0.017s	5.294

```
62.  + 1-----------<        do e = 1,EE
63.  + 1 2---------<          do k = 1,N+1
64.  + 1 2 3-------<          do j = 1,N+1
65.    1 2 3 V-----<          do i = 1,N+1
66.    1 2 3 V                  tmp1_r = 0.0
67.    1 2 3 V                  tmp2_r = 0.0
68.    1 2 3 V                  tmp3_r = 0.0
69.    1 2 3 V                  tmp1_s = 0.0
70.    1 2 3 V                  tmp2_s = 0.0
71.    1 2 3 V                  tmp3_s = 0.0
72.    1 2 3 V                  tmp1_t = 0.0
73.    1 2 3 V                  tmp2_t = 0.0
74.    1 2 3 V                  tmp3_t = 0.0
75.    1 2 3 V 5---<            do l=1,N+1
76.    1 2 3 V 5                  tmp1_r = tmp1_r + D(i,l) * u1(l,j,k,e)
77.    1 2 3 V 5                  tmp2_r = tmp2_r + D(i,l) * u2(l,j,k,e)
78.    1 2 3 V 5                  tmp3_r = tmp3_r + D(i,l) * u3(l,j,k,e)
79.    1 2 3 V 5                  tmp1_s = tmp1_s + D(j,l) * u1(i,l,k,e)
80.    1 2 3 V 5                  tmp2_s = tmp2_s + D(j,l) * u2(i,l,k,e)
81.    1 2 3 V 5                  tmp3_s = tmp3_s + D(j,l) * u3(i,l,k,e)
82.    1 2 3 V 5                  tmp1_t = tmp1_t + D(k,l) * u1(i,j,l,e)
83.    1 2 3 V 5                  tmp2_t = tmp2_t + D(k,l) * u2(i,j,l,e)
84.    1 2 3 V 5                  tmp3_t = tmp3_t + D(k,l) * u3(i,j,l,e)
85.    1 2 3 V 5--->            enddo
86.    1 2 3 V                  u1r(i,j,k,e) = u1r(i,j,k,e) +tmp1_r
87.    1 2 3 V                  u2r(i,j,k,e) = u2r(i,j,k,e) +tmp2_r
88.    1 2 3 V                  u3r(i,j,k,e) = u3r(i,j,k,e) +tmp3_r
89.    1 2 3 V                  u1s(i,j,k,e) = u1s(i,j,k,e) +tmp1_s
90.    1 2 3 V                  u2s(i,j,k,e) = u2s(i,j,k,e) +tmp2_s
91.    1 2 3 V                  u3s(i,j,k,e) = u3s(i,j,k,e) +tmp3_s
92.    1 2 3 V                  u1t(i,j,k,e) = u1t(i,j,k,e) +tmp1_t
93.    1 2 3 V                  u2t(i,j,k,e) = u2t(i,j,k,e) +tmp2_t
94.    1 2 3 V                  u3t(i,j,k,e) = u3t(i,j,k,e) +tmp3_t
95.    1 2 3 V--->>>          enddo ; enddo ; enddo
```

Excerpt 7.13.2 Optimized loop-nest from the NEK-CEM code with improved vectorization.

7.14 EXERCISES

7.1 Enumerate areas where the compiler has to introduce overhead in order to vectorize a loop.

7.2 Under what conditions might a vectorized loop run at the same speed or slower than the scalar version?

7.3 Consider the restructured version of loops 41091 below. Can you create a better rewrite?

```
 98.+1 2----------<        DO 41091 K = KA, KE, -1
 99.+1 2 3--------<          DO 41091 J = JA, JE
100. 1 2 3 Vr2----<            DO 41091 I = IA, IE
101. 1 2 3 Vr2                   AA(I,K,L,J) = AA(I,K,L,J)-BB(I,J,1,K)*AA(I,K+1,L,1)
102. 1 2 3 Vr2              *    - BB(I,J,2,K)*AA(I,K+1,L,2)-BB(I,J,3,K)*AA(I,K+1,L,3)
103. 1 2 3 Vr2              *    - BB(I,J,4,K)*AA(I,K+1,L,4)-BB(I,J,5,K)*AA(I,K+1,L,5)
104. 1 2 3 Vr2-->>> 41091 CONTINUE
```

7.4 Consider the restructured version of loops 4502* below. Can you create a better rewrite?

```
 74.     1 Vr2----------<        DO 45021 I = 1,N
 75.     1 Vr2                    F(I) = A(I) + .5
 76.     1 Vr2----------> 45021 CONTINUE
 77.     1
 78.   + 1 f------------<        DO 45022 J = 1, 10
 79.     1 f Vr2--------<         DO 45022 I = 1, N
 80.     1 f Vr2                   D(I,J) = B(J) * F(I)
 81.     1 f Vr2------->> 45022 CONTINUE
 82.     1
 83.     1 iV-----------<        DO 45023 K = 1, 5
 84.   + 1 iV fi--------<         DO 45023 J = 1, 10
 85.   + 1 iV fi ir4----<          DO 45023 I = 1, N
 86.     1 iV fi ir4                C(K,I,J) = D(I,J) * E(K)
 87.     1 iV fi ir4-->>> 45023 CONTINUE
```

7.5 Consider the restructured version of loops 4701* below. Can you create a better rewrite?

```
 97.     1               C      THE RESTRUCTURED
 98.   + 1 2----------<         DO 47016   K = 2, N - 1
 99.   + 1 2 f--------<          DO 47013  J = 2, 3
100.     1 2 f fVr2---<           DO 47013 I = 2, N
101.     1 2 f fVr2                A(I,J) = (1. - PX - PY - PZ) * B(I,J,K)
102.     1 2 f fVr2         1             + .5 * PX * ( B(I+1,J,K) + B(I-1,J,K) )
103.     1 2 f fVr2         2             + .5 * PY * ( B(I,J+1,K) + B(I,J-1,K) )
104.     1 2 f fVr2         3             + .5 * PZ * ( B(I,J,K+1) + B(I,J,K-1) )
105.     1 2 f fVr2-->> 47013 CONTINUE
106.     1 2
107.     1 2                     IF (K .EQ. 2) THEN
108.     1 2
109.   + 1 2 f--------<           DO 47014  J =2, 3
110.     1 2 f f------<            DO 47014 I =2, N
111.     1 2 f f                    C(I,J)    = A(I,J)
112.     1 2 f f----->> 47014     CONTINUE
113.     1 2
114.     1 2                     ELSE
115.     1 2
116.   + 1 2 f--------<           DO 47015  J = 2, 3
117.     1 2 f f------<            DO 47015 I = 2, N
118.     1 2 f f                    B(I,J,K-1)  = C(I,J)
119.     1 2 f f                    C(I,J)      = A(I,J)
120.     1 2 f f----->> 47015     CONTINUE
121.     1 2
122.     1 2                     ENDIF
123.     1 2
124.     1 2----------> 47016 CONTINUE
125.     1                        K = N
126.   + 1 ir2----------<         DO 47017 I = 2, N
127.   + 1 ir2 iw-------<          DO 47017  J = 2, 3
128.     1 ir2 iw                   A(I,J) = (1. - PX - PY - PZ) * B(I,J,K)
129.     1 ir2 iw             1            + .5 * PX * ( B(I+1,J,K) + B(I-1,J,K) )
130.     1 ir2 iw             2            + .5 * PY * ( B(I,J+1,K) + B(I,J-1,K) )
131.     1 ir2 iw             3            + .5 * PZ * ( B(I,J,K+1) + B(I,J,K-1) )
132.     1 ir2 iw                   B(I,J,K)    = A(I,J)
133.     1 ir2 iw                   B(I,J,K-1)  = C(I,J)
134.     1 ir2 iw                   C(I,J)      = A(I,J)
135.     1 ir2 iw------>> 47017 CONTINUE
```

CHAPTER 8

Hybridization of an Application

CONTENTS

8.1 FOREWORD BY JOHN LEVESQUE

For 10 years, the Center of Excellence at ORNL thrived. We were part of a couple Gordon Bell awards, and the XT line was extremely successful. With Ungaro at the helm and the best HPC architecture going, Cray Inc. became very successful. In 2011, I published my second book on multicore MPP systems, targeting the Cray XT. The title of the book was *High Performance Computers – Programming and Applications*. My co-author was Gene Wagenbreth, who I met during the ILLIAC IV days and have continued to work with over the past 40 years. In 2014, ORNL selected IBM and NVIDA to supply the follow-on to Titan, and I started Cray's Supercomputing Center of Excellence for the Trinity system to be installed at Los Alamos Scientific laboratory to assist researchers in optimizing applications for Intel's new Knight's Landing – the primary impetus for writing this book.

8.2 INTRODUCTION

The reason for threading on manycore nodes is greater as more and more cores are placed on the node – and threading is required to effectively utilize a GPU. Although MPI has shown that it can perform extremely well across the physical cores on a node, there are numerous reasons why MPI should not be used on hyper-threads. So the fact remains that some amount of threading should be used on this generation of high performance computers. There are many pitfalls to threading that have extended the life of the MPI-only applications. The NUMA characteristics of the Xeon and KNL nodes can significantly degrade performance. Additionally, even with the introduction of high-bandwidth memory, none of the HPC nodes have sufficient memory bandwidth to support the computational power of the cores on the node. Therefore, threading should be performed in such a way that the threads can share the available bandwidth. In this chapter we will consider OpenMP threading and discuss how best to utilize OpenMP directives to thread a previously unthreaded application. There is one tremendous advantage to threading that cannot be overlooked. In complex multiphysics codes that perform adaptive mesh refinement, work can be redistributed across the threads much easier than distributing work across MPI tasks. When work is to be distributed between MPI tasks, data must be moved, but when redistributing work across threads, data does not have to be moved.

8.3 THE NODE'S NUMA ARCHITECTURE

As we have seen in previous chapters, all Xeon and Knights systems will have significant NUMA issues on the node. As more and more cores are placed on the node, the cache coherency and the memory hierarchy will introduce latencies and bandwidth limitations that reduce the sustained performance of an application. As we have seen, all-MPI addresses some of these issues by forcing locality and simplifying the memory hierarchy within the MPI task. In summary, running MPI on all cores of a multi/manycore node gives good performance.

1. All MPI forces locality. Each MPI task allocates/deallocates within the NUMA region it is running on.

2. All MPI allows tasks to run asynchronously which allows for excellent sharing of memory bandwidth – a limited resource on the node.

Can OpenMP be implemented to mimic All-MPI? By making sure that a thread allocated as much of the memory that it uses, locality issues can be addressed. By having large parallel regions to allow the threads to run asynchronously, memory bandwidth can be shared across the threads as it is shared across the MPI tasks in an all-MPI scenario.

When an application allocates an array, the array is not fully allocated until it is initialized. The processor that initializes the array, referred to as "first touch", will allocate the memory for the array in its closest memory. Often when OpenMP is incorporated into an application, no attention is made to which core actually initializes the array. If the initialization loop is not threaded, then the master thread will allocate the memory as it is being initialized in that loop. The best technique for addressing the NUMA issues is to initialize the data in a parallel loop that employs the threads the same way they would be employed in the computational section of the program. For example, in a three-dimensional decomposition where threads would be used on the layers of the grid, the initialization of all of the grid variables should be performed in a parallel loop where the threads initialize the layers that they would be working on.

To ensure a uniform memory allocation, the grid storage for a layer should be a multiple of a memory page. It is not always possible to parallelize all computational routines with the same decomposition. If this is the case, then some of the computational routines will have to be parallelized with a different decomposition, and consequently the arrays will probably come from a distant memory. The most important aspect of utilizing the NUMA architecture of a node is to ensure that the hybrid application has the MPI tasks and threads within the MPI allocated correctly on the node. For example, on the KNL running in sub-NUMA clustering 4 (SNC4) mode, an application should use at least four MPI tasks, one within each SNC4 region. Then the threads for each MPI task should be allocated within each of the quadrants. Do not assume that this happens automatically. There are commands one can use to print out the affinity of each thread being used. When using Moab/Torque on a Cray XC40 with KNL, a good approach is to use the environment variable and aprun command presented in Excerpt 8.3.1.

```
setenv OMP_PLACES cores
setenv OMP_NUM_THREADS 4
aprun -n 64 -N 4 -d 16 -cc depth compute.x
```

Excerpt 8.3.1 Reasonable default run settings with Moab/Torque on KNL.

This will allocate the 4 MPI tasks on each node 16 cores apart so each MPI task will be able to allocate 15 additional threads close to it. There are numerous ways to achieve the correct spacing and binding and unfortunately they are quite different depending upon what work load manager is used and what vendor has supplied the system. The principal point is that getting the correct thread allocation and binding is extremely important for the best OpenMP performance.

8.4 FIRST TOUCH IN THE HIMENO BENCHMARK

Himeno will be used to illustrate the strategy for ensuring locality within a NUMA node. Himeno is a simple benchmark available on the web that computes a Jacobi iterative solver on a three-dimensional grid [24]. The benchmark is quite simple; it only contains a couple of initialization loops, a major stencil loop, and a copy loop, all of which should be parallelized. First, the major computational loops are presented in Excerpt 8.4.1. These are quite simple to parallelize; we have a reduction on WGOSA and a couple of private variables. The major arrays that are used in these loops are a, b, c, wrk1, wrk2, and p.

```
!$OMP PARALLEL DO PRIVATE(SO,SS) REDUCTION(+:WGOSA)
      DO K=2,kmax-1
         DO J=2,jmax-1
            DO I=2,imax-1
               SO=a(I,J,K,1)*p(I+1,J,K)+a(I,J,K,2)*p(I,J+1,K)
  1                  +a(I,J,K,3)*p(I,J,K+1)
  2                  +b(I,J,K,1)*(p(I+1,J+1,K)-p(I+1,J-1,K)
  3                  -p(I-1,J+1,K)+p(I-1,J-1,K))
  4                  +b(I,J,K,2)*(p(I,J+1,K+1)-p(I,J-1,K+1)
  5                  -p(I,J+1,K-1)+p(I,J-1,K-1))
  6                  +b(I,J,K,3)*(p(I+1,J,K+1)-p(I-1,J,K+1)
  7                  -p(I+1,J,K-1)+p(I-1,J,K-1))
  8                  +c(I,J,K,1)*p(I-1,J,K)+c(I,J,K,2)*p(I,J-1,K)
  9                  +c(I,J,K,3)*p(I,J,K-1)+wrk1(I,J,K)
               SS=(SO*a(I,J,K,4)-p(I,J,K))*bnd(I,J,K)
               WGOSA=WGOSA+SS*SS
               wrk2(I,J,K)=p(I,J,K)+OMEGA *SS
      enddo ; enddo ; enddo
!$OMP PARALLEL DO
      DO K=2,kmax-1
         DO J=2,jmax-1
            DO I=2,imax-1
               p(I,J,K)=wrk2(I,J,K)
      enddo ; enddo ; enddo
```

Excerpt 8.4.1 Major computation loops in Himeno.

Now we need to find where these arrays are allocated; the initialization code is presented in Excerpt 8.4.2. If this loop is not parallelized, then thread 0 would allocate all the arrays within the memories closest to it. With this parallelization, each thread will allocate its layers within its closest memories.

```
!$OMP PARALLEL DO
      do k=1,mkmax
         do j=1,mjmax
            do i=1,mimax
               a(i,j,k,1:4) = 0.0
               b(i,j,k,1:3) = 0.0
               c(i,j,k,1:3) = 0.0
               p(i,j,k)     = 0.0
               wrk1(i,j,k)  = 0.0
               wrk2(i,j,k)  = 0.0
               bnd(i,j,k)   = 0.0
      enddo ; enddo ; enddo
```

Excerpt 8.4.2 Array initialization loops in Himeno.

The next pitfall to OpenMP threading is having low granularity in a parallel region. The lower the granularity, the more OpenMP overhead will be exposed and the threads will essentially execute in lock step in a small parallel region. When the threads need operands from memory, all the threads will be requesting fetches for the operands at the same time which results in a very heavy demand on the memory bandwidth. Higher granularity allows the threads to get out of lock-step and share the bandwidth over a longer time. When memory requests are spread out, the likelihood of hitting the memory bandwidth limit is smaller. The higher the granularity, the more the threads can get out of order and share the memory bandwidth. The following set of loops are from a benchmark called BGW. Excerpt 8.4.3 shows one of the major loops contained within the BGW benchmark.

```
139.    1 M--------< !$OMP  parallel do default(none)                                 &
140.    1 M          !$OMP&  private(ig,igmax,mygpvar1,my_igp)                         &
141.    1 M          !$OMP&  shared(leftvector,matngmatmgpd,n1,ncouls,ngpown,rightvector)
142.  + 1 M m------<          do my_igp = 1, ngpown
143.    1 M m                   if (my_igp .gt. ncouls .or. my_igp .le. 0) cycle
144.    1 M m
145.    1 M m                   igmax=ncouls
146.    1 M m
147.    1 M m                   mygpvar1 = CONJG(leftvector(my_igp,n1))
148.    1 M m
149.    1 M m Vr2--<            do ig = 1, igmax
150.    1 M m Vr2                 matngmatmgpD(ig,my_igp) = rightvector(ig,n1) * mygpvar1
151.    1 M m Vr2-->            enddo
152.    1 M m----->>          enddo
```

Excerpt 8.4.3 Original parallelized major loop-nest in BGW.

The compiler listing illustrates that the innermost loop is vectorized and unrolled by 2. In this case the outermost loop is 100 and the innermost loop is 800. How might we optimize this and thread it using OpenMP? First, there are some inefficiencies in the code. It is unclear as to the function of the `if(···) cycle` construct. This would cause a problem if we want to parallelize the outermost loop. We can eliminate the `cycle` by modifying the bounds of the outermost loop; simply change the outer loop to `do my_igp = 1, min(ngpown,ncouls)`. Next we substitute `ncouls` for `igmax` and replace `mygpvar1` with the expression `CONJG(leftvector(my_igp,n1))`. The resulting code is presented in Excerpt 8.4.4.

```
137.    1 M----------< !$OMP PARALLEL DO
138.    1 M imV------<       do ig = 1, ncouls
139.  + 1 M imV ir4--<         do my_igp = 1, min(ngpown,ncouls)
140.    1 M imV ir4              matngmatmgpD(ig,my_igp) = rightvector(ig,n1) * &
141.    1 M imV ir4                             ICONJG(leftvector(my_igp,n1))
142.    1 M imV ir4-->         enddo
143.    1 M imV----->>       enddo
```

Excerpt 8.4.4 Restructured parallelized major loop-nest in BGW.

We now switch the inner and outer loops, and parallelize the loop on `ig`. The compiler in this case is smart enough to also vectorize on the `ig` loop since it is the contiguous index. This gives us excellent granularity and the best performance. The restructured version runs three times faster than the original OpenMP version on 16 threads on a Haswell system.

The next kernel from this benchmark in Excerpt 8.4.5 is much more complicated, and several issues can be seen. The outermost loop is data dependent and the straightforward approach would be to parallelize the loop on 6000. Can we do better than this? Notice that there is some array syntax in the loop on line 221 and 222, and examination determines that the computation is vectorized on `iw` as well.

```
202.  + 1 2----------<         do ifreq=1,nFreq                              (240)
203.    1 2
204.    1 2                        schDt = schDt_array(ifreq)
205.    1 2
206.    1 2                        cedifft_zb = dFreqGrid(ifreq)
207.    1 2                        cedifft_coh = CMPLX(cedifft_zb,0D0)- dFreqBrd(ifreq)
208.    1 2
209.    1 2                        if (ifreq .ne. 1) then
210.    1 2                          cedifft_zb_right = cedifft_zb
211.    1 2                          cedifft_zb_left = dFreqGrid(ifreq-1)
212.    1 2                          schDt_right = schDt
213.    1 2                          schDt_left = schDt_array(ifreq-1)
214.    1 2                          schDt_avg = 0.5D0 * ( schDt_right + schDt_left )
215.    1 2                          schDt_lin = schDt_right - schDt_left
216.    1 2                          schDt_lin2 = schDt_lin/(cedifft_zb_right-cedifft_zb_left)
217.    1 2                        endif
218.    1 2
219.    1 2             ! The below two lines are for sigma1 and sigma3
220.    1 2              if (ifreq .ne. nFreq) then
221.    1 2 fVr2----<>   schDi(:) = schDi(:) - CMPLX(0.d0,pref(ifreq)) * schDt / ( wxi(:)-cedifft_coh)
222.    1 2 f-------<>   schDi_corb(:) = schDi_corb(:) - CMPLX(0.d0,pref(ifreq)) * schDt / ( wxi(:)-cedifft_cor)
223.    1 2              endif
224.    1 2             if (ifreq .ne. 1) then
226.    1 2 M--------< !$OMP  parallel do default(none)                                &
227.    1 2 M          !$OMP&   private (intfact,iw,schdt_lin3)                          &
228.    1 2 M          !$OMP&   shared  (cedifft_zb_left,cedifft_zb_right,flag_occ,nfreqeval,   &
229.    1 2 M          !$OMP&            prefactor,sch2di,schdi_cor,schdt_avg,schdt_left,       &
230.    1 2 M          !$OMP&            schdt_lin,schdt_lin2,wxi)
231.    1 2 M mV-----<               do iw = 1, nfreqeval                           (6000)
232.    1 2 M mV       !These lines are for sigma2
233.    1 2 M mV                       intfact=abs((wxi(iw)-cedifft_zb_right)/(wxi(iw)-cedifft_zb_left))
234.    1 2 M mV                       if (intfact .lt. 1d-4) intfact = 1d-4
235.    1 2 M mV                       if (intfact .gt. 1d4) intfact = 1d4
236.    1 2 M mV                       intfact = -log(intfact)
237.    1 2 M mV                       sch2Di(iw) = sch2Di(iw) - CMPLX(0.d0,prefactor) * schDt_avg * intfact
238.    1 2 M mV       !These lines are for sigma4
239.    1 2 M mV                       if (flag_occ) then
240.    1 2 M mV                         intfact=abs((wxi(iw)+cedifft_zb_right)/(wxi(iw)+cedifft_zb_left))
241.    1 2 M mV                         if (intfact .lt. 1d-4) intfact = 1d-4
242.    1 2 M mV                         if (intfact .gt. 1d4) intfact = 1d4
243.    1 2 M mV                         intfact = log(intfact)
244.    1 2 M mV                         schDt_lin3 = (schDt_left + schDt_lin2*(-wxi(iw)-  &
                                                      cedifft_zb_left))*intfact
245.    1 2 M mV                       else
246.    1 2 M mV                         schDt_lin3 = (schDt_left + schDt_lin2*(wxi(iw)-   &
                                                      cedifft_zb_left))*intfact
247.    1 2 M mV                       endif
248.    1 2 M mV                       schDt_lin3 = schDt_lin3 + schDt_lin
249.    1 2 M mV                       schDi_cor(iw) = schDi_cor(iw) - CMPLX(0.d0,prefactor) * schDt_lin3
250.    1 2 M mV---->>                enddo
251.    1 2                         endif
252.    1 2---------->         enddo
```

Excerpt 8.4.5 Original more complicated parallelized loop-nest in BGW.

The same approach that was used in the simple kernel can be used here. The innermost loop will be pulled outside the outer loop. That way we may be able to get vectorization and parallelization on the `iw` loop. The restructured code is presented in Excerpt 8.4.6.

Now all the threads compute the code that was between the loops, and those operands will be held in registers for the innermost loop. Once again, the compiler vectorizes and parallelizes on the outermost loop which results in much better performance than the original parallel version.

In both of these kernels we have successfully increased the size of the parallel region, giving the threads more latitude to get out of sync and share the memory bandwidth. This version ran eight times faster than the original threaded version with 16 threads on a Haswell node. We did not address NUMA issues in these examples – since they are kernelized versions of a real application this was not considered.

```
193.    1 M----------< !$OMP  parallel do default(none)                                  &
194.    1 M            !$OMP&   private (cedifft_coh,cedifft_zb,cedifft_zb_left,         &
195.    1 M            !$OMP&            cedifft_zb_right,ifreq,intfact,iw,schdt,schdt_avg,    &
196.    1 M            !$OMP&            schdt_left,schdt_lin,schdt_lin2,schdt_lin3,           &
197.    1 M            !$OMP&            schdt_right)                                          &
198.    1 M            !$OMP&   shared  (cedifft_cor,dfreqbrd,dfreqgrid,flag_occ,nfreq,        &
199.    1 M            !$OMP&            nfreqeval,pref,prefactor,sch2di,schdi,schdi_cor,      &
200.    1 M            !$OMP&            schdi_corb,schdt_array,wxi)
201.    1 M mV-------<        do iw = 1, nfreqeval
202.    1 M mV 4-----<          do ifreq=1,nFreq
203.    1 M mV 4
204.    1 M mV 4                  schDt = schDt_array(ifreq)
205.    1 M mV 4
206.    1 M mV 4                  cedifft_zb = dFreqGrid(ifreq)
207.    1 M mV 4                  cedifft_coh = CMPLX(cedifft_zb,0D0)- dFreqBrd(ifreq)
208.    1 M mV 4
209.    1 M mV 4                  if (ifreq .ne. 1) then
210.    1 M mV 4                    cedifft_zb_right = cedifft_zb
211.    1 M mV 4                    cedifft_zb_left = dFreqGrid(ifreq-1)
212.    1 M mV 4                    schDt_right = schDt
213.    1 M mV 4                    schDt_left = schDt_array(ifreq-1)
214.    1 M mV 4                    schDt_avg = 0.5D0 * ( schDt_right + schDt_left )
215.    1 M mV 4                    schDt_lin = schDt_right - schDt_left
216.    1 M mV 4                    schDt_lin2 = schDt_lin/(cedifft_zb_right-cedifft_zb_left)
217.    1 M mV 4                  endif
218.    1 M mV 4
219.    1 M mV 4       ! The below two lines are for sigma1 and sigma3
220.    1 M mV 4             if (ifreq .ne. nFreq) then
221.    1 M mV 4               schDi(iw) = schDi(iw) - CMPLX(0.d0,pref(ifreq)) * schDt / ( wxi(iw)-cedifft_coh)
222.    1 M mV 4               schDi_corb(iw) = schDi_corb(iw) - CMPLX(0.d0,pref(ifreq)) * schDt / &
                                                                      ( wxi(iw)-cedifft_cor)
223.    1 M mV 4                  endif
224.    1 M mV 4                  if(ifreq.ne.1)then
225.    1 M mV 4       !These lines are for sigma2
226.    1 M mV 4                      intfact=abs((wxi(iw)-cedifft_zb_right)/(wxi(iw)-cedifft_zb_left))
227.    1 M mV 4                      if (intfact .lt. 1d-4) intfact = 1d-4
228.    1 M mV 4                      if (intfact .gt. 1d4) intfact = 1d4
229.    1 M mV 4                      intfact = -log(intfact)
230.    1 M mV 4                      sch2Di(iw) = sch2Di(iw) - CMPLX(0.d0,prefactor) * schDt_avg * intfact
231.    1 M mV 4       !These lines are for sigma4
232.    1 M mV 4                      if (flag_occ) then
233.    1 M mV 4                        intfact=abs((wxi(iw)+cedifft_zb_right)/(wxi(iw)+cedifft_zb_left))
234.    1 M mV 4                        if (intfact .lt. 1d-4) intfact = 1d-4
235.    1 M mV 4                        if (intfact .gt. 1d4) intfact = 1d4
236.    1 M mV 4                        intfact = log(intfact)
237.    1 M mV 4                        schDt_lin3 = (schDt_left + schDt_lin2*(-wxi(iw)-  &
                                                                        cedifft_zb_left))*intfact
238.    1 M mV 4                      else
239.    1 M mV 4                        schDt_lin3 = (schDt_left + schDt_lin2*(wxi(iw)-   &
                                                                         cedifft_zb_left))*intfact
240.    1 M mV 4                      endif
241.    1 M mV 4                      schDt_lin3 = schDt_lin3 + schDt_lin
242.    1 M mV 4                      schDi_cor(iw) = schDi_cor(iw) - CMPLX(0.d0,prefactor) * schDt_lin3
243.    1 M mV 4                  endif
244.    1 M mV 4--->>>        enddo ; enddo
```

Excerpt 8.4.6 Restructured more complicated parallelized loop-nest in BGW.

8.5 IDENTIFYING WHICH LOOPS TO THREAD

The ultimate method for adding OpenMP threading is to investigate the SPMD approach discussed later in this section. The SPMD approach can require significant refactoring of an application and in some cases may not be appropriate. Using OpenMP at a high level can produce good results, especially when employed within a NUMA region on the node. Employing OpenMP at a high level gives the threads the ability to get out of sync and effectively share available bandwidth. When utilizing OpenMP at a high level, there can be significant work required to examine the scoping of the high-level loop. This is especially the case if the parallel loop contains a significant number of subroutine and function calls. We will investigate the parallelization of VH1 to illustrate the issues high-level OpenMP can encounter and how one can overcome those issues.

TABLE 8.5.1 Call tree profile from VH1 showing loop statistics.

```
Table:  Function Calltree View

  Time% |        Time |       Calls |Calltree
        |             |             | PE=HIDE

 100.0% | 194.481232 | 83189234.5 |Total
|--------------------------------------------------------
| 100.0% | 194.481188 | 83189034.5 |vhone_
||-------------------------------------------------------
||  25.8% |  50.165489 | 13882484.0 |sweepy_
|||------------------------------------------------------
3||  20.7% |  40.195080 | 13854848.0 |sweepy_.LOOP.1.li.32
4||        |            |            | sweepy_.LOOP.2.li.33
5||        |            |            |  ppmlr_
||||||---------------------------------------------------
6|||||    7.9% |  15.416847 |  5306112.0 |remap_
|||||||--------------------------------------------------
7||||||    4.8% |   9.274438 |  3537408.0 |parabola_
7||||||    2.3% |   4.454351 |   589568.0 |remap_(exclusive)
|||||||==================================================
6|||||    5.1% |   9.878556 |   589568.0 |riemann_
6|||||    2.4% |   4.706512 |  1768704.0 |parabola_
6|||||    2.1% |   4.062440 |  2947840.0 |evolve_
6|||||    1.0% |   1.980040 |  1179136.0 |states_
||||||===================================================
3||    2.3% |   4.550284 |      9212.0 |mpi_alltoall
3||    1.5% |   3.011031 |      9212.0 |sweepy_(exclusive)
3||    1.2% |   2.409094 |      9212.0 |mpi_alltoall_(sync)
```

The first approach to high-level threading is to locate the best potential OpenMP loops within the application. This can be achieved by obtaining a call tree profile which includes loop statistics. Table 8.5.1 shows such a profile for the VH1 application on a representative input case. The call tree shows one of four separate sweep routines. In the main routine, VH1 calls all four of the sweep routines, and in total they utilize over 95% of the total execution time. Within each sweep routine, there is a double-nested loop and within the double-nested loop there is a significant call tree. Not shown is an additional inner loop that we can use for vectorization. In order to employ OpenMP on the outermost loop, a scoping analysis must be performed to determine if it is safe to parallelize and what if any refactorings are required to ensure that the parallelization is safe. When the loop in question contains so many calls, the scoping can be quite tedious.

Within this loop there are numerous issues that must be resolved. First, we will concentrate on the identification of private variables. Private variables require that all the threads have their own local copy. Private variables are scalars or arrays where the variable is set each pass through the loop prior to being used. This is trivial for scalars. However, it is a significant issue when an array which is not indexed by the parallel loop is utilized. For example, consider the code in Except 8.5.1 from the `parabola` routine.

```
  do n = nmin-2, nmax+1
    diffa(n) = a(n+1) - a(n)
  enddo
!  Equation 1.7 of C&W
!    da(j) = D1 * (a(j+1) - a(j)) + D2 * (a(j) - a(j-1))
  do n = nmin-1, nmax+1
   da(n) = para(n,4) * diffa(n) + para(n,5) * diffa(n-1)
   da(n) = sign( min(abs(da(n)), 2.0*abs(diffa(n-1)), 2.0*abs(diffa(n))), da(n) )
  enddo
```

Excerpt 8.5.1 Example from VH1 with array variable with private scope.

The question that must be answered to ensure that there is not a race condition with respect to `diffa`: "are all the elements of the array set in the same iteration of the parallel loop prior to being used?" Notice in the first loop `diffa` is set with a range of `nmin-2` to `nmax+1`. In the second loop, `diffa(n)` and `diffa(n-1)` are used. `diffa(n-1)` will reference `diffa(nmin-2)` through `diffa(nmax)` which is set in the first loop. `diffa(n)` references a range that is also included in the first loop. This example is quite easy, and we can safely scope `diffa` as private. The next example within the `evolve` routine in Excerpt 8.5.3 is much more difficult.

```
! grid position evolution

do n = nmin-3, nmax + 4
  dm    (n) = r(n) * dvol1(n)
  dtbdm(n) = dt / dm(n)
  xa1   (n) = xa(n)
  xa    (n) = xa(n) + dt * umid(n) / radius
  upmid(n) = umid(n) * pmid(n)
enddo

xa1(nmin-4) = xa(nmin-4)
xa1(nmax+5) = xa(nmax+5)

do n = nmin-4, nmax+5
  xa2(n)    = xa1(n) + 0.5*dx(n)
  dx (n)    = xa(n+1) - xa(n)
  xa3(n)    = xa (n) + 0.5*dx(n)
enddo
```

Excerpt 8.5.2 Example code from VH1 needing scoping.

In `evolve` we have uses of `xa`. Notice that `evolve` updates `xa` from `nmin-3` to `nmax+4` and then uses `xa(nmin-4)` to `nmax+6`. Since all of the values that are used in the second loop in `evolve` are not set in the first loop, they better be set somewhere else. Indeed in the **sweep** routine that calls **ppmlr** which calls **evolve**, **xa** is initialized from 7 to **nmax+6**, as can be seen in Excerpt 8.5.3. Once again **xa** can be scoped as private.

```
! nmin    = 7
! nmax    = imax + 6
!
do i = 1, imax+6
     n = i + 6
     r  (n) = zro(i,j,k)
     p  (n) = zpr(i,j,k)
     u  (n) = zux(i,j,k)
     v  (n) = zuy(i,j,k)
     w  (n) = zuz(i,j,k)
     f  (n) = zfl(i,j,k)

     xa0(n) = zxa(i)
     dx0(n) = zdx(i)
     xa (n) = zxa(i)
     dx (n) = zdx(i)
     p  (n) = max(smallp,p(n))
enddo
```

Excerpt 8.5.3 Example VH1 initialization code.

As this example illustrates, scoping private arrays can be very tedious. However, it is necessary to avoid the most difficult bug to identify in threaded code – a race condition that frequently gives the right answers but may occasionally give wrong answers. In this particular loop, there are over 20 variables that must be analyzed as discussed to ensure correct OpenMP.

```
hdt = 0.5*dt

do n = nmin-4, nmax+4
  Cdtdx (n) = sqrt(gam*p(n)/r(n))/(dx(n)*radius)
  svel      = max(svel,Cdtdx(n))
  Cdtdx (n) = Cdtdx(n)*hdt
  fCdtdx(n) = 1. - fourthd*Cdtdx(n)
enddo
```

Excerpt 8.5.4 Example code from VH1 with reduction.

Another example where there really is a dependency is found within the **state** routine that is called from **ppmlr**. This code is presented in Excerpt 8.5.4. In this code sequence, there is a reduction being performed on **svel**; the loop is finding the maximum sound speed to compute the timestep for the next cycle. If one thread is updating **svel** while another is fetching **svel**, we have a race condition and will potentially get incorrect answers. The solution is to introduce a critical region to ensure that only one thread performs the update at the same time, as seen in Excerpt 8.5.5.

Unfortunately, this is not very efficient since the critical region is larger than it needs to be. If we place the critical region within the loop, the loop will not vectorize. So a simple refactoring should be performed, as presented in Excerpt 8.5.6.

```
hdt = 0.5*dt

!$omp critical
do n = nmin-4, nmax+4
  Cdtdx (n) = sqrt(gam*p(n)/r(n))/(dx(n)*radius)
  svel      = max(svel,Cdtdx(n))
  Cdtdx (n) = Cdtdx(n)*hdt
  fCdtdx(n) = 1. - fourthd*Cdtdx(n)
enddo
!$omp end critical
```

Excerpt 8.5.5 Example code from VH1 with reduction and critical region.

```
hdt   = 0.5*dt
svel0 = 0.0

do n = nmin-4, nmax+4
  Cdtdx (n) = sqrt(gam*p(n)/r(n))/(dx(n)*radius)
  svel0(n)      = max(svel(n),Cdtdx(n))
  Cdtdx (n) = Cdtdx(n)*hdt
  fCdtdx(n) = 1. - fourthd*Cdtdx(n)
enddo

!$omp critical
do n = nmin-4, nmax +4
  Svel = max(svel0(n),svel)
enddo
!$omp end critical
```

Excerpt 8.5.6 Restructured VH1 code with reduction and critical region.

Now we have shortened our critical region, and we are getting vectorization on the principal loop. We could not use the reduction clause for this loop because the OpenMP directives are way up the call chain and reduction can only be used when the reduction operator is within the routine that contains the parallel loop. This brings up the next issue. Once all the variables have been scoped, only the variables that are visible in the routine that has the parallel loop can be scoped with directives. Down the call chain, the variables that are private must be local variables, so that each thread will receive a copy of the variable when the routines are called. All shared variables (e.g., `svel`) must be in a global space such as a common block or module or passed down the call chain.

Actually, in this application we have an issue with scoping down the call chain. Many of the variables that we would like to scope as private are contained within a module, as seen in Excerpt 8.5.7. Many of these variables should be private. However, since they are contained within a module they have global scope. This will result in a race condition, since all of the arrays in the module will be shared by all threads, and a significant amount of overwriting will occur. There is a special OpenMP directive for just such a case, shown in Excerpt 8.5.8. When placed in the module, this directive will set aside a copy of all of these variables for each thread. Once all of the variables within the parallel loop are scoped, the directive can be applied to the parallel loop.

```
! module sweeps
! Data structures used in 1D sweeps, dimensioned maxsweep  (set in sweepsize.mod)
!-----------------------------------------------------------------------
use sweepsize

integer :: nmin, nmax, ngeom, nleft, nright        ! number of first and last real zone
real,dimension(maxsweep) :: r, p, e, q, u, v, w    ! fluid variables
real,dimension(maxsweep) :: xa, xa0, dx, dx0, dvol ! coordinate values
real,dimension(maxsweep) :: f, flat                ! flattening parameter
real,dimension(maxsweep,5) :: para                 ! parabolic interpolation coefficient
real :: radius, theta, stheta
```

Excerpt 8.5.7 VH1 code showing global variables declared in a module.

```
!$omp threadprivate(dvol,dx,dx0,e,f,flat,p,para,q,r,  &
                    radius,theta,stheta,u,v,w,xa,xa0)
```

Excerpt 8.5.8 OpenMP threadprivate directive for module variables.

```
!      Now Loop over each column...
!$OMP  parallel do default(none)                                          &
!$OMP&   private (i,j,k,m,n)                                              &
!$OMP&   shared  (gamm,isz,js,ks,mypey,mypez,ngeomz,nleftz,npez,nrightz, &
!$OMP&            recv1,send1,zdz,zxc,zyc,zza)
do j = 1, js
 do i = 1, isz
   radius = zxc(i+mypez*isz)
   theta  = zyc(j+mypey*js)
   stheta = sin(theta)
   radius = radius * stheta
```

Excerpt 8.5.9 VH1 parallel code with all variables scoped.

Notice that none of the private variables that are contained within the module are scoped in Excerpt 8.5.9. An OpenMP syntax error will be given whenever a variable that is mentioned in a `threadprivate` directive also occurs in the `private` clause of the `parallel do` directive.

We will have other high level threading examples and show performance for this implementation in the applications chapter.

8.6 SPMD OPENMP

Conceptually, SPMD OpenMP is very simple. One introduces a high-level parallel region and then proceeds to manage the threads as one might manage MPI tasks. The implementor would then be responsible for work sharing and synchronization. With all of the details left for the implementor, there is a great deal of flexibility in employing the threads to divide the work amongst the threads. Alan Wallcraft of the Naval Research Laboratory was one of the earliest implementors of SPMD OpenMP in his HYCOM ocean model [29]. More recently, Jacob Paulsen of the Danish Meteorological Institute employed SPMD OpenMP in the ocean model used to monitor the oceans around Denmark. Jacob's work was performed when he moved his ocean model to the Knight's Corner (KNC) system [16].

Most recently Peter Mendygral has employed SPMD OpenMP in his development of WOMBAT [21], an astrophysics application. All of these implementations had the same design criteria. They required more flexibility in assigning work to the threads. With respect to Paulsen's application, he needed better heuristics for dividing his very irregular grid evenly across the threads. For Mendygral's implementation, he needed a good method for redistributing the work within the node to handle his adaptive mesh refinement. Recently, Mendygral has incorporated one-sided MPI-3 using thread-safe communication. This allows the threads to perform their own communication and avoids significant synchronization between threads.

SPMD OpenMP simply implies introducing a high level `!$OMP PARALLEL` region and then ensuring that the scoping of variables down the call chain are consistent with the language standard. Those variables down the call chain that should be shared by all threads should have global scope. Threads also need private data and private variables should be allocated by each thread when the routine/function is called.

The user is also responsible for synchronizing the threads when synchronization is necessary. For example, prior to communication using MPI that is not threaded, or I/O, the threads must be synchronized at a barrier, have the master thread perform the operation, and then continue. Additionally there could be some data race conditions that must be synchronized around – for example, a summation of all the energy in the problem.

We will use the simple Himeno benchmark to illustrate the implementation of SPMD OpenMP. While it is not a good example from a performance standpoint, it is good to illustrate the technique. First, we incorporate a high level parallel region in the main routine presented in Excerpt 8.6.1. We also do all the output on the master thread.

Next, we work down the call chain and will be distributing the work across the threads with a routine called `get_bounds`, seen in Excerpt 8.6.2. This routine uses a simple chunking scheme. However, there is complete flexibility in allocating the loop iterations across the threads. For example, it would not be unreasonable to have internal timers on each of the threads to occasionally check if load-imbalance occurs due to more complex computation in a given area (e.g., shock front). Fewer iterations can then be allocated to those threads that are taking longer for each iteration. This is one of the most powerful aspects of SPMD OpenMP. This is where and how load imbalance can be handled, and how irregular decompositions of the domain across the threads can be handled. While OpenMP gives us some options for handling load imbalance such as SCHEDULE(GUIDED), the possibilities are unlimited with this approach.

```
!$OMP PARALLEL private(cpu0,cpu1,cpu)
C     Initializing matrixes
      call initmt
!      write(*,*) ' mimax=',mimax,' mjmax=',mjmax,' mkmax=',mkmax
!      write(*,*) ' imax=',imax,' jmax=',jmax,' kmax=',kmax
C     Start  measuring
c
      cpu0=sec_timer()
C
C     Jacobi iteration
      call jacobi(nn,gosa)
C
      cpu1= sec_timer()-cpu0
      cpu = cpu1
!$OMP MASTER
      write(*,*) 'cpu : ',cpu,'sec.'
        nflop=(kmax-2)*(jmax-2)*(imax-2)*34
        if(cpu1.ne.0.0) xmflops2=nflop/cpu*1.0e-6*float(nn)
        write(*,*) ' Loop executed for ',nn,' times'
        write(*,*) ' Gosa :',gosa
        write(*,*) ' MFLOPS measured :',xmflops2
        score=xmflops2/32.27
        write(*,*) ' Score based on MMX Pentium 200MHz :',score
!$OMP END MASTER
!$OMP END PARALLEL
```

Excerpt 8.6.1 Himeno example code with high-level parallel region.

```
      subroutine get_bounds(innlo,innhi,nlo,nhi)
      integer innlo,innhi,nlo,nhi
      integer num_threads,my_thread_id
#ifdef _OPENMP
      my_thread_id = omp_get_thread_num()
      num_threads = omp_get_num_threads()
      nlo = innlo +((innhi-innlo+1)/num_threads+1)*(my_thread_id)
      nhi = min(innhi,nlo+(innhi-innlo+1)/num_threads)
#else
      nlo = innlo
      nhi = innhi
#endif
      end
```

Excerpt 8.6.2 The getbounds subroutine used with SPMD OpenMP.

In the initialization routines of Himeno we have a call to get_bounds as shown in Excerpt 8.6.3. It is important that the bounds being passed into get_bounds are the same on the initialization loop as there are on the computational loop seen in Excerpt 8.6.4.

Notice that we maintained the same get_bounds limits in the computational routine as we did in the initialization routine. It is important that each thread work only on the data that it initialized – this will take care of any NUMA issues. We also had to deal with the reduction on GOSA within a critical region. Then we have the master thread perform the communication.

```
call get_bounds(1,kmax,nlo,nhi)
do k=nlo,nhi
  do j=1,jmax
    do i=1,imax
      a(i,j,k,1:3) = 1.0
      a(i,j,k,4)   = 1.0/6.0
      b(i,j,k,1:3) = 0.0
      c(i,j,k,1:3) = 1.0
      p(i,j,k)     = float((k-1)*(k-1))/float((kmax-1)*(kmax-1))
      wrk1(i,j,k)  = 0.0
      bnd(i,j,k)   = 1.0
    enddo
  enddo
enddo
```

Excerpt 8.6.3 Himeno initialization code with getbounds subroutine call.

A much better implementation would include modifying the communication to have each of the threads perform the communication they required. One situation that occurs often when using SPMD is the need for a global value down the call chain. If a variable is not in the calling sequence, in a common block or module, then it must be allocated when calling the routine. For example, in this case `WGOSA` was a local variable. Since we want `WGOSA` to gather information across all the threads being used, it must be shared. There are two accepted ways to generate a global variable. First one could simply use the allocation: `REAL, SAVE :: WGOSA`. This would allocate a shared variable, and all calling threads would have the memory address for the variable. The other approach taken in Excerpt 8.6.5 is safer (some compilers might not like the above).

The first thread into the routine would allocate the variable and then copy the pointer to all the other threads. The variable could also be de-allocated at the end of the routine as seen in Excerpt 8.6.6.

There are two very successful implementations of SPMD OpenMP which are discussed here. One took an existing application and refactored it to incorporate a higher-level threading and the second used SPMD OpenMP from the early stages of design. Both of these efforts can give us good insight on how one might best incorporate SPMD OpenMP into their own applications. The two are Jacob Weismann Poulson and Per Berg's HBM code modernization work in preparing for Knight's Corner and KNL as discussed in "High Performance Parallelism Pearls", and the second is the work performed for the development of WOMBAT in a codesign project between the University of Minnesota and Cray's Programming environment group spearheaded by Peter Mendygral.

From a high level view of the approach, both efforts were similar. A high level parallel region was introduced and then the threads were controlled by the application. Figure 8.6.1 shows how the threads in WOMBAT are initiated to first allocate the data they would be using and second to obtain work to perform.

```
      call get_bounds(1,kmax,nlo,nhi)
      do 900 loop=1,nn
      my_gosa=0.0
      wgosa=0.0
!$OMP BARRIER
        DO 100 K=max(2,nlo),min(kmax-1,nhi)  ! Original loop went from 2, kmax-1
        DO 100 J=2,jmax-1
        DO 100 I=2,imax-1
          S0=a(I,J,K,1)*p(I+1,J,K)+a(I,J,K,2)*p(I,J+1,K)
     1      +a(I,J,K,3)*p(I,J,K+1)
     3      +b(I,J,K,1)*(p(I+1,J+1,K)-p(I+1,J-1,K)
     *                  -p(I-1,J+1,K)+p(I-1,J-1,K))
     4      +b(I,J,K,2)*(p(I,J+1,K+1)-p(I,J-1,K+1)
     *                  -p(I,J+1,K-1)+p(I,J-1,K-1))
     5      +b(I,J,K,3)*(p(I+1,J,K+1)-p(I-1,J,K+1)
     *                  -p(I+1,J,K-1)+p(I-1,J,K-1))
     6      +c(I,J,K,1)*p(I-1,J,K)+c(I,J,K,2)*p(I,J-1,K)
     *      +c(I,J,K,3)*p(I,J,K-1)+wrk1(I,J,K)
          SS=(S0*a(I,J,K,4)-p(I,J,K))*bnd(I,J,K)
          my_GOSA=my_GOSA+SS*SS
          wrk2(I,J,K)=p(I,J,K)+OMEGA *SS
  100   CONTINUE
!$OMP CRITICAL
        WGOSA =WGOSA + my_GOSA
!$OMP END CRITICAL
        DO 200 K=max(2,nlo),min(kmax-1,nhi)  ! Original loop went from 2, kmax-1
        DO 200 J=2,jmax-1
        DO 200 I=2,imax-1
          p(I,J,K)=wrk2(I,J,K)
  200   CONTINUE
!$OMP BARRIER
!$OMP MASTER
          call sendp(ndx,ndy,ndz)
          call mpi_allreduce(wgosa,
     >                       gosa,
     >                       1,
     >                       mpi_real4,
     >                       mpi_sum,
     >                       mpi_comm_world,
     >                       ierr)
!$OMP END MASTER
!$OMP BARRIER
900    Continue
```

Excerpt 8.6.4 Himeno computational code with getbounds subroutine call.

```
          Subroutine Jacobi
          Real, pointer :: wgosa

!$omp single
          allocate(gosa)
!$omp end single copyprivate(gosa)
!$OMP BARRIER
```

Excerpt 8.6.5 Example single allocation code from Himeno.

In the HBM code, the threaded region and execution was controlled with OpenMP directives like those shown in Excerpt 8.6.7.

```
!$OMP MASTER
deallocate(gosa)
!$OMP END MASTER
```

Excerpt 8.6.6 Example master de-allocation code from Himeno.

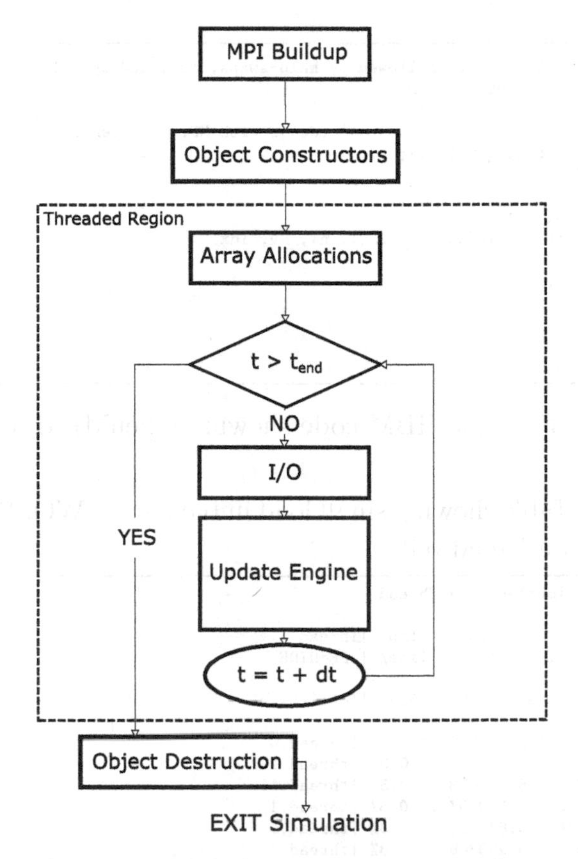

FIGURE 8.6.1 Example parallel structure of WOMBAT.

In both of these approaches, it is extremely important that the threads allocate the memory that they will be working on. This approach is trying to mimic MPI in its allocation of data local to the processor.

The two applications handle work distribution to the threads much differently. The HBM work uses a very dynamic work distribution scheme because it has to deal with an extremely sparse and irregular grid which covers the ocean structure around Denmark. Using the calls to `domp_get_domain` was the only way the developers could load-balance the computation across the threads. None of the load management directives available within OpenMP were capable of performing a well-balanced decomposition. On the other hand, WOMBAT was able to employ `!$OMP SCHEDULE(GUIDED)` very successfully.

WOMBAT's computational regions were very high granularity and any overhead from the OpenMP schedule operation was minimal. Table 8.6.1 shows statistics from a run using WOMBAT on a Broadwell node using 16 threads. As can be seen in the table, the load imbalance across the threads is extremely low.

```
! Time loop (set of calls followed by halo-swaps, followed by call
!$OMP PARALLEL DEFAULT(SHARED)
call foo(...); call bar(...); ...
call halo_update(...)                    !deals with MPI and OpenMP
call baz(:::); call quux(...); ...
!$OMP END PARALLEL

subroutine foo(...)
  call domp_get_domain(kh, 1, iw2_l, nl, nu, idx)
  do iw = nl, nu
     i = ind(1,iw)
     j = ind(2,iw)
     ...
  enddo
end subroutine foo
```

Excerpt 8.6.7 Example HBM code showing OpenMP directives.

TABLE 8.6.1 Table showing small load imbalance in WOMBAT code with 16 threads on Broadwell.

```
Table:  Load Imbalance by Thread

    Max. Time |      Imb. |  Imb. |Thread
              |      Time | Time% | PE=HIDE

 1,984.323706 | 9.329682 |   0.5% |Total
|-----------------------------------------------
| 1,989.650768 | 5.327063 |   0.3% |thread.0
| 1,980.591854 | 5.106072 |   0.3% |thread.12
| 1,980.834773 | 6.029658 |   0.3% |thread.11
| 1,980.104003 | 5.312097 |   0.3% |thread.1
| 1,980.598986 | 5.841221 |   0.3% |thread.5
| 1,980.808981 | 6.263348 |   0.3% |thread.4
| 1,980.815364 | 6.397444 |   0.3% |thread.2
| 1,980.431098 | 6.014599 |   0.3% |thread.13
| 1,980.844801 | 6.657376 |   0.3% |thread.10
| 1,980.065661 | 5.903723 |   0.3% |thread.15
| 1,980.116907 | 5.964561 |   0.3% |thread.3
| 1,980.411986 | 6.278547 |   0.3% |thread.7
| 1,979.766712 | 5.754217 |   0.3% |thread.6
| 1,980.437314 | 6.513650 |   0.3% |thread.9
| 1,979.490818 | 5.586776 |   0.3% |thread.8
| 1,980.123545 | 6.238835 |   0.3% |thread.14
```

The next area of concern when employing SPMD is to minimize the synchronization between the threads. Once again, we are trying to mimic MPI in this area. WOMBAT takes minimization of synchronization to extreme, by having each thread perform its own messaging with MPI RDMA, recently available in the MPI standard. In this approach, each thread puts its data into a neighbor's memory. When neighbor threads arrive at a point where they require that data, they will get the message and proceed. Figure 8.6.2 illustrates how the RMA boundary exchange works.

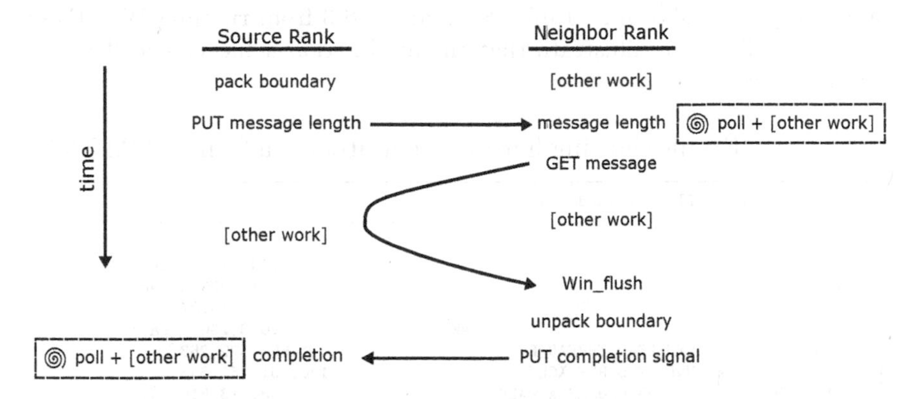

FIGURE 8.6.2 Overview of RMA boundary exchange in WOMBAT.

The work on HBM currently does not have threads perform their own communication. However, it is currently under consideration. Thread-safe MPI is only recently performant enough to be considered. The WOMBAT project has had the benefit that collaboration with the MPI developers has excellerated the development of not only a thread-safe MPI, but an extremely efficient thread-safe MPI that minimizes locks to achieve very good scaling. With the publicity that WOMBAT is making, their effort will encourage others with this requirement to refactor their applications to employ thread-safe MPI.

TABLE 8.6.2 Profile from a run of WOMBAT on Broadwell.

Time%	Time	Imb.Time	Imb. Time%	Calls	Group Function
100.0%	1,984.3233	--	--	11,869,942	Total
98.7%	1,958.7276	--	--	7,062,742	USER
14.4%	284.8705	41.7829	13.3%	1,078,200	compute_fluxes1d$mhdtvd
13.5%	268.2578	16.1907	5.9%	11,419	compute_zplane_xyfluxes3d$mhdtvd
12.9%	256.9482	34.6122	12.3%	1,078,200	compute_eigenvecs1d$mhdtvd
10.3%	203.7733	59.3891	23.4%	1	wombat_.REGION@li.63
9.6%	191.1935	19.0673	9.4%	11,419	compute_zfluxes3d$mhdtvd
4.3%	85.4743	27.5733	25.3%	3,823	halfzupdate_states3d$mhdtvd
4.1%	81.8938	25.5782	24.7%	3,823	halfyupdate_states3d$mhdtvd
3.9%	77.6199	7.9754	9.7%	1,644,167	compute_speeds_eigenvals1d$mhdtvd
3.8%	76.0699	28.4578	28.3%	3,823	halfxupdate_states3d$mhdtvd
3.0%	59.0208	10.2304	15.3%	15,293	compute_corner_emf3d$mhdtvd
[...]					
1.0%	20.064271	--	--	4,807,176	MPI

The HBM work is well-discussed in the aforementioned book and its performance on first Knight's Corner and then Knight's Landing shows excellent scaling using SPMD OpenMP. The WOMBAT work also scales extremely well and the profiles shown in Tables 8.6.2 and 8.6.3 from running WOMBAT on a Broadwell system illustrate that the application achieves excellent overall performance.

TABLE 8.6.3 Profile showing hardware counter data from WOMBAT run.

```
USER / compute_fluxes1d$mod_mhdtvd_
-----------------------------------------------------------------------------
Time%                                                        14.4%
Time                                                    284.870540 secs
Imb. Time                                                41.782990 secs
Imb. Time%                                                   13.3%
Calls                                 0.004M/sec       1,078,200.9 calls
CPU_CLK_THREAD_UNHALTED:THREAD_P                 3,011,693,522,253
CPU_CLK_THREAD_UNHALTED:REF_XCLK                   130,261,961,848
DTLB_LOAD_MISSES:MISS_CAUSES_A_WALK                     48,313,810
DTLB_STORE_MISSES:MISS_CAUSES_A_WALK                     5,142,036
L1D:REPLACEMENT                                    538,845,212,475
L2_RQSTS:ALL_DEMAND_DATA_RD                        534,082,251,116
L2_RQSTS:DEMAND_DATA_RD_HIT                         31,721,661,616
MEM_UOPS_RETIRED:ALL_LOADS                       2,171,019,773,819
FP_ARITH:SCALAR_DOUBLE                             202,264,157,103
FP_ARITH:128B_PACKED_DOUBLE                         58,990,798,259
FP_ARITH:256B_PACKED_DOUBLE                        850,729,439,694
FP_ARITH:SCALAR_SINGLE                                      527.59
CPU_CLK                                2.31GHz
HW FP Ops / User time             13,069.668M/sec    3,723,163,512,924 ops  15.5%peak(DP)
Total SP ops                           1.852 /sec               527.59 ops
Total DP ops                      13,069.668M/sec    3,723,163,512,396 ops
Computational intensity                1.24 ops/cycle             1.71 ops/ref
MFLOPS (aggregate)               352,881.05M/sec
TLB utilization                  40,613.33 refs/miss              79.32 avg uses
D1 cache hit,miss ratios              75.2% hits                  24.8% misses
D1 cache utilization (misses)          4.03 refs/miss              0.50 avg hits
D2 cache hit,miss ratio                6.8% hits                  93.2% misses
D1+D2 cache hit,miss ratio            76.9% hits                  23.1% misses
D1+D2 cache utilization                4.32 refs/miss              0.54 avg hits
D2 to D1 bandwidth             114,430.203MiB/sec   34,181,264,071,403 bytes
Average Time per Call                                         0.000264 secs
CrayPat Overhead : Time                1.2%
```

Notice that the MPI usage is well-controlled at 1.0% of the total execution time. These statistics were gathered from a run on 27 MPI tasks across 13 nodes of Broadwell, two MPI tasks per node and 16 threads per MPI task. We can ensure that the top-level routines in this profile are indeed running extremely well by examining the hardware counters seen in Table 8.6.3.

Notice that the routine is achieving 15.5% of peak on the Broadwell system. There is some caching issues, while the TLB utilization is very good. The pros of the SPMD approach are that there is much less overhead from the OpenMP implementation, scoping is handled by the language standard, and you tend to use much fewer directives. The negative part of SPMD is it is more difficult to implement in an application. The user must really understand threading and the implications of how the language handles global and local storage.

In conclusion, to scale well on many/multicore systems, application developers must develop efficient threading by:

1. Paying attention to NUMA regions.
2. Avoiding overhead caused with too much synchronization.
3. Avoiding load-imbalance.

On some systems all-MPI will perform very well and will out-perform poorly implemented OpenMP. Unfortunately, SPMD OpenMP is not performance portable to hosted accelerators.

8.7 EXERCISES

8.1 What are the major causes of inefficiencies when using OpenMP?

8.2 What are the major reasons why MPI across all the processors on a node runs well?

8.3 What are some examples where OpenMP on the node might be better than using MPI on the node?

8.4 When employing SPMD OpenMP, what are the major issues to be addressed to ensure correctness down the call chain?

CHAPTER 9

Porting Entire Applications

CONTENTS

9.1 FOREWORD BY JOHN LEVESQUE

Working with the Tri-Labs – Los Alamos National Laboratory (LANL), Sandia National Laboratory (SNL), and Lawrence Livermore National Laboratory (LLNL) – was significantly different from the work I had been doing for the previous 20 years. While there were fewer applications, those applications were extremely large multiphysics, multifunctional codes that were being used to solve very difficult high-energy physics problems. I was fortunate to have Jim Schwarzmeier working with me. While Jim had retired from Cray after

35 years, he was working part-time specifically to assist the Trinity center of excellence (CoE) in understanding the principal applications and to design approaches to be used for porting and optimizing the applications for the Knight's Landing (KNL) system. The first year and a half of working with these applications was spent identifying the important problems and what parts of these enormous codes needed to be refactored for parallelization and vectorization. The biggest surprise that really aided in the work was that running MPI across all the processors on the node worked extremely well. This meant that many of the applications which did not have threading could use MPI-only without significant modifications. The expected issue was that the scalar performance of the KNL was poor, so vectorization was critical to ensure that the application ran better on KNL than on the Xeon systems. The other aspect of the Tri-Lab work was that a majority of the applications were written in C++ with only a few in Fortran. These very large C++ frameworks presented a significant challenge for the available compilers and tools. Pretty much everything was broken, and a lot of work was expended to identify the bugs and get them fixed. When the applications are so large and convoluted, the importance of profiling tools cannot be overstated. Well, this book was finished when the Trinity CoE project was in full swing, and we still have a lot of work to do.

9.2 INTRODUCTION

This chapter looks at several applications, concentrating on approaches that can be used to make them more amenable to vectorization and/or parallelization. These applications may be smaller than many large production codes. However, they have problems whose solutions can be applied to their larger counterparts. This section will concentrate on performance portable implementations where possible. Due to differences in compilers' capabilities, one rewrite may not be best for all compilers.

9.3 SPEC OPENMP BENCHMARKS

The first group of applications that will be targeted are one node OpenMP applications running on the latest Intel Xeon chip, Broadwell, and Intel's Knight's Landing chip. These applications from the SPEC OpenMP suite [25] were analyzed in our previous book, *High Performance Computing: Programming and Applications* [19].

9.3.1 WUPWISE

The first example we will examine is WUPWISE (an acronym for "Wuppertal Wilson Fermion Solver"), a program in the area of lattice gauge theory (quantum chromodynamics). It is very well threaded with OpenMP and should scale well on both of the systems being investigated.

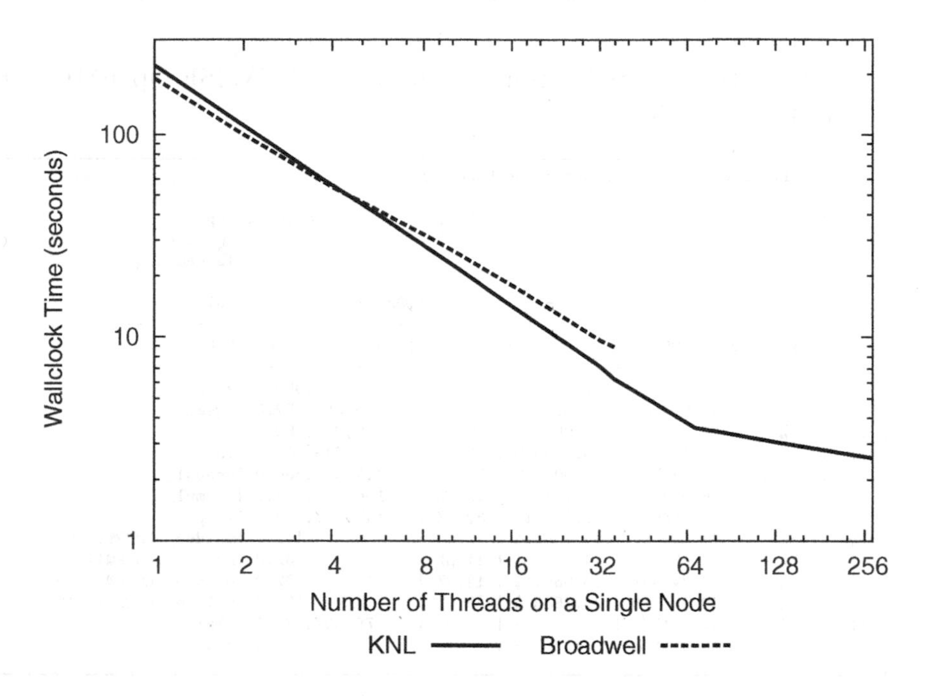

FIGURE 9.3.1 Performance of WUPWISE OpenMP code from SPEC on KNL and Broadwell.

As can be seen in Figure 9.3.1, Broadwell performs better than KNL on a small number of threads, and KNL does better at 8 and above and outperforms Broadwell by a factor of 2 when all the cores on the node are used. Can we improve the performance on both Broadwell and KNL? We can get some clue looking at the profile display of the execution of WUPWISE, presented in Table 9.3.1.

Notice that most of the time is spent in the BLASroutines. Interestingly, these routines are not being used out of the supplied vendor libraries. One would think that those might be preferred. The OpenMP loops are those four listed down the profile in `muldeo` and `muldoe` – the times given are exclusive of the routines being called within the parallel loops. Notice the very large number of times the BLAS routines are called, they undoubtedly are being called from within loops. Next we look at the call tree profile seen in Table 9.3.2 that shows loops and routines in the call tree. One other thing that needs to be examined is the load imbalance in the OpenMP loops and the routines called from within those loops.

There appears to be a double-nested loop in the `muldoe` routine and some loops down the call tree within the BLAS routines. Why aren't the vendor libraries being used here? Looking at the loop profiles presented in Table 9.3.3 we see that the loops within the BLAS routines are very small – all have a trip count of 3 or 4.

TABLE 9.3.1 Tracing profile from execution of WUPWISE OpenMP code from SPEC on KNL.

Table: Profile by Function Group and Function

Time%	Time	Imb. Time	Imb. Time%	Calls	Group / Function / Thread=HIDE
100.0%	2,051.482358	--	--	1,931,744,544.0	Total
100.0%	2,051.358903	--	--	1,931,743,415.0	USER
21.9%	449.659691	72.686583	18.5%	414,973,952.0	zaxpy_
16.7%	342.979609	19.602823	6.5%	106,954,752.0	zgemm_
15.3%	314.133816	17.950839	6.5%	106,954,752.0	gammul_
10.9%	222.939625	76.635060	39.3%	427,819,008.0	lsame_
10.6%	218.288047	71.002115	37.2%	414,974,356.0	dcabs1_
5.7%	116.333872	20.131279	19.8%	106,954,752.0	su3mul_
3.9%	80.544619	22.822715	32.4%	147,587,072.0	zcopy_
2.3%	47.495487	15.335494	12.6%	102.0	muldeo_.LOOP@li.68
2.3%	47.486578	14.059371	11.8%	102.0	muldoe_.LOOP@li.69
2.3%	47.344991	15.858112	12.9%	102.0	muldeo_.LOOP@li.106
2.3%	46.955918	15.753806	13.0%	102.0	muldoe_.LOOP@li.105
2.0%	40.101547	--	--	75,497,472.0	dlarnd_
1.9%	39.433170	--	--	1.0	rndcnf_

This is why the vendor libraries are not being used. The library calls are optimized for larger matrices/vectors, and in this case the overhead of calling the optimized library will result in the entire application running slower. So how can we fix the performance on KNL (which may also help Broadwell)? The first idea, which would be quite a bit of work, would be to try to vectorize on the second loop within the parallel region that has 16 iterations. There is some question as to whether all the routines within this loop can be inlined so that the loop can be vectorized.

Since we have so much load imbalance in the OpenMP loops and the routines called from within those loops, we should try using `schedule(guided)` on the `parallel do` directives to dynamically allocate work to the threads. The default for OpenMP is that there is a static allocation of work that divides the iterations up in even chunks across the threads. On the other hand, with `schedule(guided)` the chunk size starts out smaller than the static chunk size and then decreases to better handle load imbalance between iterations. This modification helped KNL even on smaller thread counts, achieving an overall speedup of 1.5. The schedule does not help Broadwell, and KNL outperforms Broadwell by a factor of 3 to 4 on the node. Figure 9.3.2 shows some additional performance gain by using all the hyper-threads on the node.

One last attempt at getting additional performance was to inhibit vectorization on the loops of length 3 to 4. Letting the compiler vectorize the short loops turned out to be slightly faster than running them in scalar mode. Here again the decision to incorporate **`schedule(guided)`** was due to the large amount of load imbalance that was identified in the profile of the original run.

TABLE 9.3.2 Call tree profile from WUPWISE SPEC code.

```
Table:  Function Calltree View

  Time% |        Time |         Calls |Calltree

 100.0% | 5,984.845196 |            -- |Total
|------------------------------------------------------------
| 100.0% | 5,984.845196 |          2.0 |wupwise_
||-----------------------------------------------------------
||  88.5% | 5,293.868281 |           -- |wupwise_.LOOP.1.li.146
|||----------------------------------------------------------
3||  88.3% | 5,286.100567 |         20.0 |matmul_
||||---------------------------------------------------------
4|||  44.2% | 2,643.305340 |         40.0 |muldoe_
|||||--------------------------------------------------------
5||||  20.7% | 1,241.802032 |            -- |muldoe_.LOOP.3.li.106
6||||         |              |             | muldoe_.LOOP.4.li.120
|||||||------------------------------------------------------
7||||||   9.8% |   585.064690 |  41,943,040.0 |gammul_
||||||||-----------------------------------------------------
8|||||||   5.6% |   332.822052 | 251,658,240.0 |zaxpy_
|||||||||----------------------------------------------------
9||||||||   4.9% |   291.787426 | 251,658,240.0 |zaxpy_(exclusive)
9||||||||   0.7% |    41.034626 |            -- |zaxpy_.LOOP.2.li.30
10|||||||   0.7% |    41.034626 | 125,829,120.0 | dcabs1_
|||||||||====================================================
8|||||||   2.2% |   133.218835 |  41,943,040.0 |gammul_(exclusive)
8|||||||   1.7% |    99.196599 | 104,857,600.0 |zcopy_
8|||||||   0.3% |    19.827204 |  20,971,520.0 |zscal_
||||||||=====================================================
7||||||   8.7% |   519.666796 |  41,943,040.0 |su3mul_
||||||||-----------------------------------------------------
8|||||||   7.9% |   471.486566 |  83,886,080.0 |zgemm_
|||||||||----------------------------------------------------
9||||||||   7.0% |   418.986269 |  83,886,080.0 |zgemm_(exclusive)
[...]
||||||||=====================================================
7||||||   2.3% |   137.070546 |  83,886,080.0 |zaxpy_
||||||||-----------------------------------------------------
8|||||||   2.1% |   124.192722 |  83,886,080.0 |zaxpy_(exclusive)
[...]
```

TABLE 9.3.3 Loop profile from WUPWISE SPEC code with trip counts.

Table: Inclusive and Exclusive Time in Loops (from -hprofile_generate)

Loop Incl Time%	Loop Incl Time	Time Loop Adj.	Loop Hit	Loop Trips Avg	Trips Min	Trips Max	Function=/.LOOP[.]
87.7%	7,533.334	0.00176	1	10	10	10	wupwise_.LOOP.1.li.146
24.2%	2,079.083	0.15513	22	32,768	32,768	32,768	muldoe_.LOOP.3.li.106
24.2%	2,078.928	94.21005	720,896	16	16	16	muldoe_.LOOP.4.li.120
24.2%	2,078.285	0.17135	22	32,768	32,768	32,768	muldeo_.LOOP.3.li.106
24.2%	2,078.114	94.77508	720,896	16	16	16	muldeo_.LOOP.4.li.120
24.0%	2,057.699	0.12994	22	32,768	32,768	32,768	muldoe_.LOOP.1.li.69
24.0%	2,057.569	97.50991	720,896	16	16	16	muldoe_.LOOP.2.li.80
24.0%	2,057.254	0.14945	22	32,768	32,768	32,768	muldeo_.LOOP.1.li.69
24.0%	2,057.105	95.01804	720,896	16	16	16	muldeo_.LOOP.2.li.80
13.3%	1,145.988	497.69464	719,323,136	5	3	12	zaxpy_.LOOP.2.li.30
2.4%	205.836	76.49852	92,274,688	4	4	4	zgemm_.LOOP.05.li.236
2.1%	181.998	41.41109	1	75,497k	75,497k	75,497k	rndcnf_.LOOP.2.li.52
2.0%	169.337	69.86807	92,274,688	4	4	4	zgemm_.LOOP.10.li.259

[...]

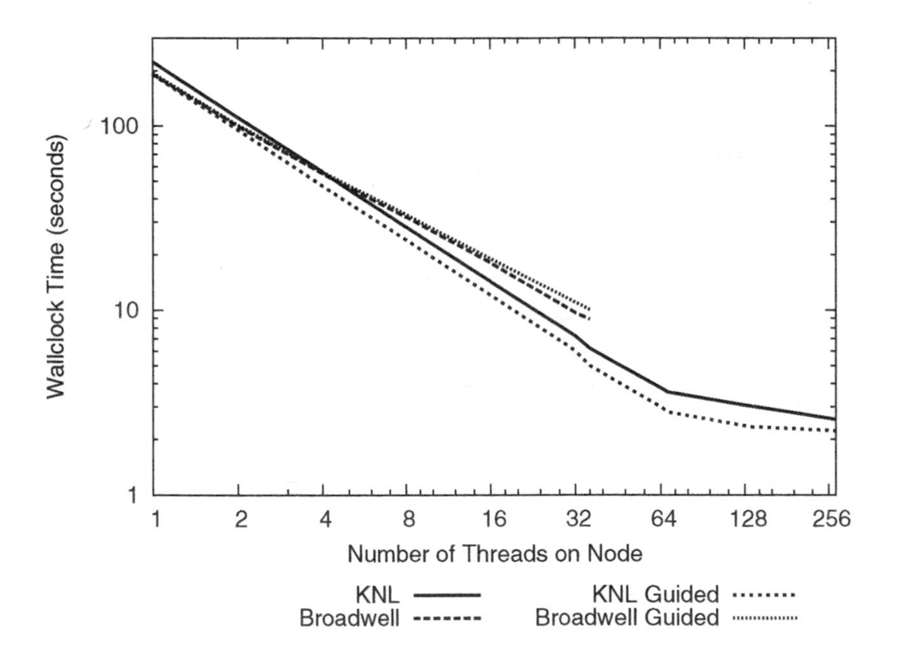

FIGURE 9.3.2 Performance of WUPWISE OpenMP code from SPEC on KNL and Broadwell with guided schedule.

9.3.2 MGRID

MGRID is a multigrid solver that seems to be a knock-off of the NASA Parallel Benchmark MG application. One of the difficulties of the multigrid solvers is that the bounds on the major loops change within the steps of the multigrid approach. The loop bounds can change by an order of magnitude. The upper bound `n` for the major loops varies from 6 to 258. This application is well threaded and can only be improved by doing some tiling of the outer two loops in the major two routines. Table 9.3.4 shows the profile of the original run on 68 threads on KNL. The loop from `resid` presented in Excerpt 9.3.1 can be tiled using the technique discussed earlier.

TABLE 9.3.4 Profile from original MGRID SPEC code.

```
Table:  Profile by Function

  Samp% |  Samp | Imb.  |  Imb.  |Group
        |       | Samp  | Samp%  | Function
        |       |       |        |  Thread=HIDE

 100.0% | 569.0 |    -- |     -- |Total
|-----------------------------------------------------------------
|  78.7% | 448.0 |    -- |     -- |USER
||----------------------------------------------------------------
||  39.2% | 223.0 | 10.6 |  4.8% |resid_.LOOP@li.364
||  18.6% | 106.0 | 14.4 | 13.3% |psinv_.LOOP@li.407
||   7.4% |  42.0 |  8.9 | 19.2% |rprj3_.LOOP@li.321
||   4.9% |  28.0 |  8.2 | 23.7% |interp_.LOOP@li.269
||   4.4% |  25.0 |  9.5 | 29.1% |interp_.LOOP@li.235
||   1.4% |   8.0 |  7.7 | 43.2% |comm3_.LOOP@li.180
||================================================================
|  14.8% |  84.0 |    -- |     -- |ETC
||----------------------------------------------------------------
||   8.8% |  50.0 | 23.5 | 94.0% |fullscan_barrier_list]
||   3.0% |  17.0 |  5.4 | 32.4% |__cray_memset_KNL
||   1.1% |   6.0 |    -- |     -- |_STOP2
||================================================================
|   4.9% |  28.0 |    -- |     -- |OMP
||----------------------------------------------------------------
||   4.2% |  24.0 |    -- |     -- |_cray$mt_start_two_code_parallel
[...]
```

In the tiled loop seen in Excerpt 9.3.2 we initially choose `ic` to be 8 and `iic` to be 16. Within a tile the total space that is utilized by this size tile is less than the level-1 cache of 32 KB. Since the loop references three arrays, the total space taken by a tile is 8*8*16*3=24 KB. Tiling should be tweaked a little to find the best combination, and the tile size with `ic` = 8 and `iic` =64 performed the best. The increased vector length delivered better performance.

The performance gain we get is not exceptional, only about 10%. Both the original and tiled versions perform better on KNL than on Broadwell when all the cores are utilized as seen in Figure 9.3.3.

```
        DO i3 = 2, n-1, 1
         DO i2 = 2, (-1)+n, 1
          DO i1 = 2, (-1)+n, 1
           r(i1, i2, i3) = v(i1, i2, i3)+(-a(0))*u(i1, i2, i3)+(-a(1))*(u(
     *i1, i2, (-1)+i3)+u(i1, i2, 1+i3)+u(i1, (-1)+i2, i3)+u(i1, 1+i2, i3
     *)+u((-1)+i1, i2, i3)+u(1+i1, i2, i3))+(-a(2))*(u(i1, (-1)+i2, (-1)
     *+i3)+u(i1, (-1)+i2, 1+i3)+u(i1, 1+i2, (-1)+i3)+u(i1, 1+i2, 1+i3)+u
     *((-1)+i1, i2, (-1)+i3)+u((-1)+i1, i2, 1+i3)+u((-1)+i1, (-1)+i2, i3
     *)+u((-1)+i1, 1+i2, i3)+u(1+i1, i2, (-1)+i3)+u(1+i1, i2, 1+i3)+u(1+
     *i1, (-1)+i2, i3)+u(1+i1, 1+i2, i3))+(-a(3))*(u((-1)+i1, (-1)+i2, (
     *-1)+i3)+u((-1)+i1, (-1)+i2, 1+i3)+u((-1)+i1, 1+i2, (-1)+i3)+u((-1)
     *+i1, 1+i2, 1+i3)+u(1+i1, (-1)+i2, (-1)+i3)+u(1+i1, (-1)+i2, 1+i3)+
     *u(1+i1, 1+i2, (-1)+i3)+u(1+i1, 1+i2, 1+i3))
        ENDDO ; ENDDO ; ENDDO
```

Excerpt 9.3.1 Original loop-nest from SPEC OMP's MGRID component.

```
        do iii = 1, n/ic*n/ic
         i3s = max(2,((iii-1)/(n/ic))*ic+1)
         i2s = max(2,mod(iii-1,n/ic)*ic + 1 )
         i3e = min(i3s+ic-1,n-1)
         i2e = min(i2s+ic-1,n-1)
        do i3=i3s,i3e
         do i2 = i2s,i2e
           do ii1 = 1, n, iic
           ii1s = max(2,ii1 )
           ii1e = min(n,ii1s+iic-1)
          DO i1 = ii1s, min(n-1,ii1e)
           r(i1, i2, i3) = v(i1, i2, i3)+(-a(0))*u(i1, i2, i3)+(-a(1))*(u(
     *i1, i2, (-1)+i3)+u(i1, i2, 1+i3)+u(i1, (-1)+i2, i3)+u(i1, 1+12, i3
     *)+u((-1)+i1, i2, i3)+u(1+i1, i2, i3))+(-a(2))*(u(i1, (-1)+i2, (-1)
     *+i3)+u(i1, (-1)+i2, 1+i3)+u(i1, 1+i2, (-1)+i3)+u(i1, 1+i2, 1+i3)+u
     *((-1)+i1, i2, (-1)+i3)+u((-1)+i1, i2, 1+i3)+u((-1)+i1, (-1)+i2, i3
     *)+u((-1)+i1, 1+i2, i3)+u(1+i1, i2, (-1)+i3)+u(1+i1, i2, 1+i3)+u(1+
     *i1, (-1)+i2, i3)+u(1+i1, 1+i2, i3))+(-a(3))*(u((-1)+i1, (-1)+i2, (
     *-1)+i3)+u((-1)+i1, (-1)+i2, 1+i3)+u((-1)+i1, 1+i2, (-1)+i3)+u((-1)
     *+i1, 1+i2, 1+i3)+u(1+i1, (-1)+i2, (-1)+i3)+u(1+i1, (-1)+i2, 1+i3)+
     *u(1+i1, 1+i2, (-1)+i3)+u(1+i1, 1+i2, 1+i3))
          ENDDO
         ENDDO
        ENDDO ; ENDDO ; ENDDO
```

Excerpt 9.3.2 Tiled loop-nest from SPEC OMP's MGRID component.

Notice that the performance on Broadwell scales well until the number of threads crosses two sockets at around 16. At this point, the NUMA effects of the Broadwell cache significantly degrade performance. On KNL the best performance is achieved when one thread is used on each core. When hyper-threads are utilized, the performance degrades.

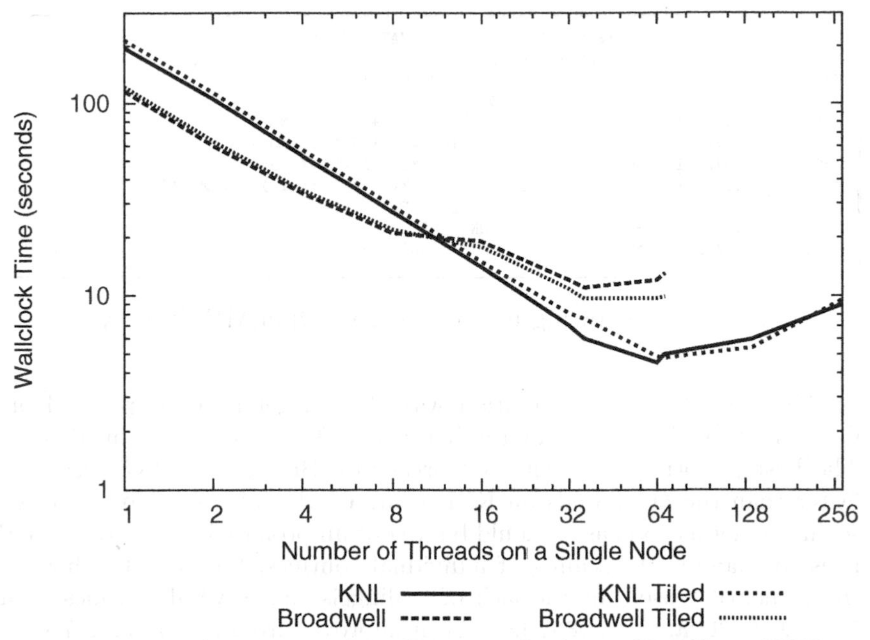

FIGURE 9.3.3 Performance of MGRID code from SPEC on KNL and Broadwell.

9.3.3 GALGEL

GALGEL is used to analyze oscillating convection instabilities in fluids [13]. It is another application that relies heavily on LAPACK and uses a Fortran version of the standard Linpack with modifications added to insert OpenMP directives. Most of the LAPACK routines are called by the Fortran 90 intrinsic MATMUL. MATMUL is used to perform both matrix multiplication as well as vector-matrix multiply. This practice is not recommended for several reasons. First, the vendors typically have much better optimized versions of LAPACK in their library suite and using MATMUL directly is probably relying too much on the compiler to do the right thing. For example, does the compiler recognize that the operation being called in the MATMUL is indeed a matrix multiply or would it be better to call the vector-matrix product instead of matrix multiply. Additionally, the application uses OpenMP within the LAPACK routines for all calls. It turns out that some of the calls within LAPACK are very small and could benefit from a selective OpenMP directive such as those in Excerpt 9.3.3. This OpenMP `IF` clause on the `PARALLEL DO` specifies that the loop should only run in parallel if `M` is greater than 32. Otherwise the overhead of the OpenMP loop would result in the parallel version running slower than the non-parallel version of the loop.

```
734.       M------------< !$OMP PARALLEL DO PRIVATE(JX,TEMP,I,J)
735.       M              !$OMP+IF(m.ge.32)
736.       M imVr2------<                    DO 50, I = 1, M
737.     + M imVr2 i----<              DO 60, J = 1, N
738.       M imVr2 i                      JX = JX0 + (J-1)*INCX
739.       M imVr2 i        !              IF( X( JX ).NE.ZERO )THEN
740.       M imVr2 i                         TEMP = ALPHA*X( JX )
741.       M imVr2 i                            Y( I ) = Y( I ) + TEMP*A( I, J )
742.       M imVr2 i        !              END IF
743.       M imVr2 i---->    60        CONTINUE
744.       M imVr2----->>    50              CONTINUE
```

Excerpt 9.3.3 Code showing use of selective OpenMP directives.

This OpenMP implementation was able to get a speedup of 14 on 64 OpenMP threads on KNL and a factor of 4.76 on 36 threads on Broadwell. The best performance is with 36 threads on Broadwell – 48.5 seconds was faster than the 66 seconds on KNL using 64 threads. To examine why the scaling is not as good as it should be we obtain profiles for all of the OpenMP runs to examine the scaling of individual routines. Figure 9.3.4 shows that the principal reason for the lack of scaling is that several routines such as `DLARFX`, `DHSEQR`, and `DGEMV` have been improved from a total of 157 seconds on 1 thread to 90 seconds on 64 threads. Unfortunately, they are not scaling well due to very low granularity and being bandwidth-limited.

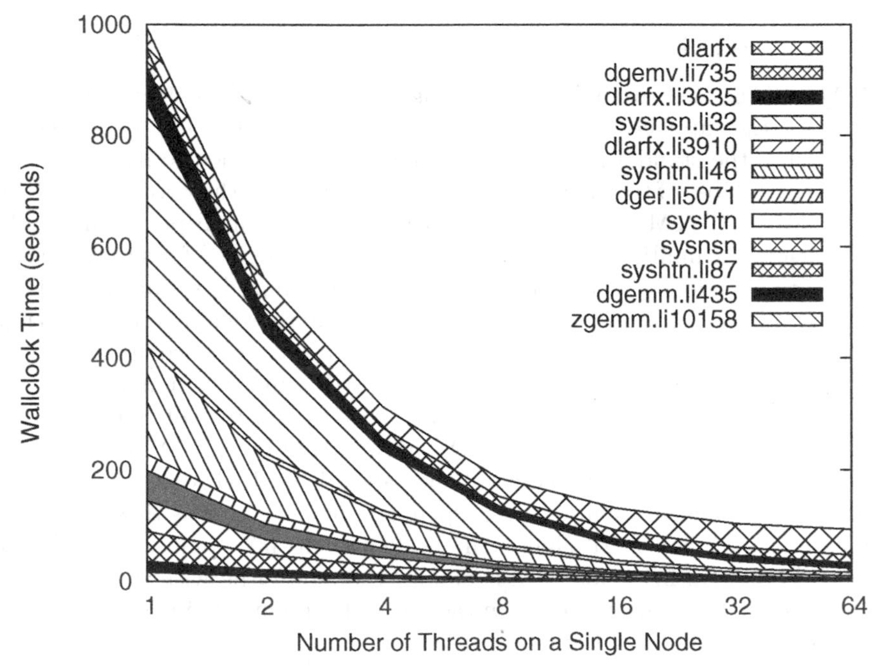

FIGURE 9.3.4 Performance of specific routines from SPEC's GALGEL.

The application needs to be rewritten to utilize optimized high level DHSEQR routine within LAPACK instead of the modified Fortran versions that are being used. Figure 9.3.5 shows the final run illustrating the scaling of the OpenMP on Broadwell and KNL.

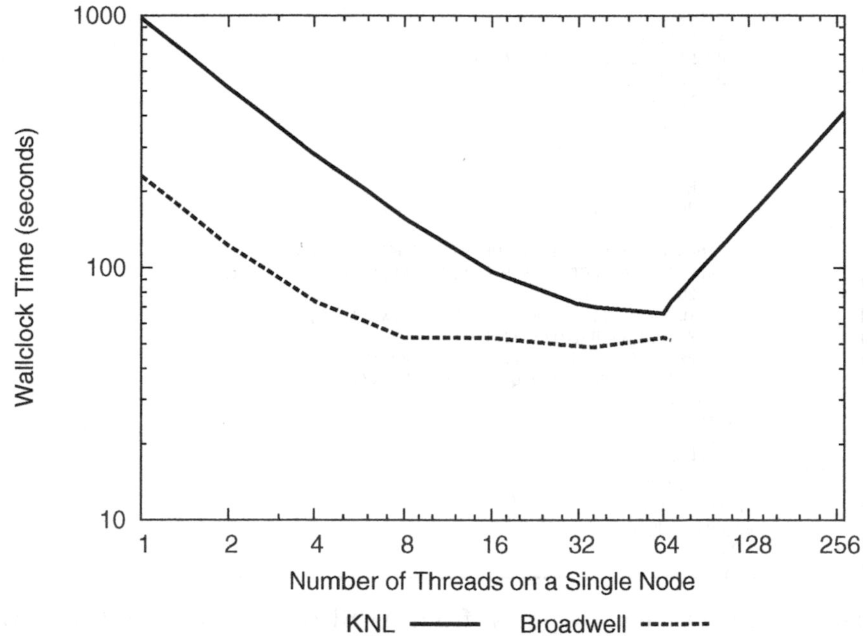

FIGURE 9.3.5 Performance of GALGEL code on KNL and Broadwell.

9.3.4 APSI

APSI is a mesoscale hydrodynamic model. This application is threaded at a very high level and it scales fairly well. Just a few issues we should point out that may not be obvious. In examining the looping structure within `wcont` presented in Excerpt 9.3.4, we see a use of a shared array to hold information to be summed outside the loop. The arrays `WWIND1` and `WSQ1` are dimensioned by the number of threads, and each OpenMP thread updates its part of the array. Each thread is running on a different core and each time the array is updated, the cache line containing that array must be fetched to its level-1 cache. Since there can only be one cache line in each core's level-1 cache containing the array, there is significant thrashing between accessing of the arrays in question. The simple solution is to dimension each array large enough that each core is accessing a different cache line. For example, if we dimension the arrays `WWIND1(32,NUMTHREADS)` and `WSQ1(32,NUMTHREADS)` and access the second dimension in the loop such as `WWIND1(1,I)` and `WSQ1(1,I)`, each reference will be on a different cache line.

```
C     DO 25 I=1,NUMTHREADS
         WWIND1(I)=0.0
         WSQ1(I)=0.0
 25   CONTINUE
!$OMP PARALLEL PRIVATE(I,K,DV,TOPOW,HELPA1,HELP1,AN1,BN1,CN1,MY_CPU_ID)
       MY_CPU_ID = OMP_GET_THREAD_NUM() + 1
!$OMP DO
      DO 30 J=1,NY
         DO 40 I=1,NX
            HELP1(1)=0.0D0 ; HELP1(NZ)=0.0D0
            DO 10 K=2,NZTOP
               IF(NY.EQ.1) THEN
                  DV=0.0D0
                           ELSE
                  DV=DVDY(I,J,K)
               ENDIF
               HELP1(K)=FILZ(K)*(DUDX(I,J,K)+DV)
 10         CONTINUE
C      SOLVE IMPLICITLY FOR THE W FOR EACH VERTICAL LAYER
            CALL DWDZ(NZ,ZET,HVAR,HELP1,HELPA1,AN1,BN1,CN1,ITY)
            DO 20 K=2,NZTOP
               TOPOW=UX(I,J,K)*EX(I,J)+VY(I,J,K)*EY(I,J)
               WZ(I,J,K)=HELP1(K)+TOPOW
               WWIND1(MY_CPU_ID)=WWIND1(MY_CPU_ID)+WZ(I,J,K)
               WSQ1(MY_CPU_ID)=WSQ1(MY_CPU_ID)+WZ(I,J,K)**2
 20         CONTINUE
 40      CONTINUE
 30   CONTINUE
!$OMP END PARALLEL DO
      DO 35 I=1,NUMTHREADS
         WWIND=WWIND+WWIND1(I)
         WSQ=WSQ+WSQ1(I)
35    CONTINUE
```

Excerpt 9.3.4 Original loop-nest from APSI component of SPEC OMP.

TABLE 9.3.5 Profile from original APSI SPEC code.

```
Table:  Profile by Function

  Samp% |     Samp | Imb.  |  Imb.  |Group
        |          | Samp  | Samp%  | Function
 100.0% | 2,976.0 |    -- |     -- |Total
|------------------------------------------------------------------------
|  85.1% | 2,532.0 |    -- |     -- |USER
||-----------------------------------------------------------------------
||  17.9% |   532.0 | 47.9 |   8.9% |rfftb1_
||  13.3% |   395.0 | 43.3 |  11.1% |rfftf1_
||   8.3% |   246.0 | 28.9 |  11.9% |dvdtz_.LOOP@li.1921
||   7.9% |   236.0 | 18.6 |   8.0% |dudtz_.LOOP@li.1749
||   7.5% |   222.0 | 17.6 |   7.5% |wcont_.LOOP@li.2033
||   6.2% |   185.0 |  7.7 |   4.2% |dkzmh_.LOOP@li.6579
||   6.0% |   179.0 | 15.1 |   8.6% |dtdtz_.LOOP@li.1531
||   5.7% |   171.0 | 18.2 |  10.8% |dcdtz_.LOOP@li.1390
||   5.4% |   160.0 | 32.0 |  20.3% |dctdy_
||   1.4% |    41.0 |  5.1 |  12.7% |dkzmh_.LOOP@li.6685
||   1.1% |    32.0 | 13.6 |  28.3% |dpdy_
||=======================================================================
|   8.8% |   261.0 |    -- |     -- |OMP
||-----------------------------------------------------------------------
||   8.2% |   245.0 |    -- |     -- |_cray$mt_barrier_part__prime_wait_others
```

```
5659.    iVr2--------<         DO       108       J=2,IPPH
5660.    iVr2                     JC = IPP2-J
5661.    iVr2                     J2 = J+J
5662. + iVr2 i------<             DO 107 K=1,L1
5663.    iVr2 i           CSPEC       CH(1,K,J) = CC(IDO,J2-2,K)+CC(IDO,J2-2,K)
5664.    iVr2 i                       CH(IX_CH(1,K),J) = CC(IX_CC(IDO,J2-2,K))+
5665.    iVr2 i               &                           CC(IX_CC(IDO,J2-2,K))
5666.    iVr2 i           CSPEC       CH(1,K,JC) = CC(1,J2-1,K)+CC(1,J2-1,K)
5667.    iVr2 i                       CH(IX_CH(1,K),JC) = CC(IX_CC(1,J2-1,K))+
5668.    iVr2 i               &                            CC(IX_CC(1,J2-1,K))
5669.    iVr2 i------>    107     CONTINUE
5670.    iVr2-------->    108 CONTINUE
```

Excerpt 9.3.5 Original strided loop-nest from SPEC's APSI component.

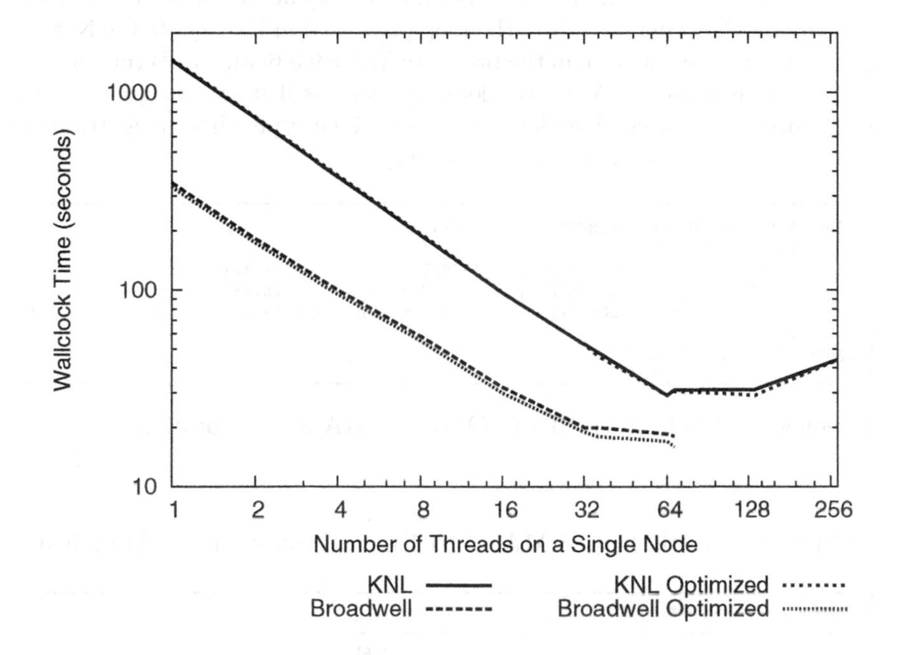

FIGURE 9.3.6 Performance of SPEC APSI code on KNL and Broadwell.

This change was also made in the **dvdtz** loop on line 1915 and the **dudtz** loop on line 1742. The results did not really make much difference since a small amount of time was spent in the loops that were modified. Table 9.3.5 shows the profile from the best run on KNL on 64 threads.

The major bottleneck for KNL are several routines called from within the high-level parallel loop. The loop iteration counts are only 8, and there are strides in the vectorized loops. Consider the loop presented in Excerpt 9.3.5 from the **radbg** routine. The compiler chooses to vectorize on the `J` loop which strides through the arrays. The `IX` references within the loop are statement functions that define which element to fetch, `IX_CH(I,J) = (J-1)*IDO + I` for example. These statement functions are inlined and do not inhibit vectorization. It does not look like the strides can be eliminated, and they prevent KNL from using its powerful vector instructions. For this reason, the performance results in Figure 9.3.6 show Broadwell beating KNL even though KNL gets close at 64 threads.

9.3.5 FMA3D

FMA3D is a crash simulation program using the finite element method [18]. This is another well-threaded application, and while KNL and Broadwell scale well, the high-level routines in the profile are not vectorizable. The MATERIAL_41_INTEGRATION routine does not even have a loop in it. The OMP regions that are not scaling have very bad memory access issues which destroys the OpenMP scaling. Consider the loop presented in Excerpt 9.3.6. Notice the poor TLB utilization seen in the profile in Table 9.3.6; any TLB reference/miss below 512 is suspect. While the loop appears as if it is contiguous in memory, since we are accessing elements of a derived type, there is a stride of the number of elements within the derived type.

```
!$OMP PARALLEL DO DEFAULT(SHARED) PRIVATE(N)
      DO N = 1,NUMRT
        MOTION(N)%Ax = NODE(N)%Minv * (FORCE(N)%Xext-FORCE(N)%Xint)
        MOTION(N)%Ay = NODE(N)%Minv * (FORCE(N)%Yext-FORCE(N)%Yint)
        MOTION(N)%Az = NODE(N)%Minv * (FORCE(N)%Zext-FORCE(N)%Zint)
      ENDDO
!$OMP END PARALLEL DO
```

Excerpt 9.3.6 Code from SPEC OMP's FMA3D component.

TABLE 9.3.6 Profile from FMA3D SPEC code showing TLB utilization.

```
USER / solve_.LOOP@li.329
------------------------------------------------------------------------
  Time%                                          4.5%
  Time                                      12.197115 secs
  Imb.Time                                   0.092292 secs
  Imb.Time%                                      1.0%
  Calls                     42.9 /sec           523.0 calls
  PAPI_L1_DCM             13.700M/sec       167144470 misses
  PAPI_TLB_DM              0.448M/sec         5460907 misses
  PAPI_L1_DCA             89.596M/sec      1093124368 refs
  PAPI_FP_OPS             52.777M/sec       643917600 ops
  User time (approx)      12.201 secs    32941756956 cycles  100.0%Time
  Average Time per Call                      0.023321 sec
  CrayPat Overhead : Time   0.0%
  HW FP Ops / User time   52.777M/sec       643917600 ops    0.5%peak(DP)
  HW FP Ops / WCT         52.777M/sec
  Computational intensity   0.02 ops/cycle     0.59 ops/ref
  MFLOPS (aggregate)       52.78M/sec
  TLB utilization         200.17 refs/miss    0.391 avg uses
  D1 cache hit,miss ratios  84.7% hits         15.3% misses
  D1 cache utilization (M)   6.54 refs/miss    0.817 avg uses
```

As can be seen in Excerpt 9.3.7, rather than the arrays being dimensioned within the derived type, the derived type is dimensioned. This results in each of the arrays having a stride of 12 elements, which is hurting both TLB and cache utilization. Performance results comparing KNL and Broadwell are presented in Figure 9.3.7.

```
TYPE :: motion_type
  REAL(KIND(0D0))  Px       ! Initial x-position
  REAL(KIND(0D0))  Py       ! Initial y-position
  REAL(KIND(0D0))  Pz       ! Initial z-position
  REAL(KIND(0D0))  Ux       ! X displacement
  REAL(KIND(0D0))  Uy       ! Y displacement
  REAL(KIND(0D0))  Uz       ! Z displacement
  REAL(KIND(0D0))  Vx       ! X velocity
  REAL(KIND(0D0))  Vy       ! Y velocity
  REAL(KIND(0D0))  Vz       ! Z velocity
  REAL(KIND(0D0))  Ax       ! X acceleration
  REAL(KIND(0D0))  Ay       ! Y acceleration
  REAL(KIND(0D0))  Az       ! Z acceleration
END TYPE
TYPE (motion_type), DIMENSION(:), ALLOCATABLE :: MOTION
```

Excerpt 9.3.7 Code with derived type from SPEC's FMA3D component.

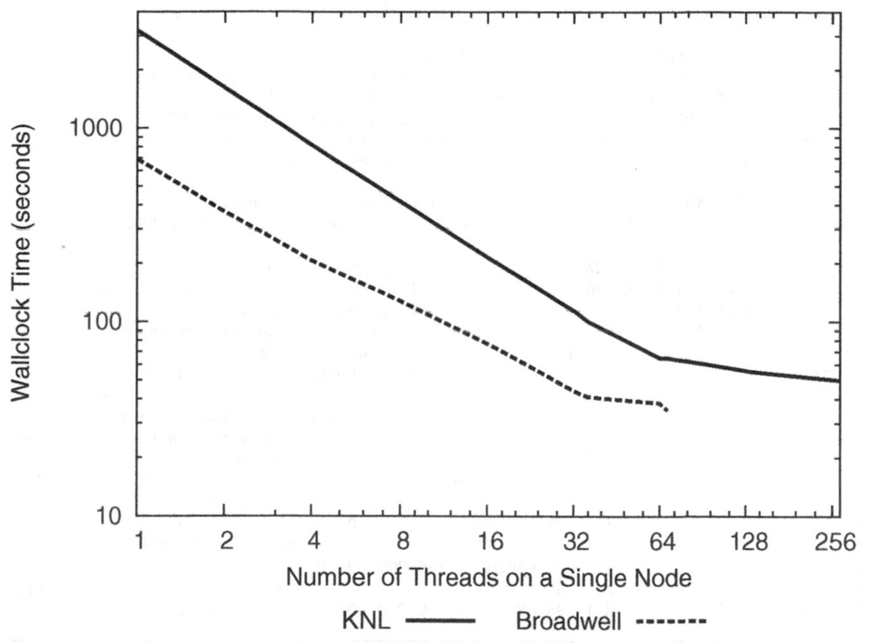

FIGURE 9.3.7 Performance of FMA3D code from SPEC on KNL and Broadwell.

9.3.6 AMMP

The AMMP component from SPEC is a molecular dynamics code written in C. Table 9.3.7 shows a call tree profile from a 28-thread run of AMMP on Broadwell. A large portion of the runtime appears to be spent in the parallel region starting on line 595 and in OpenMP locking functions. The loops over atoms here are not good candidates for vectorization due to the lock-per-atom scheme used. There are also some conditional hazards in the most time-consuming inner-loops. As this is a C code, aliasing issues come to mind. However, most of the time-consuming loops are not terribly dense with pointer or array accesses. There is actually a lot of arithmetic on scalar variables. Just in case, the `restrict` keyword was added to a few of the more commonly used variables like `nodelist` and `nodelistt`, though no performance improvement was observed. There doesn't seem to be a lot one can readily do to futher optimize this code.

TABLE 9.3.7 Calltree profile from AMMP SPEC code.

```
Table:  Calltree View with Callsite Line Numbers

  Time% |        Time |           Calls |Calltree
        |             |                 | Thread=HIDE

 100.0% | 204.128327 |             -- |Total
|-----------------------------------------------------------------------------
|  99.9% | 203.850198 |             -- |main:ammp.c:line.143
|  99.9% | 203.837505 |             -- | read_eval_do:eval.c:line.258
|||---------------------------------------------------------------------------
3||  96.6% | 197.206290 |             -- |eval:eval.c:line.709
||||--------------------------------------------------------------------------
4|||  91.1% | 185.938666 |             -- |tpac:animate.c:line.364
5|||  90.5% | 184.710192 |             -- | u_f_nonbon:unonbon.c:line.224
||||||------------------------------------------------------------------------
6|||||  87.2% | 178.019927 |             -- |f_nonbon:vnonbon.c:line.397
7|||||  86.3% | 176.185836 |             -- | fv_update_nonbon:vnonbon.c:line.130
8|||||  84.6% | 172.605279 |             -- |  main:ammp.c:line.143
9|||||        |            |                |   read_eval_do:eval.c:line.258
10||||        |            |                |    eval:eval.c:line.709
11||||        |            |                |     tpac:animate.c:line.364
12||||        |            |                |      u_f_nonbon:unonbon.c:line.224
13||||        |            |                |       f_nonbon:vnonbon.c:line.397
14||||        |            |                |        fv_update_nonbon:vnonbon.c:line.130
15||||  84.0% | 171.393526 |             -- |         mm_fv_update_nonbon:rectmm.c:li.595
16||||  84.0% | 171.393405 |          200.0 |          mm_fv_update_nonbon.REGION@li.595
||||||||||||||||--------------------------------------------------------------
17||||||||||||||||  43.6% |  89.051268 |          200.0 |mm_fv_update_nonbon.LOOP@li.595
17||||||||||||||||  22.8% |  46.455269 |             -- |mm_fv_update_nonbon.LOOP@li.1136
18||||||||||||||||  22.8% |  46.455269 | 195,809,697.0 | omp_set_lock
17||||||||||||||||  17.5% |  35.629864 |             -- |mm_fv_update_nonbon.LOOP@li.1129
18||||||||||||||||  17.5% |  35.629864 | 195,809,697.0 | omp_unset_lock             [...]
```

One thought for AMMP is that it might be possible to get a speedup by manually vectorizing some of the arithmetic that has `x`, `y`, and `z` components. That is, while there aren't really any important loops that are amenable to automatic compiler vectorization, there is some arithmetic which is quite similer in each of the three dimensions seen in the computation. A portion of this could be vectorized, with three elements in a 4-element AVX2 vector each corresponding to one of the x, y, and z components of the computation. The initial scalar version of the code is presented in Excerpt 9.3.8.

```
for(i=0; i < nng0; i++) {          /* add the new components */
  a2 = (*atomall)[natoms*o+i];
  omp_set_lock(&(a2->lock));
  v0 = a2->px-a1px;  v1 = a2->py-a1py;  v2 = a2->pz-a1pz;
  v3 = sqrt(v0*v0 + v1*v1 + v2*v2);
  if( v3 > mxcut || inclose > NCLOSE ) {
    r0 = one/v3;  r = r0*r0;  r = r*r*r; /* r0^-6 */
    xt = a1q*a2->q*dielectric*r0;  yt = a1a*a2->a*r;  zt = a1b*a2->b*r*r;
    k = xt - yt + zt;
    xt = xt*r0;  yt = yt*r0;  zt = zt*r0;
    k1 = xt - yt*six + zt*twelve;
    xt = xt*r0;  yt = yt*r0;  zt = zt*r0;
    k2 = xt*three;  ka2 = - yt*six*eight;  kb2 =  zt*twelve*14;
    k1 = -k1;  a1VP += k;
    xt     = v0*r0;  yt     = v1*r0;  zt     = v2*r0;
    a1dpx += k1*xt;  a1dpy += k1*yt;  a1dpz += k1*zt;
    xt2    = xt*xt;  yt2    = yt*yt;  zt2    = zt*zt;
    a1qxx   -= k2*(xt2-third)+ ka2*(xt2-eightth)+ kb2*(xt2-fourteenth);
    a1qxy   -= (k2+ka2+kb2)*yt*xt;
    a1qxz   -= (k2+ka2+kb2)*zt*xt;
    a1qyy   -= k2*(yt2-third)+ ka2*(yt2-eightth)+ kb2*(yt2-fourteenth);
    a1qyz   -= (k2+ka2+kb2)*yt*zt;
    a1qzz   -= k2*(zt2-third)+ ka2*(zt2-eightth)+ kb2*(zt2-fourteenth);
    a2->dpx -= k1*xt;  a2->dpy -= k1*yt;  a2->dpz -= k1*zt;
    a2->qxx -= k2*(xt2-third)+ ka2*(xt2-eightth)+ kb2*(xt2-fourteenth);
    a2->qxy -= (k2+ka2+kb2)*yt*xt;
    a2->qxz -= (k2+ka2+kb2)*zt*xt;
    a2->qyy -= k2*(yt2-third)+ ka2*(yt2-eightth)+ kb2*(yt2-fourteenth);
    a2->qyz -= (k2+ka2+kb2)*yt*zt;
    a2->qzz -= k2*(zt2-third)+ ka2*(zt2-eightth)+ kb2*(zt2-fourteenth);
  } else {
    a1->close[inclose++] = a2;
  }
  omp_unset_lock(&(a2->lock));
} /* end of loop i */
```

Excerpt 9.3.8 Scalar code from SPEC OMP's AMMP component showing 3D computation.

At first glance, it's not clear if vectorizing this will result in much of a performance improvement. A 3-element vector isn't very long and will have an unused element due to being an odd size. There will also be some overhead as a number of vector broadcast and extract instructions will be required. The easiest way to do this would probably be to use AVX2 intrinsics in the existing C code. While not portable to different ISAs, the intrinsics should be portable across a number of different compilers.

One might spend some time trying to decide if the effort to vectorize this portion of the AMMP code would be worth it, or one could simply: "Taste it and see!" The scalar code in Excerpt 9.3.8 is converted to the vector code in Excerpts 9.3.9, 9.3.10, and 9.3.11. Some minor changes were also made to the

ATOM structure for alignment reasons, which will be described in more detail a little after the core of the vector code is covered.

The start of the vector code in Excerpt 9.3.9 declares some of the temporary variables that will be needed later. It also gets some constant vectors ready. All of this will be done with four double-precision elements per vector, so all these vector variables are declared with the `__m256d` type (one may also notice that most of the instrinsic calls have names ending in `_pd`).

Excerpt 9.3.10 shows the bulk of the vector computation. While there are a few vector loads and vector stores in this code, a fair bit of the computation can be performed out of vector registers. Additionally, some amount of instruction-level parallelism (ILP) can be exposed here in addition to the vector parallelism. In fact, the vector code presented is the result of a "first pass" vectorization effort applied to the scalar code. Some additional tuning could probably result in even better performance. Specifically, one might consider unrolling the `i` loop to help expose even more ILP, depending on how much contention there is for the atom locks (as two locks would need to be held at once).

```
#include <immintrin.h>
[...]
__m256d v_v0_v1_v2,v_xt_yt_zt;
__m256d v_k2_ka2_kb2,v_xt2_yt2_zt2;
__m256d v_r0,v_t0,v_t1,v_t2,v_k2,v_ka2,v_kb2;
__m256d v_a2dpx_a2dpy_a2dpz;
__m256d v_a2qxy_a2qyz_a2qxz;
__m256d v_a2qxx_a2qyy_a2qzz;
__m256d v_1_1_1_0          = _mm256_set_pd(0.0,  1.0,  1.0, 1.0);
__m256d v_1_n6_12          = _mm256_set_pd(0.0, 12.0, -6.0, 1.0);
__m256d v_3_8_14           = _mm256_set_pd(0.0, 14.0,  8.0, 3.0);
__m256d v_frac_3           = _mm256_set1_pd(third);
__m256d v_frac_8           = _mm256_set1_pd(eightth);
__m256d v_frac_14          = _mm256_set1_pd(fourteenth);
__m256d v_a1qxx_a1qyy_a1qzz = _mm256_set_pd(a1qxx,a1qyy,a1qzz,0.0);
__m256d v_a1dpx_a1dpy_a1dpz = _mm256_set_pd(a1dpx,a1dpy,a1dpz,0.0);
__m256d v_a1qxy_a1qyz_a1qxz = _mm256_set_pd(a1qxy,a1qyz,a1qxz,0.0);

for( i=0; i< nng0; i++) { /* add the new components */
  a2 = (*atomall)[natoms*o+i];
  omp_set_lock(&(a2->lock));
  v_v0_v1_v2 = _mm256_load_pd(&(a2->px));
  v_xt_yt_zt = _mm256_set_pd(0.0, a1pz, a1py, a1px);
  v_v0_v1_v2 = _mm256_mul_pd(v_v0_v1_v2, v_1_1_1_0 );
  v_v0_v1_v2 = _mm256_sub_pd(v_v0_v1_v2, v_xt_yt_zt );
  v_t0       = _mm256_mul_pd(v_v0_v1_v2, v_v0_v1_v2 );
  v_t0       = _mm256_hadd_pd(v_t0, _mm256_permute2f128_pd(v_t0,v_t0,1));
  v_t0       = _mm256_hadd_pd(v_t0, v_t0);
  v3         = sqrt( _mm_cvtsd_f64(_mm256_castpd256_pd128(v_t0)) );
  if( v3 > mxcut || inclose > NCLOSE ) {
    r0 = one/v3;  r = r0*r0;  r = r*r*r; /* r0^-6 */
    xt = a1q*a2->q*dielectric*r0;  yt = a1a*a2->a*r;  zt = a1b*a2->b*r*r;
    v_r0  = _mm256_set1_pd(r0);
    a1VP += xt - yt + zt;
```

Excerpt 9.3.9 Example manual vector intrinsic code for 3D AMMP computation (top).

```
v_xt_yt_zt          = _mm256_set_pd(0.0, zt, yt, xt);
v_xt_yt_zt          = _mm256_mul_pd(v_xt_yt_zt, v_r0);
v_t0                = _mm256_mul_pd(v_xt_yt_zt, v_1_n6_12);
v_t0                = _mm256_hadd_pd(v_t0, _mm256_permute2f128_pd(v_t0,v_t0,1));
v_t0                = _mm256_hadd_pd(v_t0, v_t0);
k1                  = -1.0 * _mm_cvtsd_f64(_mm256_castpd256_pd128(v_t0));
v_xt_yt_zt          = _mm256_mul_pd(v_xt_yt_zt,   v_r0);
v_k2_ka2_kb2        = _mm256_mul_pd(v_xt_yt_zt,   v_1_n6_12);
v_k2_ka2_kb2        = _mm256_mul_pd(v_k2_ka2_kb2, v_3_8_14);
v_xt_yt_zt          = _mm256_mul_pd(v_v0_v1_v2,   v_r0);
v_xt2_yt2_zt2       = _mm256_mul_pd(v_xt_yt_zt,   v_xt_yt_zt);
v_t0                = _mm256_sub_pd(v_xt2_yt2_zt2, v_frac_3);
v_t1                = _mm256_sub_pd(v_xt2_yt2_zt2, v_frac_8);
v_t2                = _mm256_sub_pd(v_xt2_yt2_zt2, v_frac_14);
v_k2                = _mm256_permute2f128_pd(v_k2_ka2_kb2, v_k2_ka2_kb2, 0);
v_ka2               = v_k2;
v_kb2               = _mm256_permute2f128_pd(v_k2_ka2_kb2, v_k2_ka2_kb2, 17);
v_k2                = _mm256_permute_pd(v_k2, 0);
v_ka2               = _mm256_permute_pd(v_ka2, 15);
v_kb2               = _mm256_permute_pd(v_kb2, 0);
v_t0                = _mm256_mul_pd(v_t0, v_k2);
v_t1                = _mm256_mul_pd(v_t1, v_ka2);
v_t2                = _mm256_mul_pd(v_t2, v_kb2);
v_t0                = _mm256_add_pd(v_t0, v_t1);
v_t0                = _mm256_add_pd(v_t0, v_t2);
v_t2                = _mm256_permute2f128_pd(v_k2_ka2_kb2, v_k2_ka2_kb2, 1);
v_t2                = _mm256_hadd_pd(v_k2_ka2_kb2, v_t2);
v_t1                = _mm256_set1_pd(k1);
v_t2                = _mm256_hadd_pd(v_t2, v_t2);
v_t2                = _mm256_mul_pd(v_t2, v_xt_yt_zt);
v_t1                = _mm256_mul_pd(v_t1, v_xt_yt_zt);
v_a2qxx_a2qyy_a2qzz = _mm256_load_pd(&(a2->qxx));
v_a1qxx_a1qyy_a1qzz = _mm256_sub_pd(v_a1qxx_a1qyy_a1qzz, v_t0);
v_a2qxx_a2qyy_a2qzz = _mm256_sub_pd(v_a2qxx_a2qyy_a2qzz, v_t0);
v_t0                = _mm256_permute2f128_pd(v_xt_yt_zt, v_xt_yt_zt, 17);
v_t0                = _mm256_shuffle_pd(v_xt_yt_zt, v_t0, 1);  // yzzx
v_t0                = _mm256_permute_pd(v_t0, 6);              // yzxz
v_t0                = _mm256_mul_pd(v_t0, v_t2);
_mm256_store_pd(&(a2->qxx),v_a2qxx_a2qyy_a2qzz);
v_a2qxy_a2qyz_a2qxz = _mm256_load_pd(&(a2->qxy));
v_a1qxy_a1qyz_a1qxz = _mm256_sub_pd(v_a1qxy_a1qyz_a1qxz, v_t0);
v_a2qxy_a2qyz_a2qxz = _mm256_sub_pd(v_a2qxy_a2qyz_a2qxz, v_t0);
v_a2dpx_a2dpy_a2dpz = _mm256_load_pd(&(a2->dpx));
v_a1dpx_a1dpy_a1dpz = _mm256_add_pd(v_a1dpx_a1dpy_a1dpz, v_t1);
v_a2dpx_a2dpy_a2dpz = _mm256_sub_pd(v_a2dpx_a2dpy_a2dpz, v_t1);
_mm256_store_pd(&(a2->qxy),v_a2qxy_a2qyz_a2qxz);
_mm256_store_pd(&(a2->dpx),v_a2dpx_a2dpy_a2dpz);
```

Excerpt 9.3.10 Example manual vector intrinsic code for 3D AMMP computation (middle).

After the bulk of the computation, some additional effort is needed to integrate the produced results with the code which consumes them. Excerpt 9.3.11 shows the end of the code section being discussed. This final code portion was vectorized as well, so that the vector results produced in earlier sections could be used directly, without having to be converted back into scalars as would have been required with the original code.

To facilitate the vectorization of the computation, some changes were made to the definition of the `ATOM` type. In particular, the vector code would really like to have all its loads and stores be aligned. To accomplish this, some of the members of the `ATOM` structure are moved around, and some extra padding is added.

```
  } else {
    a1->close[inclose++] = a2;
  }
  omp_unset_lock(&(a2->lock));
}/* end of loop i */
[...]
omp_set_lock(&(a1->lock));
a1->VP += a1VP ;
v_a1qxx_a1qyy_a1qzz = _mm256_add_pd(v_a1qxx_a1qyy_a1qzz, _mm256_load_pd(&(a1->qxx)));
v_a1dpx_a1dpy_a1dpz = _mm256_add_pd(v_a1dpx_a1dpy_a1dpz, _mm256_load_pd(&(a1->dpx)));
v_a1qxy_a1qyz_a1qxz = _mm256_add_pd(v_a1qxy_a1qyz_a1qxz, _mm256_load_pd(&(a1->qxy)));
_mm256_store_pd(&(a1->qxx),v_a1qxx_a1qyy_a1qzz);
_mm256_store_pd(&(a1->dpx),v_a1dpx_a1dpy_a1dpz);
_mm256_store_pd(&(a1->qxy),v_a1qxy_a1qyz_a1qxz);
[...]
omp_unset_lock(&(a1->lock));
```

Excerpt 9.3.11 Example manual vector intrinsic code for 3D AMMP computation (bottom).

The original definition seen in Excerpt 9.3.12 is optimized to become the new definition seen in Excerpt 9.3.13. Also note that the optimized version has placed the OpenMP lock at the end of the structure. Combined with the 64-byte alignment of the structure itself, this lock placement will ensure that the OpenMP lock is never on the same cache line as another atom structure. In fact, the new lock placement is fairly far away from most of the other heavily used member variables, and the lock now essentially lives on its own cache line (assuming 64-byte cache lines).

```
typedef struct {
  double      x,y,z,fx,fy,fz;
  double      q,a,b,mass,chi,jaa;
  double      vx,vy,vz,vw,dx,dy,dz;
  double      gx,gy,gz;
  double      VP,px,py,pz,dpx,dpy,dpz;
  double      qxx,qxy,qxz,qyy,qyz,qzz;
  void       *close[200],*excluded[32],*next;
  omp_lock_t lock;
  int         serial;
  char        active,name[9],exkind[32];
  int         dontuse;
} ATOM;
```

Excerpt 9.3.12 Code from SPEC OMP's AMMP component showing original alignment.

As can be seen in Figure 9.3.8, the optimized version with manual vectorization performs a bit better than the original on Broadwell as well as KNL at all tested thread counts. While some small improvement may come from the better placement of OpenMP locks in the `ATOM` structure, most of the performance gain comes from the vectorization changes. Overall, AMMP scales fairly well with threads and also with simultaneous multithreading (SMT) or hyper-threading (HT). In fact, the best performance on KNL is with 4-way SMT and a total of 256 threads.

```
typedef struct {
  double       x,y,z,fx,fy,fz;
  double       q,a,b,mass,chi,jaa;
  double       vx,vy,vz,vw,dx,dy,dz;
  double       gx,gy,gz,VP,pad0;
  double        px, py, pz,pad1;
  double       dpx,dpy,dpz,pad2;
  double       qxx,qyy,qzz,pad3;
  double       qxy,qyz,qxz,pad4;
  void        *close[200],*excluded[32],*next;
  int          serial;
  char         active,name[9],exkind[32];
  int          dontuse;
  omp_lock_t lock;
} ATOM __attribute__ ((aligned (64)));
```

Excerpt 9.3.13 Code from SPEC OMP's AMMP component showing improved alignment.

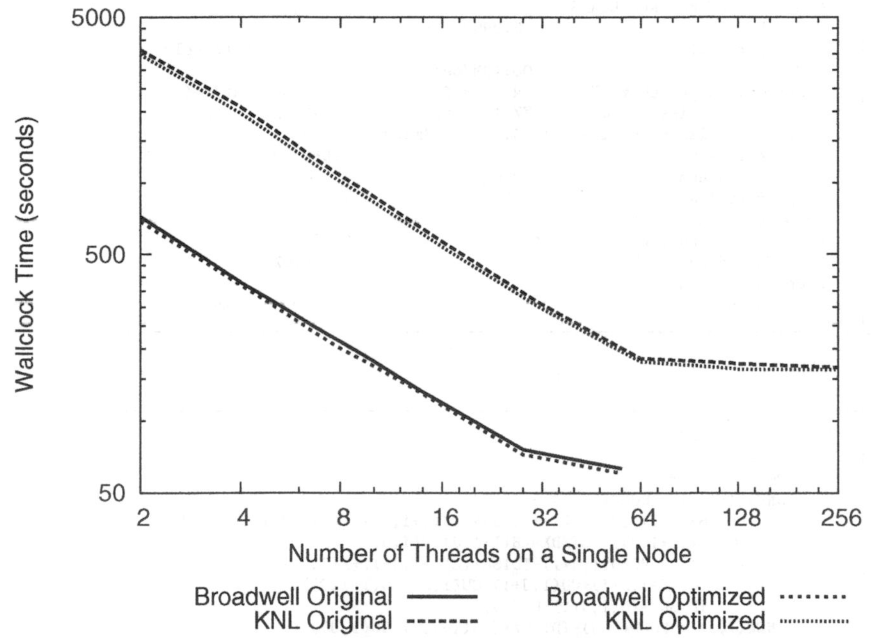

FIGURE 9.3.8 Performance of AMMP code from SPEC on KNL and Broadwell.

9.3.7 SWIM

SWIM is a small application that performs shallow water wave propagation. It is well threaded, and there is only one major problem with the application itself. The level-1 cache utilization is very poor. Table 9.3.8 shows the hardware counters from a Broadwell run on 36 threads of the `calc2` routine. Notice that the level-1 cache gets 77% hit within a cache line. Just a streaming fetch of a contiguous array would see 87.5% – a miss every 8 fetches since the cache line has 8 elements. Why is this example so poor? The reason can be seen in the major loop within the routine, presented in Excerpt 9.3.14.

TABLE 9.3.8 Profile from SWIM SPEC code showing TLB utilization.

```
  USER / calc2_
------------------------------------------------------------------------------
  Time%                                                   35.1%
  Time                                               14.538245 secs
  Imb. Time                                                 -- secs
  Imb. Time%                                                --
  Calls                           16.508 /sec            240.0 calls
  PAPI_TOT_INS                 1,703.489M/sec   24,765,733,293 instr
  CPU_CLK_THREAD_UNHALTED:THREAD_P              44,560,852,579
  CPU_CLK_THREAD_UNHALTED:REF_XCLK               1,439,994,654
  L1D:REPLACEMENT                                3,074,475,837
  MEM_UOPS_RETIRED:ALL_LOADS                    13,452,580,272
  CPU_CLK                            3.09GHz
  Instr per cycle                                         0.56 inst/cycle
  MIPS                         1,703.49M/sec
  Instructions per LD & ST       54.3% refs               1.84 inst/ref
  D1 cache hit,miss ratios       77.1% hits              22.9% misses
  D1 cache utilization (misses)  4.38 refs/miss           0.55 avg hits
  Average Time per Call                               0.060576 secs
  CrayPat Overhead : Time         0.0%
  Loop Trips Avg                                            --
  Loop Incl Time / Hit                                      -- secs
  Loop Thread Speedup                                 0.000000 secs
  Loop Thread Speedup%                                    0.0%
  Loop Incl Time%                                         0.0%
  Calls or Hits                                          240.0 calls+hits
```

```
!$OMP PARALLEL DO
      DO 200 J=1,N
      DO 200 I=1,M
      UNEW(I+1,J) = UOLD(I+1,J)+
     1    TDTS8*(Z(I+1,J+1)+Z(I+1,J))*(CV(I+1,J+1)+CV(I,J+1)+CV(I,J)
     2       +CV(I+1,J))-TDTSDX*(H(I+1,J)-H(I,J))
      VNEW(I,J+1) = VOLD(I,J+1)-TDTS8*(Z(I+1,J+1)+Z(I,J+1))
     1       *(CU(I+1,J+1)+CU(I,J+1)+CU(I,J)+CU(I+1,J))
     2       -TDTSDY*(H(I,J+1)-H(I,J))
      PNEW(I,J) = POLD(I,J)-TDTSDX*(CU(I+1,J)-CU(I,J))
     1       -TDTSDY*(CV(I,J+1)-CV(I,J))
  200 CONTINUE
```

Excerpt 9.3.14 Code from SWIM showing shared domains.

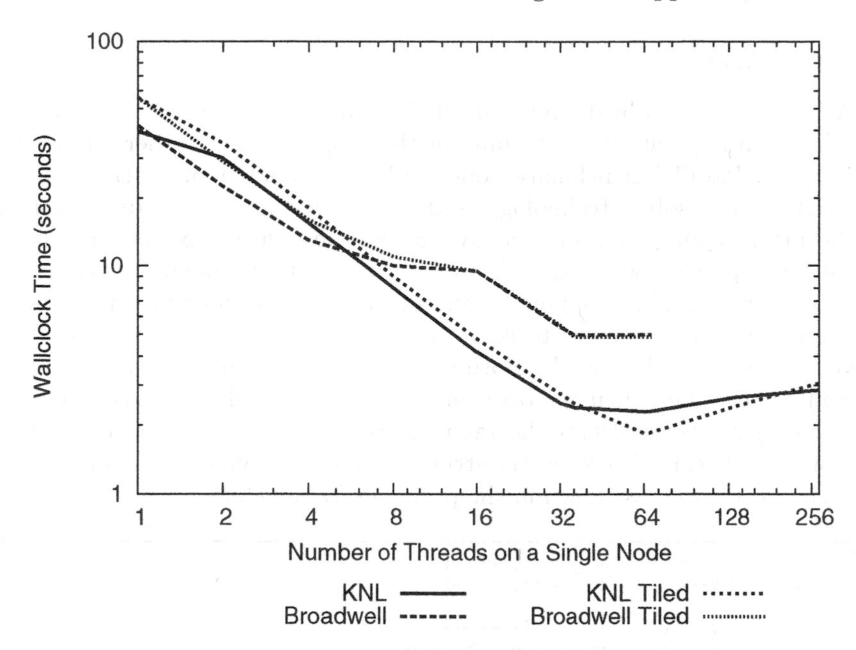

FIGURE 9.3.9 Performance of SWIM code on KNL and Broadwell.

Notice that elements are set out of their OpenMP domain. That is, the J+1 store in the second statement is storing into the domain of the next thread. Additionally, it is fetching from the other domain as well. This will result in cache thrashing, since one thread is fetching and storing out of its sphere of influence. In order to minimize that, one might consider tiling the two loops as seen in Excerpt 9.3.15. Performance results seen in Figure 9.3.9 show that this does improve on KNL at 64 threads. However, it doesn't improve on Broadwell.

```
!$OMP PARALLEL DO PRIVATE(iii,i3s,i3e,i2s,i2e,j,i)
      do iii = 1, m/ic*n/ic
       i3s = max(1,((iii-1)/(n/ic))*ic+1)
       i2s = max(1,mod(iii-1,m/ic)*ic+1)
       i3e = min(i3s+ic-1,n) ; i2e = min(i2s+ic-1,m)
      do J=i3s,i3e
       do I = i2s,i2e
      UNEW(I+1,J) = UOLD(I+1,J)+
     1    TDTS8*(Z(I+1,J+1)+Z(I+1,J))*(CV(I+1,J+1)+CV(I,J+1)+CV(I,J)
     2         +CV(I+1,J))-TDTSDX*(H(I+1,J)-H(I,J))
      VNEW(I,J+1) = VOLD(I,J+1)-TDTS8*(Z(I+1,J+1)+Z(I,J+1))
     1         *(CU(I+1,J+1)+CU(I,J+1)+CU(I,J)+CU(I+1,J))
     2         -TDTSDY*(H(I,J+1)-H(I,J))
      PNEW(I,J) = POLD(I,J)-TDTSDX*(CU(I+1,J)-CU(I,J))
     1         -TDTSDY*(CV(I,J+1)-CV(I,J))
      enddo ; enddo ; enddo
```

Excerpt 9.3.15 Code from SWIM showing tiled loops.

9.3.8 APPLU

APPLU solves coupled nonlinear PDEs using the symmetric successive over-relaxation implicit time marching method [3]. This is another adaptation of a NASA Parallel Benchmark code – LU. The main issue with most of the NPB tests of solver technology is that they use very poor dimensioning of the primary and temporary arrays, as can be seen in Excerpt 9.3.16. Since each grid point is working with a 5×5 matrix, they dimension the arrays as `u(5,5,nx,ny,nz)`. When one is vectorizing the innermost looping structures, the choice is to either vectorize on a length of 5 or to vectorize on a longer vector length with a stride. Both of these solutions are bad for KNL, which would like to have their vectors contiguous and aligned on 64-byte boundaries. The major bottlenecks are the `jacu` and `rhs` routines, which have a vectorized looping structure. However, the stride is $5 \times 5 = 25$, which results in very poor cache utilization, as shown in the profile in Table 9.3.9.

```
36.  + 1----<        do j = j1, j0, -1
37.    1 V--<        do i = i1, i0, -1
38.    1 V
39.    1 V     c-----------------------------------------------------------------
40.    1 V     c   form the block diagonal
41.    1 V     c-----------------------------------------------------------------
42.    1 V                   tmp1 = rho_i(i,j,k)
43.    1 V                   tmp2 = tmp1 * tmp1
44.    1 V                   tmp3 = tmp1 * tmp2
45.    1 V
46.    1 V                   du(1,1,i,j) =  1.0d+00
47.    1 V          >                      + dt * 2.0d+00 * (   tx1 * dx1
48.    1 V          >                                          + ty1 * dy1
49.    1 V          >                                          + tz1 * dz1 )
50.    1 V                   du(1,2,i,j) =  0.0d+00
51.    1 V                   du(1,3,i,j) =  0.0d+00
52.    1 V                   du(1,4,i,j) =  0.0d+00
53.    1 V                   du(1,5,i,j) =  0.0d+00
54.    1 V
55.    1 V                   du(2,1,i,j) =  dt * 2.0d+00
56.    1 V          >          * ( - tx1 * r43 - ty1 - tz1 )
57.    1 V          >          * ( c34 * tmp2 * u(2,i,j,k) )
58.    1 V                   du(2,2,i,j) =  1.0d+00
59.    1 V          >         + dt * 2.0d+00 * c34 * tmp1
60.    1 V          >         * (  tx1 * r43 + ty1 + tz1 )
61.    1 V          >         + dt * 2.0d+00 * (   tx1 * dx2
62.    1 V          >                            + ty1 * dy2
63.    1 V          >                            + tz1 * dz2  )
```

Excerpt 9.3.16 Strided code from SPEC OMP's APPLU component.

A difficult but necessary rewrite of this application would try to have at least one of the grid dimensions on the inside, such as `u(nx,5,5,ny,nz)`. When performing computations that are data-dependent on the `nx` direction, there would be a stride. However, the other two sweeps would be good contiguous vectors. Figure 9.3.10 shows a performance comparison of Broadwell and KNL.

TABLE 9.3.9 Profile from APPLU SPEC code showing cache utilization.

```
USER / jacu_
-------------------------------------------------------------------------------
Time                                                   0.691986 secs
Imb. Time                                              0.090822 secs
Imb. Time%                                                11.9%
Calls                              0.006M/sec           4,360.0 calls
PAPI_TOT_INS                   1,939.524M/sec     1,342,123,883 instr
CPU_CLK_THREAD_UNHALTED:THREAD_P                  1,658,382,457
CPU_CLK_THREAD_UNHALTED:REF_XCLK                     69,092,211
L1D:REPLACEMENT                                      89,949,129
MEM_UOPS_RETIRED:ALL_LOADS                          269,881,256
CPU_CLK                             2.40GHz
Instr per cycle                                            0.81 inst/cycle
MIPS                           1,939.52M/sec
Instructions per LD & ST           20.1% refs              4.97 inst/ref
D1 cache hit,miss ratios           66.7% hits             33.3% misses
D1 cache utilization (misses)     3.00 refs/miss           0.38 avg hits
Average Time per Call                                  0.000159 secs
CrayPat Overhead : Time             0.6%
```

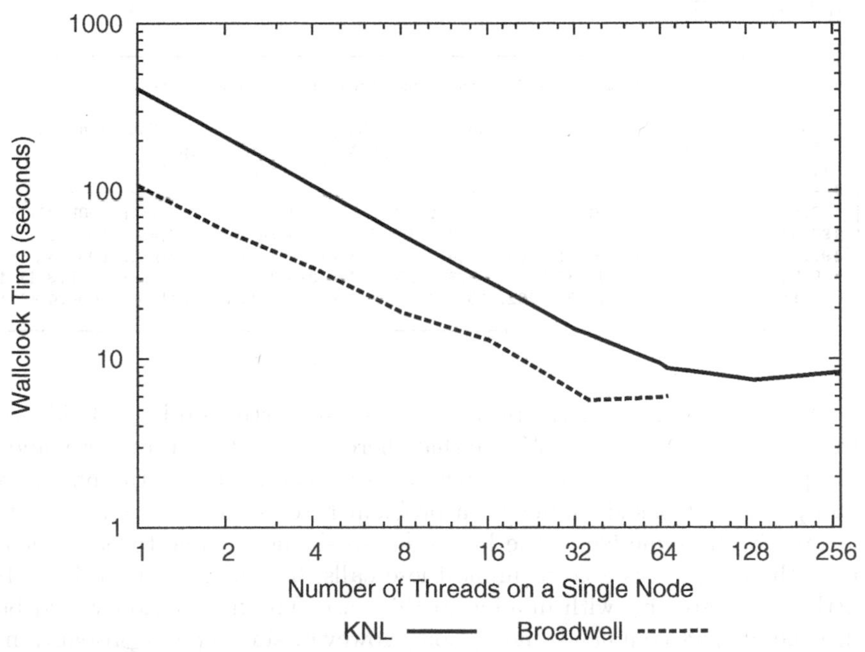

FIGURE 9.3.10 Performance of APPLU code from SPEC on KNL and Broadwell.

9.3.9 EQUAKE

EQUAKE is an earthquake model (as the name suggests) which simulates elastic wave propagation in heterogeneous media [2]. For a first step, some profiles are created with CrayPAT to find the most time-consuming functions and loops. Table 9.3.10 presents a profile from running the original EQUAKE code with 1 thread on Broadwell, while Table 9.3.11 presents a (different style of) profile with 28 threads. One can see that at low thread counts, the majority of the time is taken by the loop-nest in the `main` function, starting on line 531 (the loop on line 518 is the main timestep loop), while higher thread counts show the loop-nest in the `smvp` function starting on line 1267 to be the most time-consuming. This difference in which routines are the most time-consuming in EQUAKE at different thread counts serves as a good example that one needs to be sure to profile codes at a scale and problem size very close to what would be run in production. A profile obtained from a smaller run could be very misleading. One needs to start somewhere, and since the loop-nests starting at line 531 and at line 1267 are both mentioned as being fairly time-consuming in the 1-thread profile as well as the 28-thread profile, these two loop-nests will be selected for possible optimization efforts.

TABLE 9.3.10 Profile from EQUAKE code in SPEC OMP with 1 thread on Broadwell.

Table: Inclusive and Exclusive Time in Loops (from -hprofile_generate)

Loop Incl Time%	Loop Incl Time	Time (Loop Adj.)	Loop Hit	Loop Trips Avg	Loop Trips Min	Loop Trips Max	Function=/.LOOP[.]
99.4%	2,212.264	0.0121	1	3,334.0	3,334	3,334	main.LOOP.26.li.518
92.9%	2,066.069	8.6763	3,334	378,747.0	378,747	378,747	main.LOOP.29.li.531
92.5%	2,057.393	1,202.9260	1,262,742,498	3.0	3	3	main.LOOP.30.li.533
5.0%	110.430	19.1468	3,334	378,747.0	378,747	378,747	smvp.LOOP.3.li.1267
4.1%	91.283	91.2833	1,262,405,764	6.6	1	24	smvp.LOOP.4.li.1276

Excerpt 9.3.17 shows the primary loop-nest starting on line 531 identified from the CrayPAT profile. We see that there is no vectorization here, and the compiler is complaining about function calls. These calls to the `phi0`, `phi1`, and `phi2` functions shouldn't be a problem here, as they are small routines located in the same file as the loop-nest, so should be able to be inlined. In fact, the compiler is able to inline these calls, but needs to be told to be a little more agressive with inlining in this case. The first option would be to use the `-hipa4` command line option. However, since one is probably more concerned with vectorization of this loop nest, and inlining is a requirement for vectorization, one could add a `prefervector` pragma to get the compiler to inline the calls. However, the inlining isn't really much of a problem for

this loop-nest. After inlining is resolved with the **-hipa4** compiler flag, the compiler is complaining about aliasing, as seen in Excerpt 9.3.18.

TABLE 9.3.11 Profile from EQUAKE code in SPEC OMP with 28 threads on Broadwell (serial initialization ignored).

```
Table:  Profile by Function Group and Function

  Time% |       Time |     Imb. |  Imb. |       Calls |Group
        |            |     Time | Time% |             | Function

 100.0% | 35.775459 |       -- |    -- | 4,779,065.0 |Total
|-----------------------------------------------------------------------------
|  90.0% | 32.215304 |       -- |    -- | 4,465,505.0 |USER
||----------------------------------------------------------------------------
||  26.2% |  9.373374 | 0.002404 |  0.0% |    3,334.0 |smvp.LOOP@li.1267
||   6.0% |  2.155785 | 0.072383 |  3.4% |   10,002.0 |main.LOOP@li.533
||   4.9% |  1.763733 | 0.370949 | 14.0% |    3,334.0 |smvp.LOOP@li.1314
||   3.8% |  1.365022 | 0.084979 |  5.8% |   93,352.0 |smvp.LOOP@li.1253
||   2.3% |  0.837043 | 0.461608 | 57.2% |   96,686.0 |smvp
```

```
530.      1           #pragma omp parallel for private(j)
531.    + 1 2------<      for (i = 0; i < ARCHnodes; i++)
532.      1 2             {
533.    + 1 2 3----<        for (j = 0; j < 3; j++)
534.      1 2 3             {
535.      1 2 3               disp[disptplus][i][j] *= - Exc.dt * Exc.dt;
536.      1 2 3               disp[disptplus][i][j] += 2.0 * M[i][j] * disp[dispt][i][j] -
537.      1 2 3           (M[i][j] - Exc.dt / 2.0 * C[i][j]) * disp[disptminus][i][j] -
538.      1 2 3             Exc.dt * Exc.dt * (M23[i][j] * phi2(time) / 2.0 +
539.      1 2 3                 C23[i][j] * phi1(time) / 2.0 +
540.      1 2 3                 V23[i][j] * phi0(time) / 2.0);
541.      1 2 3               disp[disptplus][i][j] = disp[disptplus][i][j] /
542.      1 2 3                               (M[i][j] + Exc.dt / 2.0 * C[i][j]);
543.      1 2 3                 vel[i][j] = 0.5 / Exc.dt * (disp[disptplus][i][j] -
544.      1 2 3             disp[disptminus][i][j]);
545.      1 2 3---->          }
546.      1 2------>        }

CC-6262 CC: VECTOR main, File = quake.c, Line = 531
  A loop was not vectorized because it contains a call to a function on line 533.
```

Excerpt 9.3.17 Original source code from SPEC OMP's EQUAKE component with compiler markup on Broadwell.

As is common with C/C++ codes, it seems as though aliasing is going to be an issue here. Since the goal for this loop-nest is to get improved vectorization, it might be simplest to just add an `ivdep` pragma to the loop to resolve the assumed vector dependencies and be done with it. Otherwise, the **restrict** keyword can be used, but the array declarations would probably need to be modified for the following variables: `disp`, `M`, `C`, `M23`, `C23`, `V23`, and `vel`. Excerpt 9.3.19 shows the modified loop-nest with the addition of pragmas to control vectorization. This looks more promising, as the compiler is at least able to partially vectorize the outer loop. However, one can't readily determine how *much* of the loop was able to be vectorized as the compiler doesn't provide that level of detailed information.

```
530.     1           #pragma omp parallel for private(j)
531.   + 1 MmF----<     for (i = 0; i < ARCHnodes; i++)
532.     1 MmF          {
533.   + 1 MmF wF-<       for (j = 0; j < 3; j++)

CC-6290 CC: VECTOR main, File = quake.c, Line = 531
  A loop was not vectorized because a recurrence was found between "C" and "disp"
  at line 541.
```

Excerpt 9.3.18 Compiler message showing aliasing issue from SPEC OMP's EQUAKE component.

```
530.  1             #pragma omp parallel for private(j)
531.  1             #pragma ivdep
532. +1 MmVpF----<     for (i = 0; i < ARCHnodes; i++)
533.  1 MmVpF          {
534.  1 MmVpF       #pragma novector
535. +1 MmVpF wF-<       for (j = 0; j < 3; j++)
```

Excerpt 9.3.19 Partially vectorized source code from SPEC OMP's EQUAKE component with compiler markup on Broadwell.

```
530. +M-<1 2-----<     for (j = 0; j < 3; j++) {
531.  M   1 2       #pragma omp parallel for private(i)
532.  M   1 2       #pragma ivdep
533. +M   1 2 mVF-<       for (i = 0; i < ARCHnodes; i++) {
534.  M   1 2 mVF           disp[disptplus][j][i] *= - Exc.dt * Exc.dt;
535.  M   1 2 mVF I         disp[disptplus][j][i] += 2.0 * M[j][i] * disp[dispt][j][i] -
536.  M   1 2 mVF      (M[j][i] - Exc.dt / 2.0 * C[j][i]) * disp[disptminus][j][i] -
537.  M   1 2 mVF        Exc.dt * Exc.dt * (M23[j][i] * phi2(time) / 2.0 +
538.  M   1 2 mVF           C23[j][i] * phi1(time) / 2.0 +
539.  M   1 2 mVF           V23[j][i] * phi0(time) / 2.0);
540.  M   1 2 mVF           disp[disptplus][j][i] = disp[disptplus][j][i] /
541.  M   1 2 mVF                              (M[j][i] + Exc.dt / 2.0 * C[j][i]);
542.  M   1 2 mVF           vel[j][i] = 0.5 / Exc.dt * (disp[disptplus][j][i] -
543.  M   1 2 mVF        disp[disptminus][j][i]);
544.  M->1 2 mVF->       }
545.      1 2----->     }
```

Excerpt 9.3.20 Fully vectorized source code from SPEC OMP's EQUAKE component with compiler markup on Broadwell.

It turns out that the primary issue preventing vectorization here is the striding involved with the arrays used in the loop. The fastest-moving index is j which is stride-1 on the arrays, while the vector index i is the middle index and is not stride-1 when accessing the arrays. It would be better to have the vector index be stride-1 on the arrays. One can test this hypothesis without investing the time required to do a complete code update, by just changing the indexing on the existing arrays only for the loop-nest of interest, ignoring the rest of the application code. While one cannot run the code this way as it will get incorrect results or even crash, one can take a look at what the compiler does with the code. The compiler markup presented in Excerpt 9.3.20 shows that fixing the array striding problem is a viable route to acheiving full vectorization of the loop-nest.

Fixing the array striding issues in the loop-nest may be a little cumbersome in the end, considering that a fair number of arrays need to have their strides changed: `disp`, `M`, `C`, `M23`, `C23`, `V23`, and `vel`. The stride changes will affect not only the array declaration and allocation sites, but also potentially every place in the code in which these arrays are indexed. This is something to keep in mind when developing new codes. Care should be taken to ensure that the arrays used have memory layouts that are amenable to good vectorization of the most time-consuming loops. In the case of EQUAKE, making the needed changes is not unrealistic. One should note that the `disp` array needs to have its memory layout changed here, which will also impact the other time-critical loop-nest in the `smvp` function.

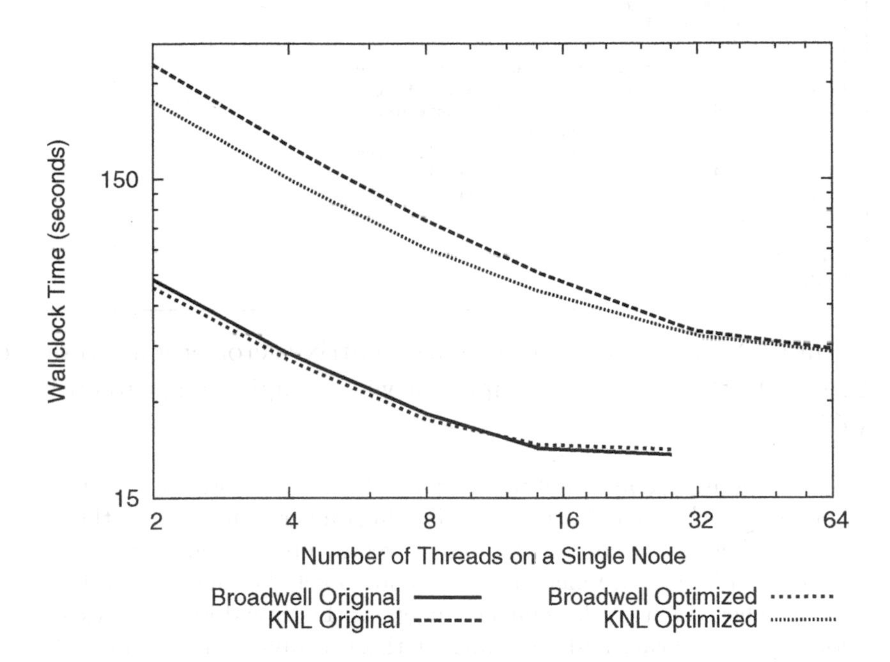

FIGURE 9.3.11 Performance of original and modified EQUAKE code from SPEC on KNL and Broadwell.

Luckily, changing the striding on the `disp` array doesn't impact the performance of `smvp` significantly. Performance of the original EQUAKE version is compared to the new optimized version in Figure 9.3.11. This figure looks reasonable considering the changes performed. At low thread counts where the loop-nest on line 533 dominates the runtime, changes to the loop result in a reasonable speedup (a factor of 1.06 for Broadwell and 1.30 for KNL). However, with more threads, the speedup vanishes (less than 5% performance difference between original and optimized versions). While the new version is always faster on KNL, it may actually be slightly slower on Broadwell with 28 threads. Overall, this shouldn't be all that surprising since the loop-nest in `main` that was just optimized here dominates the runtime primarily at lower thread counts. With some speedup attained for the loop-nest in `main`, the loop in `smvp` will be considered next.

```
1250.                void smvp(int nodes, double (*A)[3][3], int *Acol, int *Aindex,
1251.                          double **v, double **w)
1252.                {
1253.                  int i,j,tid, s, e, col;
1254.                  double sum0, sum1, sum2;
1255.
1256. +    1-----<     for(j = 0; j < numthreads; j++) {
1257.      1         #pragma omp parallel for private(i)
1258.      1 MmA-<      for(i = 0; i < nodes; i++) {
1259.      1 MmA          w2[j][i] = 0;
1260.      1 MmA->      }
1261.      1----->     }
[...]
1303.  M             #pragma omp parallel for private(j)
1304. +M   m-----<     for(i = 0; i < nodes; i++) {
1305. +M   m r4--<       for(j = 0; j < numthreads; j++) {
1306.  M   m r4            if( w2[j][i] ) {
1307.  M   m r4              w[0][i] += w1[j][i].first;
1308.  M   m r4              w[1][i] += w1[j][i].second;
1309.  M   m r4              w[2][i] += w1[j][i].third;
1310.  M   m r4            }
1311.  M   m r4-->       }
1312.  M->m----->     }
1313.                }
```

Excerpt 9.3.21 Begin and end of sparse matrix-vector product code from SPEC OMP's EQUAKE component with compiler markup on Broadwell.

The beginning and end of the `smvp` function is presented in Excerpt 9.3.21. There isn't a whole lot of exciting stuff going on here. One can see the `w2` array is cleared before the core of the matrix-vector product, and all the multiple thread-local buffers are merged at the end to get the final result. The use of `w2` is nice here, as it tracks which elements of the thread-local arrays are used, allowing full clearing and summing of those arrays to be avoided. Excerpt 9.3.22 shows the more interesting part of the computation, the core of the sparse matrix-vector product. As one can infer from the comments, the source code, and the name, the `smvp` function is computing a sparse matrix-vector product. Just to get an intuitive feel for this sparse matrix, the locations of its non-zero elements are plotted in Figure 9.3.12.

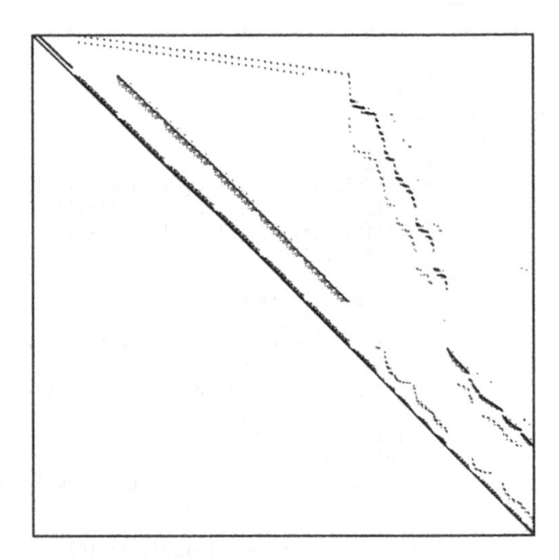

FIGURE 9.3.12 Non-zero element locations of the large 378 K×378 K sparse matrix used in the EQUAKE code from SPEC.

Reading the code for the `smvp` function shows that the matrix is stored in CSR (Compressed Sparse Row) format [10], [31]. The CSR format does have some disadvantages, particularly that it ignores dense substructures. A variation on the CSR format that is often used due to improved performance is BCSR, which utilizes blocks of contiguous elements rather than single elements. At first, it seems like performance could be improved by converting the initial matrix into the BCSR format, and then using the new version for the rest of the computation. The hope is that the cost of converting the matrix would be overcome by the benefit coming from better use of the cache and/or vectorization.

However, notice that things are a little more complex in the EQUAKE code than more common sparse matrix-vector products. In most common HPC literature and libraries, the sparse matrix-vector product being covered is the product of an $N \times M$ matrix of elements of a single primitive type with a vector of length N of elements of the same primitive type as those in the matrix (e.g., a vector of doubles and a matrix of doubles). EQUAKE is more complicated, as it performs the product of a vector of length-3 vectors with a sparse matrix of 3×3 matricies. Not only that, but the `smvp` function is actually computing *two* sparse matrix-vector products. One product with the matrix and one product with the transpose of the matrix of transposed 3×3 matricies. This makes the computation a lot more dense with a larger amount of data reuse than the simple case with vectors and matricies of scalar elements. Thus, there is not expected to be much benefit from converting from the original CSR layout to the BCSR layout. So, what can be done with the

```
1263.              #pragma omp parallel private(tid,i,s,e,col,sum0,sum1,sum2)
1264. +M-<         {
1265. +M             tid = omp_get_thread_num();
1266.  M           #pragma omp for
1267. +M  m-----<    for(i = 0; i < nodes; i++) {
1268.  M  m            s = Aindex[i];
1269.  M  m            e = Aindex[i + 1];
1270.  M  m            sum0 = A[s][0][0]*v[0][i]+ A[s][0][1]*v[1][i]+ A[s][0][2]*v[2][i];
1271.  M  m            sum1 = A[s][1][0]*v[0][i]+ A[s][1][1]*v[1][i]+ A[s][1][2]*v[2][i];
1272.  M  m            sum2 = A[s][2][0]*v[0][i]+ A[s][2][1]*v[1][i]+ A[s][2][2]*v[2][i];
1273. +M  m 2---<      while( ++s < e ) {
1274.  M  m 2            col = Acol[s];
1275.  M  m 2            sum0 += A[s][0][0]*v[0][col] + A[s][0][1]*v[1][col] +
1276.  M  m 2                    A[s][0][2]*v[2][col];
1277.  M  m 2            sum1 += A[s][1][0]*v[0][col] + A[s][1][1]*v[1][col] +
1278.  M  m 2                    A[s][1][2]*v[2][col];
1279.  M  m 2            sum2 += A[s][2][0]*v[0][col] + A[s][2][1]*v[1][col] +
1280.  M  m 2                    A[s][2][2]*v[2][col];
1281.  M  m 2            if( w2[tid][col] == 0 ) {
1282.  M  m 2              w2[tid][col] = 1;
1283.  M  m 2              w1[tid][col].first  = 0.0; w1[tid][col].second = 0.0;
1284.  M  m 2              w1[tid][col].third  = 0.0;
1285.  M  m 2            }
1286.  M  m 2            w1[tid][col].first  += A[s][0][0]*v[0][i] + A[s][1][0]*v[1][i] +
1287.  M  m 2                                   A[s][2][0]*v[2][i];
1288.  M  m 2            w1[tid][col].second += A[s][0][1]*v[0][i] + A[s][1][1]*v[1][i] +
1289.  M  m 2                                   A[s][2][1]*v[2][i];
1290.  M  m 2            w1[tid][col].third  += A[s][0][2]*v[0][i] + A[s][1][2]*v[1][i] +
1291.  M  m 2                                   A[s][2][2]*v[2][i];
1292.  M  m 2--->      }
1293.  M  m            if( w2[tid][i] == 0 ) {
1294.  M  m              w2[tid][i] = 1;
1295.  M  m              w1[tid][i].first = 0.0; w1[tid][i].second = 0.0;
1296.  M  m              w1[tid][i].third = 0.0;
1297.  M  m            }
1298.  M  m            w1[tid][i].first += sum0; w1[tid][i].second += sum1;
1299.  M  m            w1[tid][i].third += sum2;
1300.  M  m----->    }
1301.  M           }
```

Excerpt 9.3.22 Middle of sparse matrix-vector product code from SPEC OMP's EQUAKE component with compiler markup on Broadwell.

`smvp` loop-nest? There is enough freedom in the `smvp` loop-nest to rearrange many things (line numbers reference Excerpt 9.3.22):

1. The `A[s][3][3]` matrix can be reordered to be `A[3][3][s]`.
2. The conditional on line 1281 can be split out into its own inner loop.
3. Since elements of `w2` only ever contain the values 0 or 1, it can be changed from an array of 32-bit integers into an array of bytes.
4. The inner loop on line 1273 can be split in two: one with `sum+=` and one with `w1+=`. This will decouple the matrix-vector product (e.g., lines 1275 to 1280) and the transposed matrix-vector product (e.g., lines 1286 to 1291).

5. The first entry in each sparse matrix row (the diagonal of the sparse matrix) is processed a bit differently than the others (lines 1270 to 1272). The elements on the diagonal can be moved into their own array and processed contiguously with a stride of one.

Most of the easy things have already been tested and found to be wanting. There probably are ways to get more performance from this loop-nest, but at some point the developer needs to balance an improvement in application runtime against the amount of time it takes to perform the optimization.

9.3.10 ART

The ART (Adaptive Resonance Theory) component of SPEC OMP performs object recognition using a neural network [8]. Table 9.3.12 presents a calltree profile from ART, in which one can see that nearly 95% of the runtime is associated with line 1044 in the `scanner.c` source file. This line is really just a call to the `compute_values_match` function, which contains the important loop nests for the application.

TABLE 9.3.12 Basic calltree profile from ART code in SPEC OMP.

```
Table:  Calltree View with Callsite Line Numbers

  Samp% |     Samp |Calltree
        |          | Thread=HIDE

 100.0% | 4,270.0 |Total
|-----------------------------------------------------------------
|  94.2% | 4,024.0 |match:scanner.c:line.1044
|   2.1% |    91.0 |train_matchclone_21471_3:scanner.c:line.1830
|   1.7% |    73.0 |train_matchclone_21471_5:scanner.c:line.1843
```

Since the majority of the computation time is spent in the relatively small `compute_values_match` function, this is a good place to start optimizing. The function contains seven loop-nests which seem to primarily access `f1_layer`, `Y`, `busp`, and `tds`. A quick read of these loops shows that aliasing may be an issue here. For a more in-depth explanation of aliasing issues in C/C++ codes, see Section 4.9. As a first optimization, arrays should be marked with the `restrict` keyword, which is expected to help on both Broadwell and KNL. Original and modified versions of the updated source lines are presented in Excerpt 9.3.23. These small changes fix the most pressing aliasing issues, which should allow some additional vectorization as well as better scalar code generation.

The Broadwell target is considered first. Excerpt 9.3.24, showing original compiler markup, can be compared to Excerpt 9.3.25 showing compiler markup from the new source code with some of the important aliasing issues

```
// Original
double     **tds;
f1_neuron **f1_layer;
xyz        **Y;
void compute_values_match(int o, int *f1res, int *spot,  double **busp)

// Modified
double     **restrict tds;
f1_neuron **restrict f1_layer;
xyz        **restrict Y;
void compute_values_match(int o, int *f1res, int *restrict spot, double **restrict busp)
```

Excerpt 9.3.23 Original and modified source code from SPEC's ART.

removed. While there are seven loop-nests in this function, the two presented here are the only ones to have different compiler markup after fixing the aliasing. It's interesting that the compiler actually decides *not* to (partially) vectorize the loop on line 836 once it knows there are no aliasing issues, instead choosing to unroll the loop by 8. However, the compiler does (partially) vectorize the loop on line 897, which it didn't previously. All things considered, one doesn't see a large change in vectorization between the versions with and without aliasing when targeting Broadwell. However, remember that using the `restrict` keyword will help not only vector code, but scalar code as well.

```
 836.       Vpr4----------<   for (ti=0;ti<varNumF1;ti++) {
 837.       Vpr4                f1_layer[o][ti].W = f1_layer[o][ti].I[varNumCp] +
 838.       Vpr4                                    varNumA*(f1_layer[o][ti].U);
 839.       Vpr4                tnorm += f1_layer[o][ti].W * f1_layer[o][ti].W;
 840.       Vpr4---------->   }
[...]
 896.  +    1-------------<   for (ti=0;ti<varNumF1;ti++) {
 897.  +    1 r4----------<     for (tj=*spot;tj<varNumF2;tj++)
 898.       1 r4                  if ( !Y[o][tj].reset )
 899.       1 r4---------->         Y[o][tj].y += f1_layer[o][ti].P * busp[ti][tj];
 900.       1------------->   }
```

Excerpt 9.3.24 Compiler markup from SPEC OMP's original ART code targeting Broadwell.

Next, the KNL target will be considered. Excerpt 9.3.26 shows the original compiler markup which can be compared to Excerpt 9.3.27, presenting compiler markup from the new source code with some of the important aliasing issues removed. When compiling for the KNL target, one sees that aliasing was impacting the compiler's choice with respect to vectorization on the same two loops as when compiling for the Broadwell target. In this case, the compiler was able to fully vectorize the loop on line 836 once the aliasing issues were removed. While the KNL target shows clear improvement in vectorization for two of the seven loops when the aliasing is taken care of, the increase in vector coverage isn't that large overall. One again hopes more efficient scalar code will result from the optimization along with the improvement in vectorization.

```
836.  +     r8------------<     for (ti=0;ti<varNumF1;ti++) {
837.        r8                    f1_layer[o][ti].W = f1_layer[o][ti].I[varNumCp] +
838.        r8                                        varNumA*(f1_layer[o][ti].U);
839.        r8                    tnorm += f1_layer[o][ti].W * f1_layer[o][ti].W;
840.        r8------------>     }
[...]
896.  +     1-------------<     for (ti=0;ti<varNumF1;ti++) {
897.  +     1 Vpr4--------<       for (tj=*spot;tj<varNumF2;tj++)
898.        1 Vpr4                  if ( !Y[o][tj].reset )
899.        1 Vpr4-------->           Y[o][tj].y += f1_layer[o][ti].P * busp[ti][tj];
900.        1------------->     }
```

Excerpt 9.3.25 Compiler markup from modified ART code from SPEC OMP targeting Broadwell.

```
836.        Vpr6----------<     for (ti=0;ti<varNumF1;ti++) {
837.        Vpr6                  f1_layer[o][ti].W = f1_layer[o][ti].I[varNumCp] +
838.        Vpr6                                      varNumA*(f1_layer[o][ti].U);
839.        Vpr6                  tnorm += f1_layer[o][ti].W * f1_layer[o][ti].W;
840.        Vpr6---------->     }
[...]
896.  +     1-------------<     for (ti=0;ti<varNumF1;ti++) {
897.  +     1 r4----------<       for (tj=*spot;tj<varNumF2;tj++)
898.        1 r4                    if ( !Y[o][tj].reset )
899.        1 r4---------->           Y[o][tj].y += f1_layer[o][ti].P * busp[ti][tj];
900.        1------------->     }
```

Excerpt 9.3.26 Compiler markup from SPEC OMP's original ART code targeting KNL.

```
836.        Vr2-----------<     for (ti=0;ti<varNumF1;ti++) {
837.        Vr2                   f1_layer[o][ti].W = f1_layer[o][ti].I[varNumCp] +
838.        Vr2                                       varNumA*(f1_layer[o][ti].U);
839.        Vr2                   tnorm += f1_layer[o][ti].W * f1_layer[o][ti].W;
840.        Vr2----------->     }
[...]
896.  +     1-------------<     for (ti=0;ti<varNumF1;ti++) {
897.  +     1 Vpr4--------<       for (tj=*spot;tj<varNumF2;tj++)
898.        1 Vpr4                  if ( !Y[o][tj].reset )
899.        1 Vpr4-------->           Y[o][tj].y += f1_layer[o][ti].P * busp[ti][tj];
900.        1------------->     }
```

Excerpt 9.3.27 Compiler markup from modified ART code from SPEC OMP targeting KNL.

Is there anything else that can be done with the loop-nests in the `compute_values_match` function to gain more performance? Is there more that can be made to vectorize? Remember that vectorization is very important if one wants to get the most out of KNL. Consider the loop-nest presented in Excerpt 9.3.28, the only loop-nest in this function which doesn't appear to be well vectorized. From the excerpt, one can see that the only part of this loop-nest to be vectorized is the inner loop starting on line 874. Closer examination of this loop reveals it to be odd indeed. Consider that the only way the conditional inside the loop can be true is for `tj`, the loop index, to be equal to `winner[o][0]` (along with some other conditions). Thus, it isn't really a loop at all, as the only time the loop performs any operation is when the loop index has a specific value. The inner-loop can be replaced with a single `if` statement, simplifying things for the compiler. Excerpt 9.3.29 presents refactored code and compiler markup for the KNL target, as the compiler is still unable to vectorize this loop when targeting Broadwell – this is probably due to the lack of predication on Broadwell, a topic covered in more detail in the discussion of the SVE ISA in Section 10.3.1. While the primary change seen in the new code is the replacement of the old inner loop with a conditional, some other small rearrangements have also been made. From the excerpts, one can see that this refactoring allows the remaining outer loop to be fully vectorized when targeting KNL. This new version of the loop should make much better use of the vector units compared to the old loop, especially considering the fact that the old inner loop's conditional would only have been true for a single vector element. One final observation to make here is that the conditional on line 873 in Excerpt 9.3.29 can be pulled out of the loop entirely. While the compiler might be able to make this optimization automatically, one might be better off if the compiler wasn't relied upon to do so reliably.

```
869. +    1-------------<   for (ti=0;ti<varNumF1;ti++) {
870.      1                   f1_layer[o][ti].U = f1_layer[o][ti].V/oldTnorm;
871.      1
872.      1                   tsum = 0;
873.      1                   ttemp = f1_layer[o][ti].P;
874. +    1 V-----------<     for (tj=*spot;tj<varNumF2;tj++) {
875.      1 V                   if ((tj == winner[o][0])&&(Y[o][tj].y > 0))
876.      1 V                     tsum += tds[ti][tj] * varNumD;
877.      1 V----------->     }
878.      1
879.      1                   f1_layer[o][ti].P = f1_layer[o][ti].U + tsum;
880.      1
881.      1                   tnorm += f1_layer[o][ti].P * f1_layer[o][ti].P;
882.      1                   if (fabs(ttemp - f1_layer[o][ti].P) < FLOAT_CMP_TOLERANCE)
883.      1                     tresult=0;
884.      1------------->   }
```

Excerpt 9.3.28 Compiler markup from original loop-nest from ART from SPEC OMP targeting KNL.

Performance of the original and optimized versions of the ART code are presented for Broadwell and KNL in Figure 9.3.13. One can see from the figure that the code changes help both targets, but KNL sees more of an improvement than does Broadwell (speedup of 1.25 with KNL vs. 1.09 with Broadwell on the largest tested thread counts). The ART code scales fairly well with increases in the number of threads running on real CPU cores but does not improve with SMT on Broadwell. However, running 2-way SMT on KNL does provide a small benefit.

```
869.        V-------------<     for (ti=0;ti<varNumF1;ti++) {
870.        V                     f1_layer[o][ti].U = f1_layer[o][ti].V/oldTnorm;
871.        V                     ttemp = f1_layer[o][ti].P;
872.        V                     f1_layer[o][ti].P = f1_layer[o][ti].U;
873.        V                     if( (winner[o][0] >= *spot)     &&
874.        V                         (winner[o][0] < varNumF2)  &&
875.        V                         (Y[o][winner[o][0]].y > 0)     ) {
876.        V                       f1_layer[o][ti].P += tds[ti][winner[o][0]] * varNumD;
877.        V                     }
878.        V                     tnorm += f1_layer[o][ti].P * f1_layer[o][ti].P;
879.        V                     if (fabs(ttemp - f1_layer[o][ti].P) < FLOAT_CMP_TOLERANCE)
880.        V                       tresult=0;
881.        V------------->     }
```

Excerpt 9.3.29 Compiler markup from modified loop-nest from ART from SPEC OMP targeting KNL.

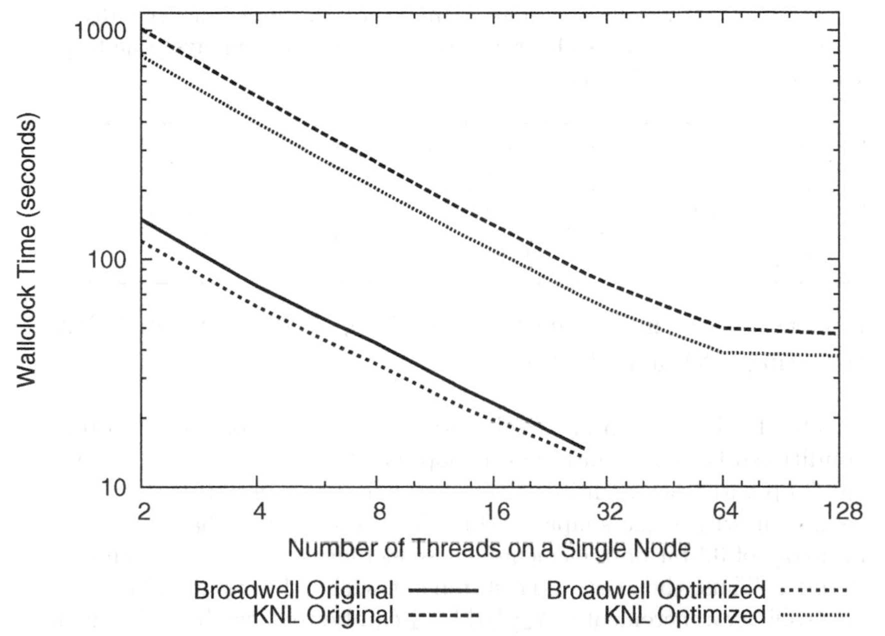

FIGURE 9.3.13 Performance of original and modified ART code from SPEC on KNL and Broadwell.

However, something doesn't feel finished here. An examination of the seven loop-nests in the `compute_values_match` function shows that most of the loops are 1D with `varNumF1` iterations, while one remaining loop is 2D with `varNumF1*varNumF2` iterations. It is this 2D loop-nest on line 896 in Excerpts 9.3.25 and 9.3.27 which is the most time-consuming. Both Broadwell and KNL are only able to partially vectorize this loop. Is there some way to improve upon this partial vectorization and to get a fully vectorized loop? Yes. A fully vector loop can be achieved with a little bit of work and an even smaller amount of clever thinking. The first thing is to observe that `Y` is an array of structs with only two members: a 32-bit integer `reset` and a 64-bit double-precision floating point `y`. Thus, when the compiler needs to access both `Y[][].y` and `Y[][].reset` as in the most time-consuming loop here, the accesses are strided and may have alignment issues due to the way the two differently sized data types are mixed. To resolve these issues, the global array `Y` is split into two arrays, each with a single primitive type. That is, `Y` is decomposed into one array of doubles `Y_y` and one array of integers `Y_reset`. This was a fairly trivial change to the ART code, though it is understandable that performing the same transformation to a full application code might be more cumbersome. Still, the point remains that the striding in the original code was an issue and should be fixed if at all possible. This is of particular importance when writing new code; using structures and objects can cause an arrays-of-structures layout in memory which does not typically vectorize as well as the structures-of-arrays layout (or just flat arrays) due to problems with striding and alignment.

```
896.  +    1--------------<   for (ti=0;ti<varNumF1;ti++) {
897.       1                 #pragma ivdep
898.       1 Vr2----------<     for (tj=*spot;tj<varNumF2;tj++)
899.       1 Vr2                  Y_y[o][tj] += f1_layer[o][ti].P * busp[ti][tj] *
900.       1 Vr2---------->                     (1.0-Y_reset[o][tj]);
901.       1-------------->   }
```

Excerpt 9.3.30 Compiler markup from final ART code from SPEC OMP targeting KNL and Broadwell.

One final tweak to allow full and efficient vectorization is to remove the conditional `if` statement from the loop. While the compiler is able to vectorize the loop with the conditional in place, it is more performant in this case to replace it with some simple arithmetic. First, realize that while `Y_reset` is an array of 32-bit data elements, these elements only ever assume the values 0 and `1`. Thus, the existing `if` statement with a `!Y_reset[o][tj]` condition controlling an increment to `Y_y[o][tj]` can be replaced by code which always performs the increment while multiplying the value to be added to `Y_y[o][tj]` by `(1.0-Y_reset[o][tj])`. The new version yields the same answer in the end, but exchanges other operations for the problematic conditional, as shown in Excerpt 9.3.30. The compiler seems to get slightly confused here, as the `ivdep` pragma probably shouldn't be required due to the previous insertion of the `restrict` keyword.

Performance of the original ART version compared to the final version produced from the second round of optimizations is presented in Figure 9.3.14. A different picture can be seen now, compared to what was seen before the most time-consuming loop was fully optimized. With the original ART code, things were slower overall but scaled better with additional threads. With the final optimized code, things are faster overall but don't scale quite as well with added threads. The optimized version is better able to saturate the SIMD units and the memory bandwidth available to the cores on a per-thread basis than the original version, so the scaling tips over at a smaller thread count. The final version shows a speedup by a factor of 2.52 on KNL and 1.15 on Broadwell.

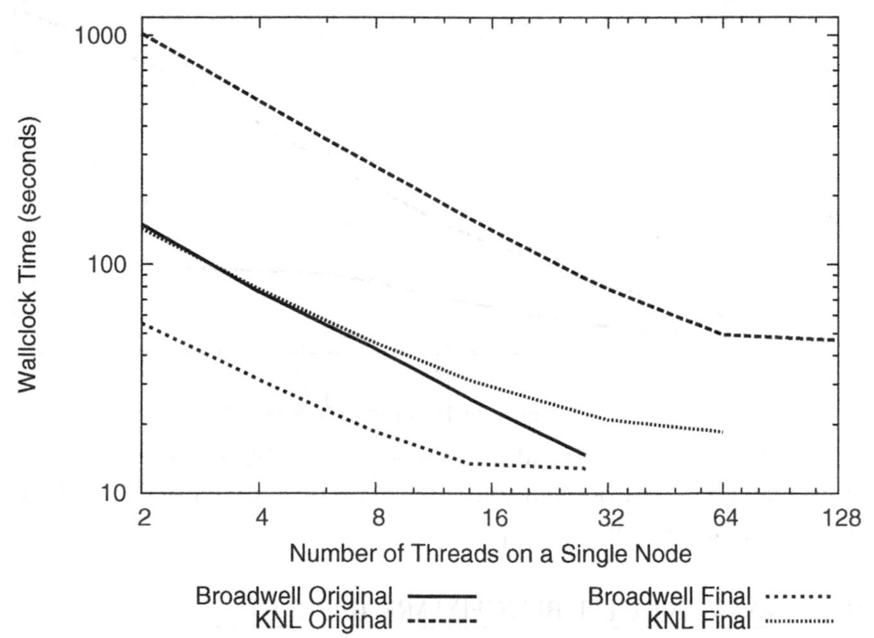

FIGURE 9.3.14 Performance of original and final ART code from SPEC on KNL and Broadwell.

The schedule used on the primary OpenMP work loop is `dynamic` in this code, which performs nearly identically to `guided`, so it doesn't appear that load-imbalance or wildly excessive synchronization overhead is an issue here. Also, a number of the arrays are allocated and initialized locally by the thread that will use them, so some thought has been given to first-touch issues. In fact, the OpenMP in the ART code seems fairly reasonable. The limiting factor for scaling to even larger thread counts will soon be the unavoidable problems associated with Amdahl's Law. Figure 9.3.15 shows the percent of ART runtime which is spent in serial portions of the benchmark as opposed to the parallel OpenMP regions, for a number of different thread counts. With nearly 70% of the computation time spent in non-threaded portions of the code, as is the case for the final 64 thread run on KNL, improvements to the existing parallel code will already result in diminishing returns. In order to scale to larger thread counts, the runtime of the currently serial computation needs to be reduced.

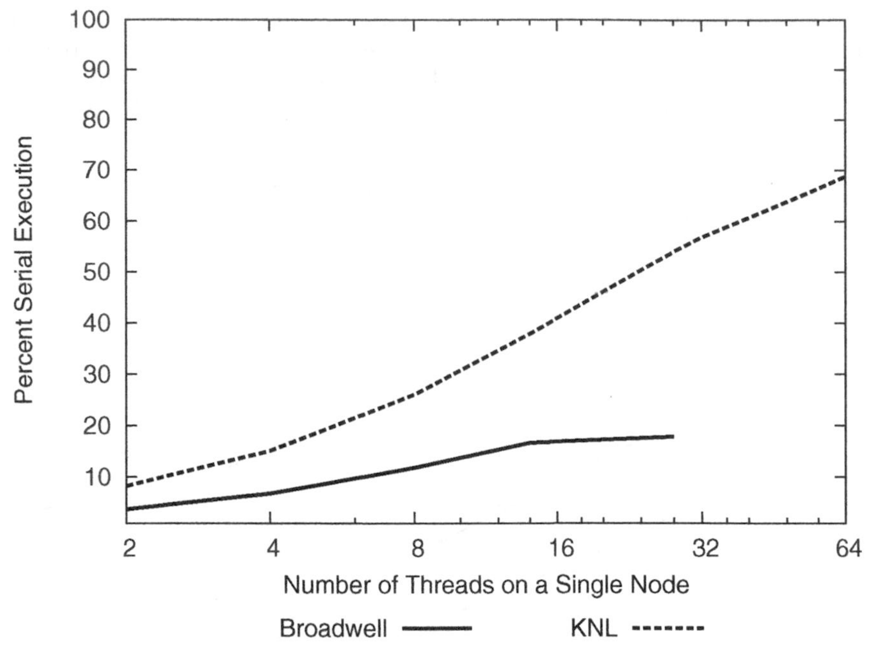

FIGURE 9.3.15 Serial percentage of runtime of ART code from SPEC on KNL and Broadwell.

9.4 NASA PARALLEL BENCHMARK (NPB) BT

The NPB suite has been around for many years and there have been many versions, from scalar, MPI, and more recently hybrid MPI/OpenMP. The MPI version of BT will be used to investigate porting an existing all MPI application to a hybrid MPI/OpenMP structure. The latest hybrid version from the NPB suite was a significant departure from the all-MPI version and an incremental approach of taking an all-MPI and moving it to a parallel vector version was not considered in that rewrite. The following incremental approach is more instructive to application developers faced with the task of porting their applications.

BT is pseudo application for a synthetic system on non-linear PDEs using a block tridiagonal solver. There are numerous sizes that can be used, in this example we will use a larger problem to illustrate the tradeoff between vectorization, parallelization, and message passing. The Class D problem is first run on 625 MPI tasks across 32 Intel Broadwell nodes, this will be our baseline test. A portion of the output from the baseline run is shown in Table 9.4.1.

Before examining the application source, we should determine the looping structure of the application. For this we will instrument the loops in the program. This instrumentation is only for loop information. Such intrusive instrumentation requires a longer execution time and obscures routine timing. Table 9.4.2 displays the principal loops in the execution.

TABLE 9.4.1 Portion of output from NPB BT baseline test on Broadwell.

```
BT Benchmark Completed.
Class           =                        D
Size            =            408x 408x 408
Iterations      =                      250
Time in seconds =                    45.28
Total processes =                      625
Compiled procs  =                      625
Mop/s total     =               1288366.50
Mop/s/process   =                  2061.39
Operation type  =           floating point
Verification    =               SUCCESSFUL
Version         =                      3.2
Compile date    =              22 Oct 2016
```

TABLE 9.4.2 Principal loops from BT instrumented for loop statistics.

```
 Loop| Loop Incl|     Time| Loop Hit| Loop| Loop| Loop|Function=/.LOOP[.]
 Incl|      Time|    (Loop|         |Trips|Trips|Trips| PE=HIDE
Time%|          |    Adj.)|         |  Avg|  Min|  Max|
|-----------------------------------------------------------------
|98.6%|233.178614| 0.000264|        1|250.0|  250|  250|mpbt_.LOOP.2.li.179
|30.1%| 71.243534| 0.009947|      251| 25.0|   25|   25|z_solve_.LOOP.1.li.32
|28.5%| 67.364757| 0.010030|      251| 25.0|   25|   25|x_solve_.LOOP.1.li.34
|28.4%| 67.072951| 0.009394|      251| 25.0|   25|   25|y_solve_.LOOP.1.li.33
|27.5%| 65.107169| 0.004856|    6,275| 16.2|   15|   17|z_solve_cell_.LOOP.1.li.413
|27.5%| 65.102314| 0.339276|  101,906| 16.2|   15|   17|z_solve_cell_.LOOP.2.li.414
|25.7%| 60.786872| 0.005421|    6,275| 16.2|   15|   17|y_solve_cell_.LOOP.1.li.413
|25.7%| 60.781451| 0.269313|  101,906| 16.2|   15|   17|y_solve_cell_.LOOP.2.li.414
|25.5%| 60.363368| 0.003114|    6,275| 16.2|   15|   17|x_solve_cell_.LOOP.1.li.418
|25.5%| 60.360253| 0.126434|  101,906| 16.2|   15|   17|x_solve_cell_.LOOP.2.li.419
|24.8%| 58.604785|13.496033|1,654,953| 16.2|   15|   17|z_solve_cell_.LOOP.6.li.709
|24.1%| 57.012239|13.307301|1,654,953| 16.2|   15|   17|y_solve_cell_.LOOP.6.li.708
|24.1%| 57.005820|13.298918|1,654,953| 16.2|   15|   17|x_solve_cell_.LOOP.5.li.703
```

The top loop which uses 98.6% of the time is the time step loop – this computation runs for 250 iterations. Below that loop we have several loops whose average iteration count is 25. This will be the first we should consider for high level parallelization. Table 9.4.3 shows the call tree display of the routines with loops interspersed. This gives a very good idea of the structure of the computation and is useful to determine which loops might be good candidates for parallelization The indentation in the call tree display gives the nesting of the calling structure. Then the inclusive time and the number of times the line item is called is displayed Upon closer examination of the call tree profile, the loops in `x`, `y`, and `z_solve` contain calls to MPI routines. That is, the parallelization of those loops would have to deal with threads performing their own communication and/or synchronizing around MPI. While this might be possible, the loops within `x`, `y`, and `z_solve_cell` have reasonable iteration counts and they do not contain MPI calls. Since the nest level of the loops in these routines are at the same level of the MPI call, the MPI call is outside of the loop. We should investigate these loops for parallelization.

TABLE 9.4.3 Call tree profile from NPB BT with interspersed loops.

```
Table:  Function Calltree View

  Time% |        Time |         Calls |Calltree
        |             |               | PE=HIDE

 100.0% | 181.612195 |            -- |Total
|-----------------------------------------------------------------------
| 100.0% | 181.612175 |          2.0 |mpbt_
||----------------------------------------------------------------------
||  98.5% | 178.959425 |            -- |mpbt_.LOOP.2.li.179
|||---------------------------------------------------------------------
3||  98.2% | 178.277896 |        250.0 |adi_
||||--------------------------------------------------------------------
4|||  32.5% |  59.055764 |        500.0 |z_solve_
|||||-------------------------------------------------------------------
5||||  29.1% |  52.886429 |            -- |z_solve_.LOOP.1.li.32
||||||------------------------------------------------------------------
6|||||  25.8% |  46.820341 |    12,500.0 |z_solve_cell_
|||||||-----------------------------------------------------------------
7||||||  14.9% |  27.014988 |            -- |z_solve_cell_.LOOP.1.li.413
8||||||        |             |               | z_solve_cell_.LOOP.2.li.414
|||||||||---------------------------------------------------------------
9||||||||  14.8% |  26.929310 |            -- |z_solve_cell_.LOOP.6.li.709
||||||||||--------------------------------------------------------------
10||||||||   6.7% |  12.222413 | 53,538,732.8 |matvec_sub_
10||||||||   5.8% |  10.600108 | 53,538,732.8 |matmul_sub_
10||||||||   2.3% |   4.106789 | 26,769,366.4 |binvcrhs_
[...]
|||||||||==============================================================
7||||||  10.9% |  19.787606 |    12,500.0 |z_solve_cell_(exclusive)
7||||||   0.0% |   0.017748 |    12,500.0 |lhsabinit_
|||||||================================================================
6|||||   3.0% |   5.426147 |    12,000.0 |mpi_wait
6|||||   0.2% |   0.321180 |    12,000.0 |z_send_solve_info_
[...]
```

There are significant issues with the loops within `z_solve_cell`. Several routines are called, and scoping of the shared and private variables in the loop will be difficult. Additionally, data dependency analysis will be a challenge. Excerpt 9.4.1 presents the start of the top level loop within `z_solve_cell`.

```
 413.  + 1---------------<          do j=start(2,c),jsize
 414.  + 1 2-------------<             do i=start(1,c),isize
[...]
 424.    1 2 r2----------<                do k = start(3,c)-1, cell_size(3,c)-end(3,c)
 425.    1 2 r2                              utmp(1,k) = 1.0d0 / u(1,i,j,k,c)
 426.    1 2 r2 VsR-----<>                   utmp(2,k) = u(2,i,j,k,c)
 427.    1 2 r2                              utmp(3,k) = u(3,i,j,k,c)
 428.    1 2 r2                              utmp(4,k) = u(4,i,j,k,c)
 429.    1 2 r2                              utmp(5,k) = u(5,i,j,k,c)
 430.    1 2 r2                              utmp(6,k) = qs(i,j,k,c)
 431.    1 2 r2---------->                end do
 432.    1 2
 433.    1 2 Vp----------<                do k = start(3,c)-1, cell_size(3,c)-end(3,c)
 434.    1 2 Vp
 435.    1 2 Vp                              tmp1 = utmp(1,k)
 436.    1 2 Vp                              tmp2 = tmp1 * tmp1
 437.    1 2 Vp                              tmp3 = tmp1 * tmp2
```

Excerpt 9.4.1 Start of top level loop in NPB BT routine.

In this case we are analyzing the threading of the `j` loop. If we review our scoping rules:

1. All variables only used within the parallel loop should be shared.
2. All variables that are set and then used within the parallel loop should be private.
3. All variables that are referenced by the loop index should be shared.

The first challenge is to determine if the variable `utmp` is set and then used within the loop. The complication here is that `utmp` is an array constant, that is, the array is not referenced by the parallel index and each thread will require its own copy of the array if and only if all elements of the array are set each pass through the loop prior to being used. So `utmp(1:6,k)` is set and where is it used? Upon examination of the routine, we see that all the elements of `utmp` that are used are set prior to their use. This means that there is no loop carried dependency on `utmp`, and `utmp` should be private.

The next array constant of interest is `fjac`, shown in Excerpt 9.4.2. It is three dimensional and all three dimensions are independent of the parallel loop `j`. Once again we must investigate the uses of `fjac` and assure ourselves that the uses are covered by the sets. This is a little more difficult since the sets of `fjac` are within the loop starting on line 433, and the uses are within the loop starting on line 522. Also, `fjac` is used with `k-1`. Are all of the uses of `fjac` in this loop covered by the sets in the loop starting at 433? Since the sets start at `k = start(3,c)-1` and the `k-1` uses start at `do k = start(3,c)`, the left element has been set in the first loop. To determine if the upper loop bounds also allow the uses to be covered by the sets, we must understand the relationship between `cell_size(3,c)-end(3,c)` and `ksize-end(3,c)`. On line 409 we see `ksize = cell_size(3,c)-1`, which does indicate that all the uses are covered by the sets.

```
 409.                              ksize = cell_size(3,c)-1
[...]
 432.      1 2
 433.      1 2 Vp----------<             do k = start(3,c)-1, cell_size(3,c)-end(3,c)
 434.      1 2 Vp
 435.      1 2 Vp                           tmp1 = utmp(1,k)
 445.      1 2 Vp                           fjac(2,1,k) = - ( utmp(2,k)*utmp(4,k) )
 446.      1 2 Vp                  >             * tmp2
 447.      1 2 Vp                           fjac(2,2,k) = utmp(4,k) * tmp1
 449.      1 2 Vp                           fjac(2,4,k) = utmp(2,k) * tmp1
 452.      1 2 Vp                           fjac(3,1,k) = - ( utmp(3,k)*utmp(4,k) )
 453.      1 2 Vp                  >             * tmp2
 455.      1 2 Vp                           fjac(3,3,k) = utmp(4,k) * tmp1
 456.      1 2 Vp                           fjac(3,4,k) = utmp(3,k) * tmp1
 459.      1 2 Vp                           fjac(4,1,k) = - (utmp(4,k)*utmp(4,k) * tmp2 )
 460.      1 2 Vp                  >             + c2 * utmp(6,k)
 461.      1 2 Vp                           fjac(4,2,k) = - c2 *  utmp(2,k) * tmp1
 462.      1 2 Vp                           fjac(4,3,k) = - c2 *  utmp(3,k) * tmp1
 463.      1 2 Vp                           fjac(4,4,k) = ( 2.0d+00 - c2 )
 464.      1 2 Vp                  >             *  utmp(4,k) * tmp1
 467.      1 2 Vp                           fjac(5,1,k) = ( c2 * 2.0d0 * utmp(6,k)
 468.      1 2 Vp                  >             - c1 * ( utmp(5,k) * tmp1 ) )
 469.      1 2 Vp                  >             * ( utmp(4,k) * tmp1 )
 470.      1 2 Vp                           fjac(5,2,k) = - c2 * ( utmp(2,k)*utmp(4,k) )
 472.      1 2 Vp                           fjac(5,3,k) = - c2 * ( utmp(3,k)*utmp(4,k) )
 474.      1 2 Vp                           fjac(5,4,k) = c1 * ( utmp(5,k) * tmp1 )
 475.      1 2 Vp                  >             - c2 * ( utmp(6,k)
 476.      1 2 Vp                  >             + utmp(4,k)*utmp(4,k) * tmp2 )
 477.      1 2 Vp                           fjac(5,5,k) = c1 * utmp(4,k) * tmp1
[...]
 527.      1 2 Vp                           lhsa(1,1,k) = - tmp2 * fjac(1,1,k-1)
 528.      1 2 Vp                  >                      - tmp1 * njac(1,1,k-1)
```

Excerpt 9.4.2 NPB BT routine with fjac variable.

There is a slight problem with these arrays that we would like to scope as private. They are in a COMMON block which means they are global variables. Given the COMMON block, we need to either make the variables local – it turns out that the common block was being used to save memory – or we can use the !$OMP THREADPRIVATE directive. We chose to make the variables local.

This is without a doubt the most difficult part of the scoping analysis. Array constants are used heavily in many applications and to thread a loop that contains array constants, the use-set analysis depends upon which elements are referenced and the loop bounds controlling the references. A second issue with array constants is found in this example. One possibility is that not all of the elements of an array constant may be set within the loop, and all elements may be used. If this is the case, first value setting must be used to initialize each thread with the master thread's value entering the parallel loop. This is the case with lhsa and lhsb arrays. Prior to the parallel loop there is a call to lhsabinit, and lhsa and lhsb are passed to this routine. Within the routine, those two arrays are initialized. Some of the elements in those initialized arrays are not set within the parallel loop. Therefore, private arrays must be set from the master thread when initialized. Overlooking first value getting is a typical cause of a race condition that can give incorrect answers.

```
!$OMP PARALLEL DO
!$OMP&firstprivate(lhsa, lhsb)
!$OMP&      private(j,i,k,utmp,tmp1,tmp2,tmp3,fjac,njac)
      do j=start(2,c),jsize
```

Excerpt 9.4.3 OpenMP directives with scoping for NPB BT loop.

Fortunately, the routines called from with the parallel loop are quite simple and the scoping and data dependencies can be resolved. There are many array constants that must be scoped accurately to thread this loop. Additionally, `x_solve_cell` and `y_solve_cell` are similar to `z_solve_cell`, so those three routines can be threaded with a similar analysis. The resultant OpenMP directives for `z_solve_cell` are presented in Excerpt 9.4.3. Variables `j`, `i`, and `k`, which are all loop indices, do not necessarily need to be scoped. However, it is probably good practice to scope everything. We then apply this same analysis to `x_solve_cell` and `y_solve_cell`. Running this OpenMP version on Broadwell did not give much improvement: 43 seconds compared to the original 45 seconds. One of the issues with the BT benchmark is that the number of MPI tasks must be a square. The runs on Broadwell were run on 36 core Broadwell nodes using 32 MPI tasks/node. The threading was done across two hyper-threads. When we run on KNL, we will probably see better performance.

TABLE 9.4.4 Portion of output from NPB BT baseline test on KNL.

```
BT Benchmark Completed.
Class           =                        D
Size            =           408x 408x 408
Iterations      =                      250
Time in seconds =                    95.64
Total processes =                      625
Compiled procs  =                      625
Mop/s total     =                609958.10
Mop/s/process   =                   975.93
Operation type  =           floating point
Verification    =               SUCCESSFUL
Version         =                      3.2
Compile date    =              22 Oct 2016
```

To compare Broadwell to KNL, we want to use the same number of nodes – 20. This will give us 1360 cores on KNL. KNL also has 4 hyper-threads/core. We will first run our baseline on KNL to compare to Broadwell. Table 9.4.4 contains output from our baseline run on KNL. This is quite a bit slower than Broadwell; we will now try threading and use up to 8 threads/MPI task. Running the OpenMP version on KNL shows a good improvement for 2 threads. However, when we ran with hyper-threads (SMT) the performance was lost, as seen in Table 9.4.5. Broadwell is still faster by a factor of 1.5. However, we have not examined vectorization of this application. KNL will improve with vectorization. Actually Broadwell will as well but not as much. Excerpt 9.4.4 of the listing of the loop in `z_solve_cell` shows some vectorization. However, the routines being called from within the loop are not being vectorized.

TABLE 9.4.5 NPB BT performance on KNL with 2, 4, and 8 threads.

Total Threads	SMT Threads	Time (seconds)
2	1	59.23
4	2	61.12
8	4	71.90

```
712. + M m 3 4-----<            do k=kstart+first,ksize-last
713.   M m 3 4
714.   M m 3 4        c-----------------------------------------------------------------
715.   M m 3 4        c      subtract A*lhs_vector(k-1) from lhs_vector(k)
716.   M m 3 4        c
717.   M m 3 4        c      rhs(k) = rhs(k) - A*rhs(k-1)
718.   M m 3 4        c-----------------------------------------------------------------
719. + M m 3 4                    call matvec_sub(lhsa(1,1,k),
720.   M m 3 4             >                          rhs(1,i,j,k-1,c),rhs(1,i,j,k,c))
721.   M m 3 4
722.   M m 3 4        c-----------------------------------------------------------------
723.   M m 3 4        c      B(k) = B(k) - C(k-1)*A(k)
724.   M m 3 4        c      call matmul_sub(aa,i,j,k,c,cc,i,j,k-1,c,bb,i,j,k,c)
725.   M m 3 4        c-----------------------------------------------------------------
726. + M m 3 4                    call matmul_sub(lhsa(1,1,k),
727.   M m 3 4             >                          lhsc(1,1,i,j,k-1,c),
728.   M m 3 4             >                          lhsb(1,1,k))
729.   M m 3 4
730.   M m 3 4        c-----------------------------------------------------------------
731.   M m 3 4        c      multiply c(i,j,k) by b_inverse and copy back to c
732.   M m 3 4        c      multiply rhs(i,j,1) by b_inverse(i,j,1) and copy to rhs
733.   M m 3 4        c-----------------------------------------------------------------
734. + M m 3 4                    call binvcrhs( lhsb(1,1,k),
735.   M m 3 4             >                          lhsc(1,1,i,j,k,c),
736.   M m 3 4             >                          rhs(1,i,j,k,c) )
737.   M m 3 4----->            enddo
```

Excerpt 9.4.4 Loop in NPB BT code which shows some vectorization.

How can we vectorize the computation within the calls? One way is to inline the subroutines. Ideally the compiler would inline and then vectorize. However, in this case it is asking too much of the compiler.

Several refactorings were applied to this routine with little or no improvement. The table below shows the final timings. Considered first is the original using MCDRAM as cache and also running the application completely out of MCDRAM as a separate address space. The amount of memory required per node on this example is less than 16 GB, the size of the MCDRAM. Considered next is the OpenMP version that was able to get close to a factor of two improvement. Finally, two vectorization versions are considered: one where the subroutines were inlined and one additional attempt to vectorize on the

I index, hoping to avoid some of the striding caused by addressing on the first two indices (5,5). The runtime performance of these versions is shown in Table 9.4.6.

TABLE 9.4.6 NPB BT refactored performance (seconds) on KNL.

	KNL (cache)	KNL (flat)	Broadwell
Original	90.87	91.14	46.6
OpenMP	58.48	59.31	43.8
OMP - Vector 1	58.34	58.64	43.9
OMP - Vector 2	59.97	60.43	47.7

We now investigate OpenMP scaling by running on more nodes. If we run on twice the number of nodes with half the number of MPI tasks/node, we can determine if more threading will help our performance. We will employ the OMP-Vector 1 version of the application. Table 9.4.7 gives the times from running on 20, 40, and 80 nodes. Two threads are used on 20 nodes, two and four threads on 40 nodes, and two, four, and eight threads on 80 nodes.

TABLE 9.4.7 NPB BT OpenMP performance (seconds) on KNL.

Threads	20 Nodes	40 Nodes	80 Nodes
2	58.48	60.33	58.68
4		43.41	42.45
8			33.55

We can gleam a lot of information from these timings. While we get a factor of 1.74 moving from 20 to 80 nodes, we are using 4 times the resources. Another clue that we get from these runs, which should have also been apparent from the runs using MCDRAM as direct memory, is that this application is not memory bandwidth limited. If it were memory bandwidth limited, we would have seen an improvement when using flat memory on KNL, and we would see a bigger improvement when employing more nodes. More nodes provide more available memory bandwidth. While we are getting a reasonable improvement with OpenMP threading, why are we not seeing a good improvement in vectorization? This application was written 15 to 20 years ago when the most important issue was cache utilization. Most of the arrays are dimensioned `(5,5,i,j,k)`. This is extremely bad for vectorization. The elements for the vectors must be contiguous within the register, which on KNL is the length of a cache line. When the compiler vectorizes a loop, these array references must be packed into a contiguous temporary. Ideally, the ultimate improvement would be to perform global index reordering on these arrays and change them to `(i,5,5,j,k)`. Performing this refactoring to an excerpt of `z_solve` illuminates a serious performance choice for this routine. First, there is a slight improvement when performing the refactoring to generate vectorization on the `I` index instead of `K` (10% improvement on

KNL). However, we were expecting a larger increase from the vectorization. What happened to generate the lower increase in performance? The answer is shown by gathering hardware counters from the two examples. First the original, and then the example where the arrays' dimensions were reordered. This array redimensioning is shown in Excerpt 9.4.5.

```
c Original:
  double precision
 >   qs   (-1:IMAX,    -1:JMAX,    -1:KMAX,     1),
 >   u    (5,          -2:IMAX+1, -2:JMAX+1,  -2:KMAX+1, 1),
 >   lhsc (5,          5,          -2:IMAX+1,  -2:JMAX+1, -2:KMAX+1, 1)

c Refactored:
  double precision
 >   qs   (-1:IMAX,    -1:JMAX,    -1:KMAX,     1),
 >   u    (-2:IMAX+1, 5,           -2:JMAX+1,  -2:KMAX+1, 1),
 >   lhsc (-2:IMAX+1, 5,           5,           -2:JMAX+1, -2:KMAX+1, 1)
```

Excerpt 9.4.5 NPB BT array redimensioning of qs, u, and lhsc.

The idea was to be able to vectorize on the innermost, contiguous index `I`. Recall that the loop structure was the same in both examples, as seen in Excerpt 9.4.6.

```
c      Original:
!$OMP PARALLEL DO
!$OMP&firstprivate(lhsa, lhsb)
!$OMP&       private(j,i,k,utmp,tmp1,tmp2,tmp3,fjac,njac)
       do j=1, jsize
          do i=1, isize
             do k = 0, ksize+1
                utmp(1,k) = 1.0d0 / u(1,i,j,k,c)
                utmp(2,k) = u(2,i,j,k,c)
                utmp(3,k) = u(3,i,j,k,c)
                utmp(4,k) = u(4,i,j,k,c)
                utmp(5,k) = u(5,i,j,k,c)
                utmp(6,k) = qs(i,j,k,c)

c      Refactored:
!$OMP PARALLEL DO
!$OMP&firstprivate(lhsa, lhsb)
!$OMP&       private(j,i,k,utmp,tmp1,tmp2,tmp3,fjac,njac)
       do j=1, jsize
          do k = 0, ksize+1
            do i=1, isize
                utmp(i,1,k) = 1.0d0 / u(i,1,j,k,c)
                utmp(i,2,k) = u(i,2,j,k,c)
                utmp(i,3,k) = u(i,3,j,k,c)
                utmp(i,4,k) = u(i,4,j,k,c)
                utmp(i,5,k) = u(i,5,j,k,c)
                utmp(i,6,k) = qs(i,j,k,c)
```

Excerpt 9.4.6 NPB BT array redimensioning loop structure.

Table 9.4.8 displays the hardware counters from the run of this excerpt of `z_solve`. Notice the excellent cache reuse in this code. The entirety of the excerpt has a great deal of reuse with the cache due to the original dimensioning. The hardware counters from the refactored version seen in Table 9.4.9 do not indicate good cache reuse, due to the amount of data required to be accessed within the `I` loop. So whatever gain we get from vectorization is lost due to inefficient cache reuse.

TABLE 9.4.8 NPB BT hardware counters from original dimensioning.

```
Time%                                       99.9%
Time                                     0.113643 secs
Imb. Time                                      -- secs
Imb. Time%                                     --
Calls                        8.799 /sec       1.0 calls
UNHALTED_CORE_CYCLES                  119,451,534
UNHALTED_REFERENCE_CYCLES             111,490,106
INSTRUCTION_RETIRED                    37,012,202
LLC_REFERENCES                          4,827,003
LLC_MISSES                                289,096
LLC cache hit,miss ratio    94.0% hits      6.0% misses
Average Time per Call                    0.113643 secs
CrayPat Overhead : Time       0.0%
```

TABLE 9.4.9 NPB BT hardware counters from refactored dimensioning.

```
Time%                                       99.9%
Time                                     0.101959 secs
Imb. Time                                      -- secs
Imb. Time%                                     --
Calls                        9.808 /sec       1.0 calls
UNHALTED_CORE_CYCLES                  100,329,092
UNHALTED_REFERENCE_CYCLES              93,640,792
INSTRUCTION_RETIRED                    31,431,210
LLC_REFERENCES                          4,520,126
LLC_MISSES                              1,941,657
LLC cache hit,miss ratio    57.0% hits     43.0% misses
Average Time per Call                    0.101959 secs
CrayPat Overhead : Time       0.0%
```

Attaining the highest effective memory bandwidth from a computation is the most important element of achieving good performance. Vectorization and threading will be limited if the computation is bandwidth limited. Memory bandwidth is *always* a limiting issue in most applications. Whenever one can effectively utilize cache to improve the utilized memory bandwidth, the better the performance will be. This is an excellent example where refactoring for better vectorization is not a good approach if it destroys cache utilization.

BT represents a suite of applications that perform computations in three different directions: X, Y, and Z. In the X sweep the computations are data-dependent upon the X direction. Likewise in the Y and Z sweeps, the computations are data dependent upon the Y and Z directions, respectively. When the computation is data dependent on a particular direction, then the OpenMP threading must be on the outermost of the other two dimensions. So for X and Y sweeps, the OpenMP would be on the Z dimension or `K` loop. Then, the vectorization needs to be on the innermost of the order two dimensions. Ideally on the Y and Z sweeps the vectorization would be on the X direction of the `I` loop. The worse case for lacking contiguous access is the X sweep where the OpenMP would be on Z (`K`) and the vectorization would be on Y (`J`). These types of applications have striding as a necessary part of the algorithm, and the refactoring process should be done to minimize as much of the striding as possible.

9.5 REFACTORING VH-1

TABLE 9.5.1 Profile of top functions in VH1.

```
Table:  Profile by Function

  Samp% |      Samp |  Imb.  |  Imb.  |Group
        |           |  Samp  | Samp%  | Function
        |           |        |        |  PE=HIDE

 100.0% | 7,771.0 |     -- |     -- |Total
|-----------------------------------------------------
|  52.7% | 4,097.9 |     -- |     -- |USER
||----------------------------------------------------
||  15.4% | 1,197.9 |  85.1 |  6.6% |parabola_
||  10.2% |   790.8 | 100.2 | 11.3% |riemann_
||   6.2% |   480.2 |  59.8 | 11.1% |sweepy_
||   6.1% |   475.5 |  63.5 | 11.8% |sweepz_
||   4.4% |   341.0 |  55.0 | 13.9% |remap_
||   2.6% |   202.7 |  50.3 | 19.9% |paraset_
||   1.9% |   151.2 |  40.8 | 21.3% |sweepx2_
||   1.6% |   125.3 |  41.7 | 25.0% |states_
||   1.6% |   123.6 |  32.4 | 20.8% |sweepx1_
||   1.5% |   115.3 |  39.7 | 25.6% |evolve_
||   1.0% |    75.9 |  30.1 | 28.4% |flatten_
||====================================================
|  43.8% | 3,399.9 |     -- |     -- |MPI
||----------------------------------------------------
||  40.0% | 3,107.2 |  54.8 |  1.7% |mpi_alltoall
||   3.5% |   270.0 | 281.0 | 51.0% |mpi_comm_split
```

VH-1 is a multidimensional ideal compressible hydrodynamics code written in FORTRAN. It is a small application that illustrates a number of interesting issues when vectorizing, threading, and even employing message passing in an application. First we instrument the application on 32 nodes of Broadwell using 32 MPI tasks/node. Table 9.5.1 shows the top level routines.

In addition to the computational routines, we see a lot of time spent in `mpi_alltoall`. `mpi_alltoall` takes more and more time as we run on more and more processors. In this case, one of the reasons to introduce good threading is to reduce the number of MPI tasks to reduce the time spent in `mpi_alltoall` – we are currently running on 1024 MPI tasks. To investigate threading we want to obtain the looping structure, presented in Table 9.5.2.

TABLE 9.5.2 Profile showing looping structure in VH1.

Table: Inclusive and Exclusive Time in Loops (from -hprofile_generate)

Loop Incl Time%	Loop Incl Time	Time (Loop Adj.)	Loop Hit	Loop Trips Avg	Loop Trips Min	Loop Trips Max	Function=/.LOOP[.] PE=HIDE
97.7%	98.184450	0.000354	1	50.0	50	50	vhone_.LOOP.2.li.219
21.8%	21.908535	0.000246	100	32.0	32	32	sweepy_.LOOP.1.li.32
21.8%	21.908289	0.042587	3,200	32.0	32	32	sweepy_.LOOP.2.li.33
19.3%	19.378648	0.000076	50	32.0	32	32	sweepz_.LOOP.05.li.48
19.3%	19.378571	0.038525	1,600	32.0	32	32	sweepz_.LOOP.06.li.49
17.3%	17.373445	12.792886	307,200	1,031.0	1,031	1,031	riemann_.LOOP.2.li.63
9.9%	9.998203	0.000165	50	32.0	32	32	sweepx2_.LOOP.1.li.28
9.9%	9.998038	0.018652	1,600	32.0	32	32	sweepx2_.LOOP.2.li.29
9.7%	9.740628	0.000049	50	32.0	32	32	sweepx1_.LOOP.1.li.28
9.7%	9.740579	0.018907	1,600	32.0	32	32	sweepx1_.LOOP.2.li.29
4.6%	4.580559	4.580559	316,723,200	12.0	12	12	riemann_.LOOP.3.li.64

Once again, the top level loop is the time stepping loop and then we have four high level loops in routines `sweepy`, `sweepz`, `sweepx1`, and `SWEEPX2` with double-nested loops. All of these loops have an average iteration count of 32. This is more evident in the call tree display of these statistics seen in Table 9.5.3.

This just displays the information down the call tree through `sweepz`. Notice that there are numerous routines called from the double-nested loop on lines 48 and 49 of `sweepz`. This also shows where a large amount of time is spent in a call to `mpi_alltoall`. So the challenge will be to analyze the outermost loop in `sweepz` to determine if it is parallelizable and to deal with scoping of the variables within the loop. In this application `sweepz`, `sweepy`, `sweepx1`, and `sweepx2` are all very similar. The analysis of `sweepy` for parallelization is discussed in Chapter 8 and a summary of issues encountered follows:

1. Numerous array constants were investigated to assure that all the uses of the elements of those arrays were covered by sets.

2. Introduction of a critical section around the reduction operation in `states`.

3. Using the `!$OMP THREADPRIVATE` directive for private variables contained within the module file.

TABLE 9.5.3 Profile from VH1 showing call tree.

```
Table:  Function Calltree View

  Time% |       Time |        Calls |Calltree
        |            |              | PE=HIDE

 100.0% | 98.594988 |            -- |Total
|-------------------------------------------------------
| 100.0% | 98.594874 |          2.0 |vhone_
||------------------------------------------------------
||  97.6% | 96.252189 |            -- |vhone_.LOOP.2.li.219
|||-----------------------------------------------------
3||  50.7% | 50.010498 |        100.0 |sweepz_
||||----------------------------------------------------
4|||  26.2% | 25.843463 |        100.0 |mpi_alltoall
4|||  16.7% | 16.420581 |            -- |sweepz_.LOOP.05.li.48
5|||         |            |              | sweepz_.LOOP.06.li.49
6|||  16.7% | 16.420581 |   102,400.0 |  ppmlr_
||||||||-----------------------------------------------
7||||||    6.4% |  6.345912 |   204,800.0 |riemann_
7||||||    5.7% |  5.647568 |   204,800.0 |remap_
||||||||-----------------------------------------------
8|||||||    3.3% |  3.274622 | 1,228,800.0 |parabola_
8|||||||    1.8% |  1.773262 |   204,800.0 |remap_(exclusive)
8|||||||    0.4% |  0.437516 |   204,800.0 |paraset_
8|||||||    0.2% |  0.162169 |   204,800.0 |volume_
||||||||===============================================
7||||||    1.7% |  1.657099 |   614,400.0 |parabola_
7||||||    1.1% |  1.059529 |   204,800.0 |evolve_
[...]
|||||||================================================
4|||   7.6% |  7.470206 |        100.0 |sweepz_(exclusive)
4|||   0.3% |  0.276248 |        100.0 |mpi_alltoall_(sync)
||||===================================================
```

After applying all the OpenMP directives, the identical problem was run with 1 and 2 threads on the Broadwell system and the results, presented in Table 9.5.4, were promising. Even the single OpenMP thread slightly outperformed the original. Now we take the two versions of VH1, the original and the OpenMP version, and run them on 32 nodes of KNL. Results are presented in Table 9.5.5.

TABLE 9.5.4 VH1 OpenMP runtime performance on Broadwell.

	Time (sec)	Hyper-Threads
Original	98	1
OpenMP x1	91	1
OpenMP x2	72	2

KNL is still somewhat slower than Broadwell. Once again we want to examine the vectorization to determine if we can improve that for KNL. If we look at the sampling profile of the KNL run we can determine where we might need some vectorization. First the routine profile, shown in Table 9.5.6, is considered. With the sampling profile, we can now look at the routines of interest, such as **riemann**, shown in Table 9.5.7. From the compiler's listing of those lines, presented in Excerpt 9.5.1, we see that the loop is not vectorized.

TABLE 9.5.5 VH1 OpenMP runtime performance on KNL.

	Time (sec)	SMT Threads
Original	166.17	1
OpenMP x1	158.45	1
OpenMP x2	96.53	2
OpenMP x4	85.88	4
OpenMP x8	90.97	8

TABLE 9.5.6 Sampling profile from VH1 on KNL.

```
Table:  Profile by Function

  Samp% |      Samp |  Imb.  |  Imb.  |Group
        |           |  Samp  | Samp%  | Function
        |           |        |        |  PE=HIDE

 100.0% | 17,730.7 |     -- |     -- |Total
|-----------------------------------------------------
|  73.5% | 13,030.2 |     -- |     -- |USER
||----------------------------------------------------
||  21.4% |  3,797.3 | 268.7 |  6.6% |riemann_
||  18.3% |  3,243.4 | 431.6 | 11.8% |parabola_
||   8.3% |  1,473.6 | 158.4 |  9.7% |sweepy_
||   5.7% |  1,012.0 | 135.0 | 11.8% |sweepz_
||   5.4% |    951.1 | 152.9 | 13.9% |remap_
```

TABLE 9.5.7 Sampling profile of riemann routine from VH1 on KNL.

```
||  21.4% |  3,797.3 |     -- |     -- |riemann_
3|        |          |        |        | dal/levesque/VH1_version1_orig/riemann.f90
||||--------------------------------------------------------------------------------
4|||   1.4% |     241.8 |  59.2 | 19.7% |line.69
4|||   3.2% |     566.6 |  71.4 | 11.2% |line.73
4|||   1.1% |     196.2 |  42.8 | 17.9% |line.75
4|||   5.1% |     911.6 | 138.4 | 13.2% |line.77
4|||   4.8% |     845.2 | 105.8 | 11.1% |line.78
```

This is due to the exit out of the inner loop. This loop is actually iterating on the zone pressure to converge to a value less than `tol`. Might it be possible to bring the `do l` loop inside the `do n` loop and gain some vectorization? This type of a situation would require that we generate an array of the converged state for each of the zones, and then once all of the zones have converged we can exit the loop testing for convergence. In such a rewrite, there will be more computation performed. However, it will be vectorized. As long as a few cells need many more iterations than the rest, we should run faster. The refactored loop is shown in Excerpt 9.5.2.

```
63. + 1----< do l = lmin, lmax
64. + 1 2--<   do n = 1, 12
65.   1 2        pmold(l) = pmid(l)
66.   1 2        wlft (l) = 1.0 + gamfac1*(pmid(l) - plft(l)) * plfti(l)
67.   1 2        wrgh (l) = 1.0 + gamfac1*(pmid(l) - prgh(l)) * prghi(l)
68.   1 2        wlft (l) = clft(l) * sqrt(wlft(l))
69.   1 2        wrgh (l) = crgh(l) * sqrt(wrgh(l))
70.   1 2        zlft (l) = 4.0 * vlft(l) * wlft(l) * wlft(l)
71.   1 2        zrgh (l) = 4.0 * vrgh(l) * wrgh(l) * wrgh(l)
72.   1 2        zlft (l) = -zlft(l) * wlft(l)/(zlft(l) - gamfac2*(pmid(l) - plft(l)))
73.   1 2        zrgh (l) =  zrgh(l) * wrgh(l)/(zrgh(l) - gamfac2*(pmid(l) - prgh(l)))
74.   1 2        umidl(l) = ulft(l) - (pmid(l) - plft(l)) / wlft(l)
75.   1 2        umidr(l) = urgh(l) + (pmid(l) - prgh(l)) / wrgh(l)
76.   1 2        pmid (l) = pmid(l) + (umidr(l) - umidl(l))*(zlft(l) * zrgh(l)) /
                            (zrgh(l)-zlft(l))
77.   1 2        pmid (l) = max(smallp,pmid(l))
78.   1 2        if (abs(pmid(l)-pmold(l))/pmid(l) < tol ) exit
79.   1 2->>   enddo ; enddo
```

Excerpt 9.5.1 Compiler listing with non-vectorized VH1 riemann loop.

```
62.   A----<> converged =.F.
63. + 1-----< do n = 1, 12
64.   1 Vr2-<   do l = lmin, lmax
65.   1 Vr2       if(.not.converged(l))then
66.   1 Vr2        pmold(l) = pmid(l)
67.   1 Vr2        wlft (l) = 1.0 + gamfac1*(pmid(l) - plft(l)) * plfti(l)
68.   1 Vr2        wrgh (l) = 1.0 + gamfac1*(pmid(l) - prgh(l)) * prghi(l)
69.   1 Vr2        wlft (l) = clft(l) * sqrt(wlft(l))
70.   1 Vr2        wrgh (l) = crgh(l) * sqrt(wrgh(l))
71.   1 Vr2        zlft (l) = 4.0 * vlft(l) * wlft(l) * wlft(l)
72.   1 Vr2        zrgh (l) = 4.0 * vrgh(l) * wrgh(l) * wrgh(l)
73.   1 Vr2        zlft (l) = -zlft(l) * wlft(l)/(zlft(l) - gamfac2*(pmid(l) - plft(l)))
74.   1 Vr2        zrgh (l) =  zrgh(l) * wrgh(l)/(zrgh(l) - gamfac2*(pmid(l) - prgh(l)))
75.   1 Vr2        umidl(l) = ulft(l) - (pmid(l) - plft(l)) / wlft(l)
76.   1 Vr2        umidr(l) = urgh(l) + (pmid(l) - prgh(l)) / wrgh(l)
77.   1 Vr2        pmid (l) = pmid(l) + (umidr(l) - umidl(l))*(zlft(l) * zrgh(l)) /
                              (zrgh(l)-zlft(l))
78.   1 Vr2        pmid (l) = max(smallp,pmid(l))
79.   1 Vr2        if (abs(pmid(l)-pmold(l))/pmid(l) < tol ) then
80.   1 Vr2        converged(l) = .T.
81.   1 Vr2        endif
82.   1 Vr2       endif
83.   1 Vr2->   enddo
84. + 1         if(all(converged(lmin:lmax)))exit
85.   1-----> enddo
```

Excerpt 9.5.2 Compiler listing showing vectorized VH1 riemann loop.

Now we see that the important loop does vectorize, and the timings are improved somewhat, getting down to 80 seconds. This is is still 10% slower than Broadwell. Since VH1 was originally written for vector machines, the remaining code vectorizes very well and may be able to see some improvement.

9.6 REFACTORING LESLIE3D

TABLE 9.6.1 Profile from Leslie3D code.

```
Table:  Profile by Function

  Samp% |     Samp | Imb.  |  Imb.  |Group
        |          | Samp  | Samp%  | Function
        |          |       |        |  PE=HIDE

 100.0% | 3,905.9 |    -- |     -- |Total
|--------------------------------------------------
|  53.0% | 2,069.9 |    -- |     -- |USER
||-------------------------------------------------
||  13.6% |   532.6 | 13.4 |  2.5% |fluxk_
||   6.9% |   270.6 | 10.4 |  3.8% |fluxj_
||   6.4% |   250.5 | 21.5 |  8.0% |fluxi_
||   5.3% |   206.7 | 16.3 |  7.4% |extrapi_
||   4.9% |   189.7 |  8.3 |  4.3% |extrapk_
||   4.5% |   173.8 | 10.2 |  5.6% |extrapj_
||   3.3% |   128.2 |  3.8 |  2.9% |grid_
||   2.1% |    83.5 |  2.5 |  2.9% |setiv_
||   1.9% |    74.5 | 10.5 | 12.5% |update_
||   1.4% |    56.4 | 14.6 | 20.9% |mpicx_
||   1.0% |    39.7 |  8.3 | 17.6% |parallel_
||=================================================
|  28.4% | 1,107.6 |    -- |     -- |MPI
[...]
||=================================================
|  18.6% |   725.2 |    -- |     -- |ETC
||-------------------------------------------------
||  18.4% |   720.1 | 16.9 |  2.3% |__cray_memset_KNL
```

Leslie3D is a "Large Eddy Simulation" code that performs finite differences on a 3D grid. Leslie3D uses a 3D decomposition, so the mesh is divided among the processors as small cubes. Table 9.6.1 is a profile for the original run on KNL. As is the case in most of these types of applications, the `FLUXI`, `FLUXJ`, and `FLUXK` typically take about the same amount of time. In this case we see that `FLUXK` is taking about twice as much, so we should look a little deeper and see if there is something abnormal about the computation within `FLUXK`. We gather hardware counters for the three flux routines (Table 9.6.2) and see that the cache utilization of `FLUXK` is significantly worse than the other two routines.

TABLE 9.6.2 Hardware counters for last-level cache from Leslie3D.

Routine	LLC Misses	LLC hit ratio
FLUXI	119,816,654	62.5%
FLUXJ	25,332,157	90.6%
FLUXK	18,992,047	92.4%

```
      DO J = 1, JCMAX
        DO K = 0, KCMAX
           QS(1:IND) = UAV(1:IND,J,K) * SKX(1:IND,J,K) +
     >                 VAV(1:IND,J,K) * SKY(1:IND,J,K) +
     >                 WAV(1:IND,J,K) * SKZ(1:IND,J,K)
           IF ( NSCHEME .EQ. 2 ) THEN
              L = K + 1 - KADD
              DO I = 1, IND
                 QSP = U(I,J,L) * SKX(I,J,K) +
     >                 V(I,J,L) * SKY(I,J,K) +
     >                 W(I,J,L) * SKZ(I,J,K)
                 QSPK = (QSP - QS(I)) * DBLE(1 - 2 * KADD)
                 IF (QSPK .GT. 0.0D+00) QS(I) = 0.5D+00 * (QS(I) + QSP)
              ENDDO
           ENDIF
           FSK(1:IND,K,1) =  QAV(1:IND,J,K,1) * QS(1:IND)
           FSK(1:IND,K,2) =  QAV(1:IND,J,K,2) * QS(1:IND) +
     >                       PAV(1:IND,J,K) * SKX(1:IND,J,K)
           FSK(1:IND,K,3) =  QAV(1:IND,J,K,3) * QS(1:IND) +
     >                       PAV(1:IND,J,K) * SKY(1:IND,J,K)
           FSK(1:IND,K,4) =  QAV(1:IND,J,K,4) * QS(1:IND) +
     >                       PAV(1:IND,J,K) * SKZ(1:IND,J,K)
           FSK(1:IND,K,5) = (QAV(1:IND,J,K,5) + PAV(1:IND,J,K)) *
     >                                                QS(1:IND)
           IF (ISGSK .EQ. 1) THEN
              FSK(1:IND,K,7) = QAV(1:IND,J,K,7) * QS(1:IND)
           ENDIF
           IF ( ICHEM .GT. 0 ) THEN
              DO L = 8, 7 + NSPECI
                 FSK(1:IND,K,L) = QAV(1:IND,J,K,L) * QS(1:IND)
              ENDDO
           ENDIF
           IF ( VISCOUS ) CALL VISCK ( 1, IND, J, K, FSK )
        ENDDO
        DO K = 1, KCMAX
           DO L = 1, 5
              DQ(1:IND,J,K,L) = DQ(1:IND,J,K,L) -
     >           DTV(1:IND,J,K) * (FSK(1:IND,K,L) - FSK(1:IND,K-1,L))
           ENDDO
           IF (ISGSK .EQ. 1) THEN
              DQ(1:IND,J,K,7) = DQ(1:IND,J,K,7) -
     >           DTV(1:IND,J,K) * (FSK(1:IND,K,7) - FSK(1:IND,K-1,7))
           ENDIF
           IF ( ICHEM .GT. 0 ) THEN
              DO L = 8, 7 + NSPECI
                 DQ(1:IND,J,K,L) = DQ(1:IND,J,K,L) -
     >            DTV(1:IND,J,K) * (FSK(1:IND,K,L) - FSK(1:IND,K-1,L))
              ENDDO
           ENDIF
      ENDDO ; ENDDO
```

Excerpt 9.6.1 Example loop-nest from original Leslie3D code.

So we have a significant issue with cache utilization within FLUXK. Looking at the routine in Excerpt 9.6.1, we determine what is causing the issue. The outermost index on most of the arrays is being used as the second looping construct in the computation. This will not only impact cache reuse, it will also impact the TLB, since we are jumping a large distance in memory each invocation of the K loop. The principal reason for having K as the second looping construct – is the innermost in the array assignments – is due to the temporary array FSK. The only way we can pull the J loop inside the K loop is to have FSK dimensioned on J as well, which unfortunately will use more memory. The good news is FSK is a local variable and its storage is released upon exiting from the routine.

In the refactoring, we add the dimension on to FSK and we can also combine the two K loops by simply testing the code within the second K loop and only execute it when K is greater or equal to 1. Excerpt 9.6.2 shows the rewrite with vectorization by the Cray compiler.

```
28.  + F-----------<        DO K = 0, KCMAX
29.  + F F---------<        DO J = 1, JCMAX
30.    F F V-------<         DO I = 1, IND
31.    F F V                     QS(I) = UAV(I,J,K) * SKX(I,J,K) +
32.    F F V               >                 VAV(I,J,K) * SKY(I,J,K) +
33.    F F V               >                 WAV(I,J,K) * SKZ(I,J,K)
34.    F F V
35.    F F V                     FSK(I,J,K,1) =  QAV(I,J,K,1) * QS(I)
36.    F F V                     FSK(I,J,K,2) =  QAV(I,J,K,2) * QS(I) +
37.    F F V               >                           PAV(I,J,K) * SKX(I,J,K)
38.    F F V                     FSK(I,J,K,3) =  QAV(I,J,K,3) * QS(I) +
39.    F F V               >                           PAV(I,J,K) * SKY(I,J,K)
40.    F F V                     FSK(I,J,K,4) =  QAV(I,J,K,4) * QS(I) +
41.    F F V               >                           PAV(I,J,K) * SKZ(I,J,K)
42.    F F V                     FSK(I,J,K,5) = (QAV(I,J,K,5) + PAV(I,J,K)) *
43.    F F V               >                                    QS(I)
44.    F F V
45.    F F V                     IF (ISGSK .EQ. 1) THEN
46.    F F V                        FSK(I,J,K,7) = QAV(I,J,K,7) * QS(I)
47.    F F V                     ENDIF
48.    F F V------->          ENDDO
49.    F F V I----<>             IF ( VISCOUS ) CALL VISCK ( 1, IND, J, K, FSK )
50.    F F                       IF(K.ge.1)THEN
51.    F F V-------<         DO I = 1, IND
52.    F F V            !dir$ unroll(5)
53.    F F V w-----<             DO L = 1, 5
54.    F F V w                      DQ(I,J,K,L) = DQ(I,J,K,L) -
55.    F F V w             >        DTV(I,J,K) * (FSK(I,J,K,L) - FSK(I,J,K-1,L))
56.    F F V w----->             ENDDO
57.    F F V
58.    F F V                     IF (ISGSK .EQ. 1) THEN
59.    F F V                        DQ(I,J,K,7) = DQ(I,J,K,7) -
60.    F F V               >         DTV(I,J,K) * (FSK(I,J,K,7) - FSK(I,J,K-1,7))
61.    F F V                     ENDIF
62.    F F V------->        ENDDO
63.    F F                       ENDIF
64.    F F-------->>        ENDDO ; ENDDO
```

Excerpt 9.6.2 Example loop-nest from optimized Leslie3D code.

Notice that we used the directive to tell the compiler to unroll the L loop – the w indicates that the loop was completely unwound and eliminated. We also changed the array assignments to good old DO loops, which really utilizes cache better. The compiler did fuse the array assignments in the original loop. However, a good programmer doesn't leave important stuff entirely up to the compiler. The restructured loops do indeed improve the cache performance from 62.5% to 91.3%, and the routine runs twice as fast as it did before. Use the hardware counters to identify issues like this when possible, so you can narrow in on poorly performing code. It's the old whack-a-mole principal: identify the hot spots and see how to improve them.

9.7 REFACTORING S3D – 2016 PRODUCTION VERSION

S3D is a 3D combustion application that exhibits very good "weak scaling". In setting up the input for S3D, the user specifies the size of the computational grid per processor. As the number of processors increases, the number of grid blocks increases accordingly. The communication in S3D is all nearest neighbor. Although S3D is a CFD code, it is not a spectral model. It solves the physical equations using a higher-order Runga Kutta algorithm which does not require global transposes as the spectral model does.

The version we will use is the production version as of Fall 2016. A previous version developed for the GPUs using OpenACC is not employed in production, so the version we will examine is all-MPI with no threading and very little vectorization. The next section will examine the project that was performed in moving the 2011 version of S3D to the Titan system at Oak Ridge National Laboratory.

Table 9.7.1 shows the sampling profile (the sample size is 1/100 of a second) run on one node of KNL and Broadwell using 64 MPI tasks on the node. While the KNL had 64 cores, hyper-threads had to be used on Broadwell to accommodate the 64 MPI tasks.

TABLE 9.7.1 Sampling profile results from S3D on KNL and Broadwell.

Routine	KNL Samples	Broadwell Samples
Total	54,496	41,689
USER	39,217	30,737
rdot	5,766	1,226
ratt	4,922	982
ratx	4,639	982
stif	3,264	515
rhsf	3,211	7,795
rdsmh	2,843	467
qssa	2,195	429
mcavis	1,794	3,675
reaction	1,483	949
diffflux	977	1,831
integrate	701	1,925
derivative x	682	451
mcadif	669	1,157
computeheat	585	861
EXP	3,641	2,072

The routines are ordered by the largest samples on KNL, our primary target. Notice that the Broadwell version takes fewer samples since the original version of S3D runs faster on Broadwell than on KNL. The top seven routines on KNL are part of the computational chemistry in S3D. From the comparison to Broadwell it is evident that these routines do not vectorize. Broadwell is outperforming KNL on the routines by a factor of 4 to 5. Our first step will be to investigate the vectorization/optimization of these important chemistry routines. The principal top level chemistry routine is **GETRATES**, which is called from **REACTION_RATE_BOUNDS**, as seen in Excerpts 9.7.1 and 9.7.2.

```
! get reaction rate from getrates and convert units
do k = kzl,kzu
    do j = jyl,jyu
      do i = ixl, ixu
       yspec(:) = yspecies(I,j,k,:)
       call getrates(pressure(i,j,k)*pconv,temp(i,j,k)*tconv,  &
                     yspec,diffusion(I,j,k)*diffconv,          &
                     max(tchem_fast,dt*timeconv),              &
                     ickwrk,rckwrk,rr_r1)
        rr_r(i,j,k,:) = rr_r1(:) * rateconv * molwt(:)
  enddo ; enddo ; enddo
```

Excerpt 9.7.1 S3D code showing call to getrates routine.

```
      SUBROUTINE GETRATES  (P, T, Y, DIFF, DT, ICKWRK, RCKWRK, WDOT)
        IMPLICIT DOUBLE PRECISION (A-H, O-Z), INTEGER (I-N)
      PARAMETER (IREAC=283,KK=52,KSS=16,KTOTAL=68)
      DIMENSION Y(*), ICKWRK(*), RCKWRK(*), WDOT(*), DIFF(*)
      DIMENSION RF(IREAC),RB(IREAC),RKLOW(12),C(KK),XQ(KSS)
C
      CALL YTCP(P, T, Y, C)
      CALL RATT(T, RF, RB, RKLOW)
      CALL RATX(T, C, RF, RB, RKLOW)
      CALL QSSA(RF, RB, XQ)
      CALL STIF(RF, RB, DIFF, DT, C)
      CALL RDOT(RF, RB, WDOT)
      END
```

Excerpt 9.7.2 S3D code showing body of getrates routine.

The **getrates.f** file is actually automatically generated and contains over 8,000 lines of computations. **EXP** and **ALOG10** are used extensively as well as **Real** raised to a **Real**. There is a lot of very important computation. To vectorize the computations within these calls, we must pull arrays of elements into the routines, which will result in loop constructs that can be vectorized. One important note about **getrates** is that it updates cell quantities without the need for communication with neighbors. This allows us to be very flexible in pulling in a vector of quantities. In the rewrite for OpenACC, a loop of 512 quantities was used, and all subroutine boundaries were eliminated. This resulted in excellent performance. However, the readability of the code suffered. This rewrite will be performed to minimize the amount of refactoring and still achieve good performance gain. Initially, we will simply pull in the innermost loop on **I** and use **#define** statements to allow one to choose between the vector and scalar version at compile time, as shown in Excerpts 9.7.3 and 9.7.4.

```
  do k = kzl,kzu
    do j = jyl,jyu
#ifdef ORIGINAL_GETRATES
      do i = ixl, ixu
        yspec(:) = yspecies(I,j,k,:)
        diff(:)  = diffusion(i,j,k,:)*diffconv
        call getrates(pressure(i,j,k)*pconv,temp(i,j,k)*tconv,  &
                      yspec,diff,max(tchem_fast,dt*timeconv),   &
                      ickwrk,rckwrk,rr_r1)
        rr_r2(i,:) = rr_r1(:)
      enddo
#else
      do i = ixl, ixu
        rr_r2(i,:) = 0.0
        yspec2(i,:) = yspecies(i, j, k, :)
        diff2(i,:)  = diffusion(i,j,k,:)*diffconv
        press(i)    = pressure(i,j,k)*pconv
        t(i)        = temp(i,j,k)*tconv
      enddo
      call V_getrates(ixl, ixu,n_species,press,t,               &
                      yspec2,diff2,max(tchem_fast,dt*timeconv), &
                      ickwrk,rckwrk,rr_r2)
#endif
      do i = ixl, ixu
        rr_r(i,j,k,:) = rr_r2(i,:) * rateconv * molwt(:)
  enddo ; enddo ; enddo
```

Excerpt 9.7.3 Optimized S3D code with vector and scalar versions.

The major difference is that we are now passing arrays down into V_GETRATES and will propagate them down into the major computational routines. Now the modification to V_GETRATES just replaces the scalar calls with the vector calls. Notice that the temporary arrays used down the call chain are all now dimensioned the length of the vector, in this case IXL:IXU. Additional memory must be used to hold temporaries that communicate between the seven computational routines. This will increase data motion and while the rewrite will perform better on KNL, Broadwell may suffer, since the improvement from vectorization on that system is not as great as KNL, and the increased data motion will tend to degrade the performance on Broadwell.

Rather than showing over 8,000 lines of changes a short snippet will illustrate how the computational routines are structured in Excerpt 9.7.5.

The refactoring of the computational routines needs to be performed in a way such that all compilers that will be potentially used can vectorize the major loop(s). For example, the rewrite in Excerpt 9.7.6 was done to ensure that combining the two loops would not inhibit vectorization by a compiler.

Combining the loops would necessitate that the compiler vectorize on the outer-most loop on I – some compilers do not perform outer-loop vectorization. The increase in performance was almost a factor of two on KNL with only a slight improvement on Broadwell, and the optimized/vectorized code runs faster on KNL than on Broadwell. Table 9.7.2 is a comparison of the original and optimized profiles on KNL and Broadwell.

```
      SUBROUTINE V_GETRATES(ixl,ixu,n_species,P,T,YI,DIFF,DT,
     +                             ICKWRK, RCKWRK, WDOT)
        IMPLICIT DOUBLE PRECISION (A-H, O-Z), INTEGER (I-N)
      PARAMETER (IREAC=283,KK=52,KSS=16,KTOTAL=68)
      DIMENSION YI(ixl:ixu,n_species), ICKWRK(*), RCKWRK(*)
      DIMENSION  WDOT(ixl:ixu,n_species), DIFF(ixl:ixu,n_species)
      DIMENSION  W(n_species), D(n_species),Y(n_species)
      DIMENSION  P(ixl:ixu), T(ixl:ixu),CI(ixl:ixu,n_species)
      DIMENSION RF(IREAC),RB(IREAC),RKLOW(12),C(n_species),XQ(KSS)
      DIMENSION RFI(ixl:ixu,IREAC),RBI(ixl:ixu,IREAC),RKLOWI(ixl:ixu,12)
C
       call V_YTCP(ixl,ixu,n_species,P, T, YI, CI)
       call V_RATT(ixl,ixu,IREAC,T ,RFI,RBI, RKLOWI)
       call V_RATX(ixl,ixu,n_species,IREAC,T ,CI ,RFI,RBI, RKLOWI)
       CALL V_QSSA(ixl,ixu,IREAC,RFI, RBI, XQ)
       CALL V_STIF(ixl,ixu,n_species,IREAC,RFI, RBI,DIFF, DT, CI)
       CALL V_RDOT(ixl,ixu,n_species,IREAC,RFI, RBI, WDOT)
       END
```

Excerpt 9.7.4 Optimized S3D code showing calls to vectorized routines.

```
      SUBROUTINE V_RATT (ixl, ixu,IREAC,T, RFI, RBI, RKLOWI)
        IMPLICIT DOUBLE PRECISION (A-H, O-Z), INTEGER (I-N)
      PARAMETER (RU=8.314510D7, RUC=RU/4.184D7, PATM=1.01325D6)
      DIMENSION EQK(283)
      DIMENSION SMH(ixl:ixu,68)
      DIMENSION RFI(ixl:ixu,IREAC),RBI(ixl:ixu,IREAC)
      DIMENSION RKLOWI(ixl:ixu,12)
      DIMENSION  EG(ixl:ixu,68), T(ixl:ixu)
      DATA SMALL/1.D-200/
C
      do i = ixl, ixu
      ALOGT = LOG(T(i))
      TI = 1.0D0/T(i)
      TI2 = TI*TI
C
      RFI(i,1) = EXP(3.52986472D1 -4.D-1*ALOGT)
      RFI(i,2) = EXP(9.75672617D0 +3.D0*ALOGT -4.13845387D3*TI)
      RFI(i,3) = EXP(1.21704455D1 +2.4D0*ALOGT -1.0597743D3*TI)
      RFI(i,4) = EXP(2.87784236D1 +5.D-1*ALOGT -5.17809951D3*TI)
      RFI(i,5) = 1.02D14
      RFI(i,6) = EXP(1.18493977D1 +1.95D0*ALOGT +6.77832851D2*TI)
      RFI(i,7) = EXP(3.29142248D1 -8.32320368D3*TI)
```

Excerpt 9.7.5 Optimized S3D code with example comutational routine.

These routines were further optimized for the Titan system at Oak Ridge National Laboratory by combining all of the chemical mechanism routines into one routine and eliminating temporary arrays. This would definitely help on this version for KNL. The refactoring of S3D for the Titan system which consisted of 15,000 nodes that had a Opteron host and an Nvidia GPU was much more extensive. That version does still exist. However, the S3D development team did not pick it up as their current version. All of the modifications for Titan did improve that version of S3D on Xeon systems and that version runs extremely well on KNL. Node for node, the KNL system outperforms Titan by a factor of two.

```
      do i = ixl, ixu
       DO N = 1, 67
          EG(i,N) = EXP(SMH(i,N))
       ENDDO
      enddo
C
      do i = ixl, ixu
       PFAC = PATM / (RU*T(i))
       PFAC2 = PFAC*PFAC
       PFAC3 = PFAC2*PFAC
C
       EQK(1)=EG(i,11)/EG(i,1)/EG(i,10)/PFAC
       EQK(14)=EG(i,18)/EG(i,1)/EG(i,17)/PFAC
       EQK(15)=EG(i,5)*EG(i,10)/EG(i,22)*PFAC
```

Excerpt 9.7.6 Optimized S3D code showing portable rewrite of computational loop.

TABLE 9.7.2 Performance comparison of S3D on KNL and Broadwell.

Routine/ Function	Original KNL Samples 1/100 Second	Optimized KNL Samples 1/100 Second	Original Broadwell Samples 1/100 Second	Optimized Broadwell Samples 1/100 Second
Total	54,496.00	32,293.80	41,688.50	40,723.60
rdot	5,765.90	2,172.50	1,226.10	1,126.20
ratt	4,921.80	1,012.00	982.4	1,759.00
ratx	4,639.00	1,959.70	982.4	876.70
rhsf	3,210.50	3,316.50	7,794.50	7,843.10
mcavis	1,794.10	1,787.90	3,675.40	3,659.50
reaction	1,482.70	876.9	949.1	515.3
diffflux	977.1	1,012.00	1,831.30	1,832.30
integrate	700.5	647	1,924.50	1,923.30
derivative x	682.2	441.9	450.7	430.5
mcadif	668.5	683.3	1,157.10	565.2
computeheat	585	515.5	861.1	873.80
EXP	3,640.70	2208	2,072.00	2,029.20

9.8 PERFORMANCE PORTABLE – S3D ON TITAN

S3D running on Titan in 2013 is an example of a truly performance portable application. Not having the time to take the current version of S3D and rewrite it for multi/manycore systems, we will discuss the work that was performed around 5 years ago in moving S3D to the Titan supercomputer at Oak Ridge National Laboratory. Rather than using performance data from that time, we will demonstrate the 2013 all MPI version of S3D and the refactored version on Broadwell and KNL, discussing how the modifications were made to port/optimize for the Nvidia GPU and how they improved the performance on current multi/manycore architectures.

The following profile was obtained on a Broadwell system running on 8 nodes utilizing 64 threads of execution and 8 nodes of KNL using 64 cores. While the particular Broadwell only had 36 cores (18 per socket), two hyper-threads were used on 32 cores (16 per socket). This does not degrade the Broadwell system and is to compare node to node to the KNL system. The following chart shows 18 of the top level computational routines that account for over 85% of the total runtime.

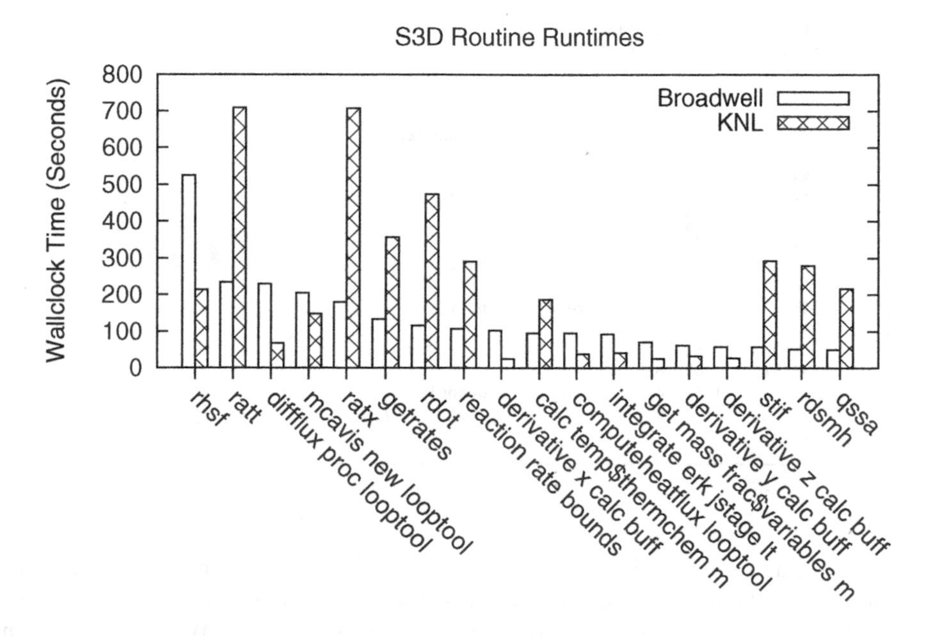

FIGURE 9.8.1 Runtime of selected routines from original S3D code.

It is obvious from the chart that the routines that account for most of the time on KNL are not vectorized. Broadwell outperforms KNL on those routines by a wide margin. Many of the top routines on Broadwell do vectorize on KNL because KNL outperforms Broadwell. From a vectorization standpoint, it is obvious the routines that we vectorized in the previous section are indeed the ones to be addressed in this version of the code. Another disturbing result from the statistics gathered on the Broadwell runs (where one can obtain excellent cache utilization statistics) shows this version of S3D has very poor cache utilization, as seen in Table 9.8.1.

We cannot obtain similar statistics on KNL, and there is no reason to expect that they would be better on KNL – they would probably be worse due to the smaller amount of low-level cache available. At first look, there is a problem with cache utilization and a problem with vectorization of important computational routines.

TABLE 9.8.1 Table showing poor cache utilization in version of S3D.

```
D1 + D2 cache utilization:

    95.3% of total execution time was spent in 15 functions with
    combined D1 and D2 cache hit ratios below the desirable minimum of
    80.0%. Cache utilization might be improved by modifying the
    alignment or stride of references to data arrays in these functions.

     D1+D2   Time%  Function
     cache           PE=HIDE
       hit
     ratio

      32.9%   30.4%  reaction_rate_bounds$chemkin_m_
      37.6%   14.5%  rhsf_
      38.1%    2.8%  stif_
      38.8%    1.7%  derivative_x_calc_buff_
      39.7%    2.8%  rdsmh_
      42.5%    2.9%  mcavis_new_looptool_
      43.9%    1.1%  derivative_y_calc_buff_
      44.9%    3.7%  rdot_
      45.1%    1.0%  derivative_z_calc_buff_
      46.3%    4.8%  ratx_
      47.0%   12.7%  ratt_
      51.5%    1.0%  mcadif_
      53.9%    1.5%  computeheatflux_looptool_
      64.6%    3.5%  diffflux_proc_looptool_
      73.5%   11.0%  MPI_WAIT
```

Looking at vectorization first is getting the cart before the horse. When moving to the multi/manycore systems, the first thing to examine is threading and organization of the data to minimize data movement. While the previous table was useful from a vectorization standpoint, Table 9.8.2 is important from a data organization/structure standpoint.

While this is not a complete call tree of the computation, it gives an idea of the data usage within the Runga Kunta loop in `INTEGRATE_ERK`. All the computation is performed within the three- and four-dimensional loops. The number in parenthesis after the loop indicates the average loop iteration count. While there may be some amount of cache reuse within a nested loop computation, there will not be any from one nested loop to the next, since the data being accessed is too large to fit into lower level caches. This particular problem size is too large to fit into last level caches on the Xeons or even within MCDRAM on KNL. Of course, one could use more nodes to reduce the working set down to fit into MCDRAM on KNL. However, that is a significant misuse of the node resources. Why not find a way to handle larger problem sizes and get cache reuse?

The first refactoring that was done on S3D in moving to the Titan system was to reorganize the looping structures to obtain better cache utilization and promote vectorization. Within the S3D computation, there are several routines that update cell quantities without requiring data from their neighbors.

TABLE 9.8.2 Calltree profile from S3D code.

```
Table:  Function Calltree View

  Time% |        Time |       Calls |Calltree
        |             |             | PE=HIDE

 100.0% | 707.862999 |          -- |Total
|-----------------------------------------------------------------------------
| 100.0% | 707.862920 |         1.0 |s3d_
||----------------------------------------------------------------------------
||  99.5% | 704.492944 |         2.0 |solve_driver_
|||---------------------------------------------------------------------------
3||  98.4% | 696.886256 |          -- |solve_driver_.LOOP.1.li.161  (10)
||||--------------------------------------------------------------------------
4|||  97.7% | 691.875123 |        20.0 |integrate_
|||||-------------------------------------------------------------------------
5||||  96.3% | 681.558040 |        20.0 |integrate_erk_jstage_lt_
||||||------------------------------------------------------------------------
6|||||  95.8% | 678.352700 |          -- |integrate_erk_jstage_lt_.LOOP.1.li.47 (6)
|||||||-----------------------------------------------------------------------
7||||||  95.7% | 677.518132 |       120.0 |rhsf_
8|||||||    0.3% |   2.007678 |       120.0 |get_mass_frac$variables_m_  (48,48,48)
8|||||||    0.0% |   0.053106 |        60.0 |get_velocity_vec_  (48,48,48)
8|||||||    0.1% |   0.580969 |       120.0 |calc_inv_avg_mol_wt$thermchem_m_  (48,48,48)
8|||||||    7.4% |  52.632581 |       120.0 |calc_temp$thermchem_m_  (48,48,48)
8|||||||    0.0% |   0.021695 |        60.0 |calc_gamma$thermchem_m_  (48,48,48)
8|||||||    0.0% |   0.019763 |        60.0 |calc_press$thermchem_m_  (48,48,48)
8|||||||    0.3% |   1.916414 |       120.0 |calc_specenth_allpts$thermchem_m_ (48,48,48)
8|||||||    2.0% |  14.032747 |       120.0 |computevectorgradient_  (48,48,48)
|||||||||---------------------------------------------------------------------
9||||||||    2.0% |  14.031220 |          -- |computevectorgradient_.LOOP.1.li.92 (52)
10|||||||    2.0% |  14.031220 |       180.0 | computescalargradient_  (48,48,48)
|||||||||||-------------------------------------------------------------------
11|||||||||    1.3% |   8.950916 |       360.0 |derivative_x_calc_
||||||||||||------------------------------------------------------------------
12||||||||||    1.2% |   8.814708 |       720.0 |MPI
||||||||----------------------------------------------------------------------
8|||||||    2.8% |  19.799414 |          -- |rhsf_.LOOP.1.li.239
9|||||||    2.8% |  19.799414 |     3,120.0 | computescalargradient5d_
8|||||||    9.7% |  68.685447 |        60.0 |getdiffusivefluxterms_
|||||||||---------------------------------------------------------------------
9||||||||    8.2% |  58.393736 |        20.0 |computecoefficients$transport_m_
[...]
|||||||||=====================================================================
8|||||||   63.3% | 448.072929 |        60.0 |reaction_rate$chemkin_m_
|||||||||---------------------------------------------------------------------
9||||||||   63.3% | 448.072003 |       120.0 |reaction_rate_bounds$chemkin_m_
||||||||||--------------------------------------------------------------------
10||||||||   58.9% | 416.871345 |          -- |reaction_rate_bounds_.LOOP.1.li.487 (48)
11||||||||          |            |             | reaction_rate_bounds_.LOOP.2.li.488 (48)
```

Other routines require differences, and these routines end up exchanging halos with their neighbors. We chose to partition the grid in two different ways for refactoring S3D. The first way was to handle cell-centered computation, where nearest-neighbor communication was not required. The three grid loops were collapsed, and a chunk size was used to strip-mine across all the routines that did not require nearest-neighbor information.

The innermost loop that went over the chunk size was vectorized, and the outer loop that went over the number of chunks was threaded as seen in Excerpt 9.8.1. Notice that the `#IFDEF` macros are used to choose between OpenACC for the attached accelerator or OpenMP for the multicore Xeons and KNL. `MS` is the chunk size, and this was parameterized to best utilize cache.

```
#ifdef GPU
!$acc parallel loop  gang  private(i,ml,mu)
#else
!$omp parallel private(i, ml, mu)
!$omp do
#endif
 do i = 1, nx*ny*nz, ms
  ml = i
  mu =  min(i+ms-1, nx*ny*nz)
  call calc_gamma_r( gamma, cpmix, avmolwt, ml, mu)
  call calc_press_r( pressure, q(1,1,1,4), temp, avmolwt, ml, mu )
  call calc_specEnth_allpts_r(temp, h_spec, ml, mu)
end do
#ifdef GPU
!$acc end parallel loop
#else
!$omp end parallel
#endif
```

Excerpt 9.8.1 S3D code refactored to better utilize cache and promote vectorization.

The second way was to keep the 3 loops, vectorizing on the innermost and threading on the outermost, as presented in Excerpt 9.8.2. This worked well. However, a much better plan for ensuring consistent memory usage would be to maintain a single decomposition of the 3 loops, so that a thread would be updating the same memory regardless of the computation it was performing. In other words, it would have been much better to utilize a SPMD OpenMP approach with S3D. OpenMP scaling on the NUMA systems like the Xeons and KNL would have been better if we had used a single decomposition across the entire computation.

What we were able to do in the rewrite of S3D for Titan was to move computation blocks and communication blocks around to maximize as much overlap of computation and communication as possible. Within each of the computation blocks, we were able to have high-granularity threading and excellent vectorization for the GPU. The threading also aided execution on the x86 systems at that time.

Next we gather statistics on the refactored application on Broadwell and KNL. Figure 9.8.2 is the same profile comparison for the refactored S3D. The total times for both versions on both systems seen in Table 9.8.3 gives a 58% speedup on Broadwell and 95% speedup on KNL; Broadwell still beats KNL. The cache analysis for the refactored version seen in Table 9.8.4 is somewhat better, but still not the best.

```
#ifdef GPU
!$acc parallel loop gang collapse(2) private(n,i,j,k)
#else
!$omp parallel do private(n,i,j,k)
#endif
  do n=1,n_spec
    do k = 1,nz
#ifdef GPU
!$acc loop vector collapse(2)
#endif
      do j = 1,ny
        do i = 1,nx
          grad_Ys(i,j,k,n,1) = 0.0
          grad_Ys(i,j,k,n,2) = 0.0
          grad_Ys(i,j,k,n,3) = 0.0
          if(i.gt.iorder/2 .and. i.le.nx-iorder/2) then
             grad_Ys(i,j,k,n,1) = scale1x(i)*(
                   aex *( yspecies(i+1,j,k,n)-yspecies(i-1,j,k,n) )   &
                 + bex *( yspecies(i+2,j,k,n)-yspecies(i-2,j,k,n) )   &
                 + cex *( yspecies(i+3,j,k,n)-yspecies(i-3,j,k,n) )   &
                 + dex *( yspecies(i+4,j,k,n)-yspecies(i-4,j,k,n) ))
          endif
          if(j.gt.iorder/2 .and. j.le.ny-iorder/2) then
             grad_Ys(i,j,k,n,2) = scale1y(j)*(
                   aey *( yspecies(i,j+1,k,n)-yspecies(i,j-1,k,n) )   &
                 + bey *( yspecies(i,j+2,k,n)-yspecies(i,j-2,k,n) )   &
                 + cey *( yspecies(i,j+3,k,n)-yspecies(i,j-3,k,n) )   &
                 + dey *( yspecies(i,j+4,k,n)-yspecies(i,j-4,k,n) ))
          endif
          if(k.gt.iorder/2 .and. k.le.nz-iorder/2) then
             grad_Ys(i,j,k,n,3) = scale1z(k)*(
                   aez *( yspecies(i,j,k+1,n)-yspecies(i,j,k-1,n) )   &
                 + bez *( yspecies(i,j,k+2,n)-yspecies(i,j,k-2,n) )   &
                 + cez *( yspecies(i,j,k+3,n)-yspecies(i,j,k-3,n) )   &
                 + dez *( yspecies(i,j,k+4,n)-yspecies(i,j,k-4,n) ))
          endif
      end do ; end do ! i, j
#ifdef GPU
!$acc end loop
#endif
  end do ; end do ! k, n
#ifdef GPU
!$acc end parallel loop
#endif
```

Excerpt 9.8.2 Second optimization applied to example S3D code.

TABLE 9.8.3 Runtime of original and refactored S3D code on Broadwell and KNL.

Version	Broadwell	KNL
Original	2457	4128
Refactored	1554	2120

We then ran the same size problem using strong scaling up to 128 nodes of Broadwell and KNL. Broadwell continues to out-perform KNL as both systems are scaling very well . This particular set of runs used all-MPI, even though S3D can do threading. Figure 9.8.3 shows those results.

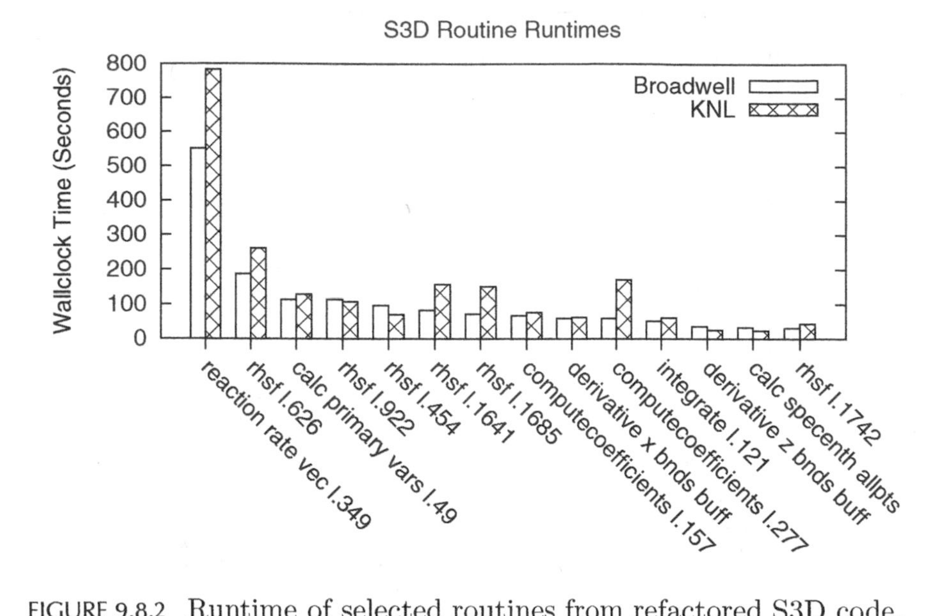

FIGURE 9.8.2 Runtime of selected routines from refactored S3D code.

TABLE 9.8.4 Cache utilization of refactored S3D code.

```
D1 + D2 cache utilization:

    46.8% of total execution time was spent in 8 functions with combined
    D1 and D2 cache hit ratios below the desirable minimum of 80.0%.
    Cache utilization might be improved by modifying the alignment or
    stride of references to data arrays in these functions.

     D1+D2    Time%  Function
     cache             PE=HIDE
       hit
     ratio

      58.1%     8.9%  rhsf_.LOOP@li.626
      67.4%    26.1%  reaction_rate_vec_.LOOP@li.349
      74.6%     1.3%  s3d_
      76.3%     1.7%  derivative_z_bnds_buff_r_
      76.4%     2.8%  derivative_x_bnds_buff_r_
      77.5%     1.6%  calc_specenth_allpts_r$thermchem_m_
      78.9%     1.9%  rhsf_.LOOP@li.1742
      79.6%     2.6%  integrate_.LOOP@li.121
```

There are two notable vectorizations that were performed in the port to the Titan system. First is the iteration loop within the CALC_TEMP routine that computes a new temperature and uses an iterative scheme to converge to a better temperature. Excerpt 9.8.3 shows the loop of interest.

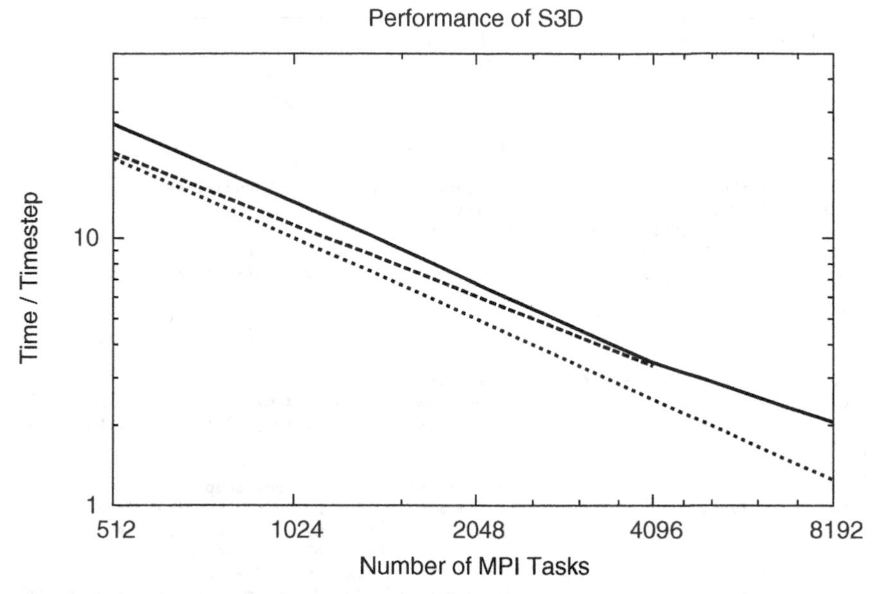

FIGURE 9.8.3 Performance of S3D on KNL and Broadwell.

```
277.  + 1----------<      do k = 1, nz
278.  + 1 2--------<         do j = 1, ny
279.  + 1 2 3------<            do i = 1, nx
281.    1 2 3                      icount = 1
282.    1 2 3                      r_gas = Ru*avmolwt(i,j,k)
283.  + 1 2 3 r8--<>               yspec(:) = ys(i, j, k, :)
285.  + 1 2 3 4----<               ITERATION: do
288.  + 1 2 3 4                       cpmix(i,j,k) =   mixCp( yspec, temp(i,j,k) )
289.  + 1 2 3 4                       enthmix      = mixEnth( yspec, temp(i,j,k) )
291.    1 2 3 4                     !-- calculate deltat, new temp
292.    1 2 3 4                     !   remember tmp1 holds the internal energy
293.    1 2 3 4           deltat = ( tmp1(i,j,k) - (enthmix-r_gas*temp(i,j,k)) )&
294.    1 2 3 4                              /( cpmix(i,j,k) - r_gas )
295.    1 2 3 4                       temp(i,j,k) = temp(i,j,k) + deltat
297.    1 2 3 4                     !-- check for convergence
298.    1 2 3 4                       if( abs(deltat) < atol ) then  ! converged
300.  + 1 2 3 4                       cpmix(i,j,k) =   mixCp( yspec, temp(i,j,k) )
301.    1 2 3 4                          exit
302.    1 2 3 4                       elseif( icount > icountmax ) then
303.    1 2 3 4                          write(6,*)'calc_temp cannot converge
304.    1 2 3 4                          write(6,*) 'for processor with rank
305.    1 2 3 4                          write(6,*) 'i=',i
306.    1 2 3 4                          write(6,*) 'j=',j
307.    1 2 3 4                          write(6,*) 'k=',k
308.    1 2 3 4                          stop
309.    1 2 3 4                       else
310.    1 2 3 4                          icount = icount + 1
311.    1 2 3 4                       endif
312.    1 2 3 4---->               enddo ITERATION
313.    1 2 3---->>>      enddo ; enddo ; enddo
```

Excerpt 9.8.3 Original example loop from S3D before vectorization.

```
305.  + 1-----------------<       ITERATION: do
306.    1                     !dir$ concurrent
307.    1                     #ifdef GPU
308.    D                     !$acc loop vector private(n,mm)
309.    1                     #endif
310.    1 fVcr2-----------<     do m = ml, mu
311.    1 fVcr2
312.    1 fVcr2                   n = int((temp(m)-temp_lobound)*invEnthInc)+1
313.    1 fVcr2                   cpmix(m) = 0.0
314.  + 1 fVcr2 fVpr2-----<       do mm=1,n_spec
315.    1 fVcr2 fVpr2                cpmix(m) = cpmix(m) +  &
316.    1 fVcr2 fVpr2                    ys(m,mm)*(cpCoef_aa(mm,n) * temp(m) + &
                                  cpCoef_bb(mm,n) )
317.    1 fVcr2 fVpr2----->       enddo
318.    1 fVcr2                   n = int((temp(m)-temp_lobound)*invEnthInc)+1
319.    1 fVcr2                   enthmix(m) = 0.0
320.    1 fVcr2 f---------<       do mm=1,n_spec
321.    1 fVcr2 f                    enthmix(m) = enthmix(m) + &
                            ys(m,mm)*(enthCoef_aa(mm,n)*temp(m) + enthCoef_bb(mm,n))
322.    1 fVcr2 f--------->       enddo
323.    1 fVcr2
324.    1 fVcr2                    !-- calculate deltat, new temp
325.    1 fVcr2                    !   remember tmp1 holds the internal energy
326.    1 fVcr2                   deltat(m) = ( tmp1(m) - (enthmix(m)-   &
                                               Ru*avmolwt(m)*temp(m)) )  &
327.    1 fVcr2                                /( cpmix(m) - Ru*avmolwt(m) )
328.    1 fVcr2                   if(iconverge(m).eq.0)temp(m) = temp(m) + deltat(m)
329.    1 fVcr2----------->     enddo
330.    1                     #ifdef GPU
331.    D                     !$acc end loop
332.    1                     #endif
333.    1                     #ifdef GPU
334.    D                     !$acc loop vector private(n,mm)
335.    1                     #endif
336.  + 1 f---------------<     do m = ml, mu
337.    1 f                        !-- check for convergence
338.    1 f                       if( abs(deltat(m)) < atol.and.iconverge(m).eq.0 )
341.    1 f                           iconverge(m) = 1
342.    1 f                           n = int((temp(m)-temp_lobound)*invEnthInc)+1
343.    1 f                           cpmix(m) = 0.0
344.  + 1 f Vpr6----------<           do mm=1,n_spec
345.    1 f Vpr6                         cpmix(m) = cpmix(m) +  &
346.    1 f Vpr6                  ys(m,mm)*(cpCoef_aa(mm,n) * temp(m) + cpCoef_bb(mm,n) )
347.    1 f Vpr6---------->           enddo
348.    1 f                       endif
349.    1 f--------------->     enddo
350.    1                     #ifdef GPU
351.    D                     !$acc end loop
352.    1                     #endif
353.  + 1                            if(all(iconverge(ml:mu).eq.1))EXIT ITERATION
354.  + 1 2---------------<          do m = ml,mu
355.    1 2                          if(iconverge(m).eq.0)then
356.    1 2                            if( icount(m) > icountmax ) then
357.    1 2                              write(6,*)'calc_temp cannot converge
358.    1 2                              write(6,*) 'for processor with rank =',myid
359.    1 2                              write(6,*) 'm=',m
360.    1 2                              stop
361.    1 2                            else
362.    1 2                              icount(m) = icount(m) + 1
363.    1 2                            endif
364.    1 2                          endif
365.    1 2--------------->          enddo
366.    1----------------->       enddo ITERATION
```

Excerpt 9.8.4 Refactored example loop from S3D after vectorization.

Notice that the grid loops are inside this routine. In the refactored version we pull in a chunk of the grid loops as discussed previously. In this routine, the compilers cannot do any optimization due to the ITERATION loop inside the grid loops. Can we pull the grid loop inside of the ITERATION loop? To do this, we would have to maintain a mask array that indicates when the temperature converges, and continue to update until all the elements within the chunk of the array converges.

Excerpt 9.8.4 is the rewrite that vectorizes nicely and delivers a good performance increase. Fortunately the number of iterations that is typically required for convergence is less than 5. If the number of iterations were much larger, then this rewrite would have to be done differently or left to run in scalar mode. Notice the use of iconverge which is set to 1 when the zone converges. To get the same answers, it is important to stop updating TEMP once a zone converges.

```
1643.+Mm--------<   SPECIES62: do n=1,n_spec-1
1644.+Mm 2------<             do k = 1, nz
1645. Mm 2        #ifdef GPU
1646. MD          !$acc loop vector collapse(2) private(diffx,diffy,diffz)
1647. Mm 2        #endif
1648.+Mm 2 3----< do j = 1, ny
1649.+Mm 2 3 r2-<  do i = 1, nx
1650. Mm 2 3 r2      diffusion(i,j,k,n) = 0.0d0 ! RG because deriv routines accumulate
1651. Mm 2 3 r2      diffx=0
1652. Mm 2 3 r2      diffy=0
1653. Mm 2 3 r2      diffz=0
1654. Mm 2 3 r2      if(i.gt.iorder/2 .and. i.le.nx-iorder/2)                         &
1655. Mm 2 3 r2      diffx = (scale1x(i)*(                                            &
                                aex *( diffFlux(i+1,j,k,n,1)-diffFlux(i-1,j,k,n,1) )  &
1656. Mm 2 3 r2                + bex *( diffFlux(i+2,j,k,n,1)-diffFlux(i-2,j,k,n,1) )  &
1657. Mm 2 3 r2                + cex *( diffFlux(i+3,j,k,n,1)-diffFlux(i-3,j,k,n,1) )  &
1658. Mm 2 3 r2                + dex *( diffFlux(i+4,j,k,n,1)-diffFlux(i-4,j,k,n,1) )))
1659. Mm 2 3 r2      if(j.gt.iorder/2 .and. j.le.ny-iorder/2)                         &
1660. Mm 2 3 r2      diffy = (scale1y(j)*(                                            &
                                aey *( diffFlux(i,j+1,k,n,2)-diffFlux(i,j-1,k,n,2) )  &
1661. Mm 2 3 r2                + bey *( diffFlux(i,j+2,k,n,2)-diffFlux(i,j-2,k,n,2) )  &
1662. Mm 2 3 r2                + cey *( diffFlux(i,j+3,k,n,2)-diffFlux(i,j-3,k,n,2) )  &
1663. Mm 2 3 r2                + dey *( diffFlux(i,j+4,k,n,2)-diffFlux(i,j-4,k,n,2) )))
1664. Mm 2 3 r2      if(k.gt.iorder/2 .and. k.le.nz-iorder/2)                         &
1665. Mm 2 3 r2      diffz = (scale1z(k)*(                                            &
                                aez *( diffFlux(i,j,k+1,n,3)-diffFlux(i,j,k-1,n,3) )  &
1666. Mm 2 3 r2                + bez *( diffFlux(i,j,k+2,n,3)-diffFlux(i,j,k-2,n,3) )  &
1667. Mm 2 3 r2                + cez *( diffFlux(i,j,k+3,n,3)-diffFlux(i,j,k-3,n,3) )  &
1668. Mm 2 3 r2                + dez *( diffFlux(i,j,k+4,n,3)-diffFlux(i,j,k-4,n,3) )))
1669. Mm 2 3 r2      diffusion(i,j,k,n) = -diffx-diffy-diffz
1670. Mm 2 3 r2->  enddo
1671. Mm 2 3----> enddo
1672. Mm 2        #ifdef GPU
1673. MD          !$acc end loop
1674. Mm 2        #endif
1675. Mm 2------>             enddo
1676. Mm-------->   enddo SPECIES62
```

Excerpt 9.8.5 Refactored RHSF loop from S3D showing combined computations.

The next example is illustrative of the idea of moving computation around to achieve good cache reuse and maximize the computational intensity. The loop presented in Excerpt 9.8.5 is from the refactored RHSF routine and it represents combining computations from several derivative routines computing the diffusion that updates the X, Y, and Z directions separately. The advantage here is that diffFlux has tremendous reuse; when the three components of the diffusion were computed separately, the diffFlux arrays had to be transversed three times.

```
    isync = 0
     do i = 1, reqcount
        call MPI_WAITANY(reqcount,req,index, stat, ierr )
if(direction(index).eq.1)then
#ifdef GPU
!$acc update device(pos_f_x_buf(:,:,:,idx(index):idx(index)+nb(index)-1)) async(isync)
#endif
endif
if(direction(index).eq.2)then
#ifdef GPU
!$acc update device(neg_f_x_buf(:,:,:,idx(index):idx(index)+nb(index)-1)) async(isync)
#endif
endif
if(direction(index).eq.3)then
#ifdef GPU
!$acc update device(pos_f_y_buf(:,:,:,idx(index):idx(index)+nb(index)-1)) async(isync)
#endif
endif
if(direction(index).eq.4)then
#ifdef GPU
!$acc update device(neg_f_y_buf(:,:,:,idx(index):idx(index)+nb(index)-1)) async(isync)
#endif
endif
if(direction(index).eq.5)then
#ifdef GPU
!$acc update device(pos_f_z_buf(:,:,:,idx(index):idx(index)+nb(index)-1)) async(isync)
#endif
endif
if(direction(index).eq.6)then
#ifdef GPU
!$acc update device(neg_f_z_buf(:,:,:,idx(index):idx(index)+nb(index)-1)) async(isync)
#endif
endif
isync=isync+1
    enddo
```

Excerpt 9.8.6 Example MPI optimization in S3D code.

Once all of the computation was improved, scaling to a higher number of nodes on Titan was desirable, so the MPI message passing had to be improved significantly. Once again, the reorganization of the computation and communication allowed us to have numerous messages in flight at the same time, and since this is a three-dimensional application, the nearest neighbors in the logical sense may not be close physically on the interconnect. First all the sends were accomplished using MPI_ISEND, and then the code for receiving the messages was rewritten in the form shown in Excerpt 9.8.6.

The interesting aspect of this logic is that the `MPI_TASK` waits for any message and then, given the header information, it puts the data directly into the GPU memory asynchronously. If S3D is running on the non-GPU platform, the data just goes into the appropriate buffer directly. This helped overlap the receives and eliminate at lot of `MPI_WAIT` time. However, to scale to the full extent of the Titan system (>15,000 nodes), we had to ensure that the three-dimensional decomposition was laid out on the 3D torus in such a way that the nearest neighbors were indeed close to each other on the network. Consequently, the weak scaling benchmarks showed perfect scaling.

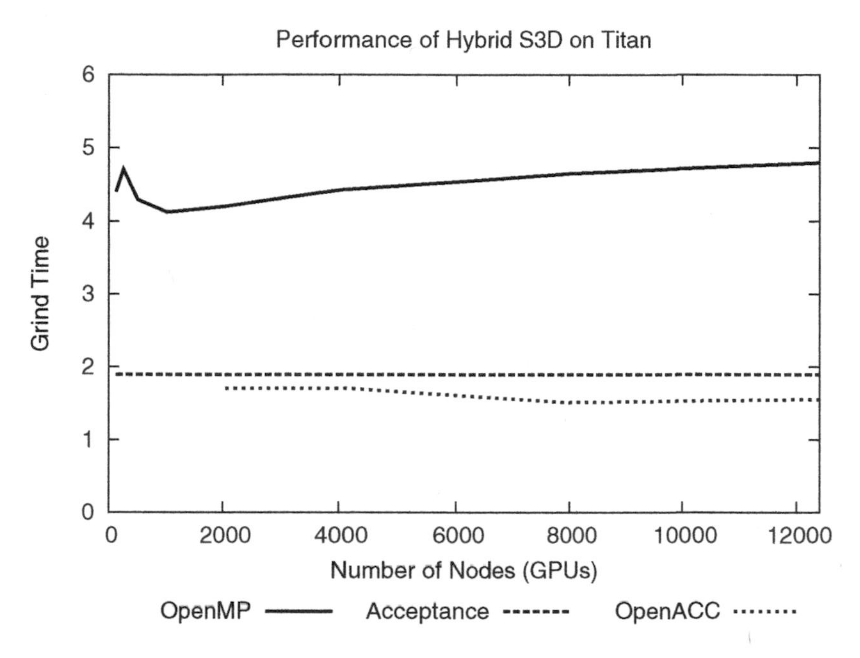

FIGURE 9.8.4 Scaling performance of hybrid S3D on KNL and Broadwell.

Figure 9.8.4 illustrates the acceptance criteria we were shooting for and both the runs using the GPU (OpenACC) and running using OpenMP on the host.

9.9 EXERCISES

9.1 Enumerate steps to be taken when approaching the task of porting/optimizing a large application to a new architecture. For each task, itemize what portions of the book address those steps.

9.2 Consider the `restrict` keyword as used in Excerpt 9.3.23. How could things be improved here?

CHAPTER 10

Future Hardware Advancements

CONTENTS

10.1 INTRODUCTION

This chapter examines potential future hardware advancements, and how the application developer may prepare for those hardware advancements. Note that this chapter contains forward-looking statements which come with some amount of speculation and uncertainty.

10.2 FUTURE X86 CPUS

While non-x86 architectures such as Power and SPARC do see some success in the HPC market, current systems primarily utilize x86 CPUs. There is some indication that this may begin to change in the future, but x86 is expected to continue to be the dominant ISA in HPC for many years. Intel currently has the majority market share in HPC, though AMD did particularly well with their Opteron line for some time, even claiming the number one spot on the TOP500 list with the DoE's Titan supercomputer in 2012. Titan still takes third place on this list as of June 2016. However, Titan is fairly old at this point, and the majority of new systems being sold use Intel CPUs. In this section, two new Intel microarchitectures will be examined, one each from Intel's Xeon and Xeon Phi product lines. Additionally, AMD's new Zen microarchitecture will be examined in more detail.

10.2.1 Intel Skylake

Skylake is expected to be built on a 14 nm manufacturing process and come with more cores than previous Intel Xeon CPUs. However, one of the most important features coming with the new Skylake chips is the addition of support for the AVX-512 instruction set (subsets F, CD, VL, BW, and DQ). Previously, this SIMD ISA was only available on the Xeon Phi line (e.g., KNL). The longer vector width supported by AVX-512 means many application developers will need to pay particularly close attention to the vectorization of their codes in order to take full advantage of the increased performance available from the new chips. Optimizing for wide vectors will no longer be applicable primarily to accelerators like GPUs and to a lesser extent the Xeon Phi; even the traditional Xeon line will require vectorization work for maximum performance. Additional microarchitecture details are provided by the Intel optimization manual [15].

10.2.2 AMD Zen

AMD largely left the HPC and server market to Intel after failing to make significant upgrades to the Opteron line of CPUs for quite some time. However, after a period of relative stagnation, AMD has announced a new microarchitecture named Zen. AMD claims Zen is a "from scratch" microarchitecture and a significant departure from previous designs. AMD announced that Zen has a 40% instructions per clock (IPC) improvement over earlier architectures and also features 2-way simultaneous multi-threading (SMT), a first for AMD CPUs. The new Zen chips are expected to be built on a 14 nm manufacturing process by Globalfoundries and/or TSMC. The Zen model targeting the server and HPC markets is expected to have up to 32 cores and support up to 8 channels of DDR4. If new hardware arrives with the expected features, AMD might just have a winning server, or even HPC, CPU. That said, there is some concern about Zen's suitability for floating-point heavy HPC codes.

Zen supports the AVX2 ISA with a vector length of 256 bits (32 bytes). However, each core only supports two 16-byte loads from and one 16-byte store to the level-1 cache per cycle. This means that there is not enough bandwidth between the core and the level-1 cache to fully feed the SIMD units at a rate of 1 FMA per clock. These load/store concerns aside, Zen is expected to come with more cores, wider SMT, and wider vectors than previous AMD designs.

10.3 FUTURE ARM CPUS

ARM is starting a significant push from the embedded and mobile space into the server and HPC markets. While still in the early stages, this push has been making slow and steady progress. ARM's historical focus on low power solutions with high energy efficiency is not a terrible fit for the HPC and server space. However, some improvement over traditional ARM designs may be needed in terms of raw performance. Keep in mind that an HPC system consists of more than just CPUs. Slower, more efficient cores means a longer time to solution, which adds overhead from the rest of the system. Additionally, strong scalar performance may be needed for codes which do not vectorize well. Such designs with powerful, deeply out-of-order cores are largely left to the ARM ecosystem and silicon design partners to create. That is, ARM designs the ISA and provides model CPU implementations, then licenses this IP to other designers and silicon partners to produce physical implementations of the ISA – ARM sells IP, not CPUs. To enable silicon vendors to create competitive HPC CPUs, ARM is designing ISA extensions targeting the HPC and server markets and working with partners to create CPUs implementing the new ISA. The newest such ISA extension is called "Scalable Vector Extension" (SVE).

10.3.1 Scalable Vector Extension

In 2016 at the Hot Chips 28 conference in Cupertino, ARM announced an extension to the ARMv8 ISA, the "Scalable Vector Extension" (SVE) [28]. This extension is an important part of ARM's efforts to expand into the HPC and server markets, providing a powerful new vectorization capability to the ARM ecosystem. This SIMD extension supplements the existing ARM NEON vector instructions, providing instructions primarily tailored to the needs of HPC applications, though many of the instructions are also valuable to the wider server market. A number of common SIMD ISA features are compared along with SVE in Table 10.3.1.

An important aspect of this new "scalable" extension is its support for vector length agnosticism – that is, properly written code will work on a large range of vector lengths without any modification. Specifically, SVE supports any vector length from 128 to 2048 bits in increments of 128 bits. This is important for the ARM ecosystem, as silicon partners design implementations for a diverse market; small embeded ARM processors may not need as large of a vector unit as HPC CPUs. Additionally, the scalable nature of SVE allows application code to be more future-proof than similar x86 code. Historically, every time the x86 ISA has been extended to support a larger SIMD width, a new set of instructions has been created to expose this functionality to users and compilers. In order to take advantage of the increase in SIMD processing power, x86 application codes need to be recompiled, even rewritten in the case of hand-tuned ASM and/or library code. In making SVE vector length agnostic, properly constructed code can automatically take advantage of new hardware providing longer vector lengths and increased performance.

TABLE 10.3.1 Key features of common SIMD ISAs.

Arch	SIMD ISA	Max Width	Predication	Gather	Scatter
x86	AVX2	256 bits		✓	
	AVX-512	512 bits	✓	✓	✓
ARM	NEON	128 bits			
	SVE	2048 bits	✓	✓	✓
Power	VMX2	128 bits			

Another important aspect of SVE is its pervasive support for predication. Predication support helps the compiler to vectorize loops containing conditional statements which would not otherwise be profitable to vectorize. That is, a large number of SVE instructions support a vector mask operand, instructing the CPU to only execute the instruction's operation (e.g., `ADD`, `MUL`, `XOR`) on those vector elements with a 1 in their associated position in the mask vector. While support for predicate masks is not unique to the SVE ISA, the ISA was designed from the beginning with predication in mind and has very good support for this feature; SVE not only provides more instruction forms accepting a predicate mask than AVX-512, but also provides many more mask registers than AVX-512 (which only provides 8: `k0 - k7`). A concrete example of the importance of predicate support is presented in Excerpts 10.3.1 and 10.3.2, showing a loop from LULESH (Livermore Unstructured Lagrangian Explicit Shock Hydrodynamics) [17].

As can be seen in Excerpt 10.3.1, the Intel Haswell does not support predication with AVX2, and the loop does not vectorize. The Cray Compiler Environment (CCE) includes information in the compiler listing output file describing what is preventing the compiler from vectorizing the loop. While this summary description does indicate there is an issue with conditional code on line 295 in the example loop, more information is available through the `explain CC-6339` command. A more detailed description of the `CC-6339` message directly points toward predicate support being the compiler's desired solution, as seen in Table 10.3.2.

Excerpt 10.3.2 shows that the example loop from LULESH is able to be vectorized by CCE on the Intel Knight's Landing with AVX-512. The eight predicate registers provided by the ISA are sufficient to avoid the potential hazards in the conditional code and allow this loop to be vectorized. In addition to the loop-mark indication that the loop was vectorized (V – vectorized), a message is also included in the compiler listing output file.

```
287.                    #pragma ivdep
288.  +    F------<       for (long i = 0 ; i < length ; ++i) {
289.       F                 Real_t vhalf = Real_t(1.) / (Real_t(1.) + compHalfStep[i]);
290.       F
291.       F                 if ( delvc[i] > Real_t(0.) ) {
292.       F                   q_new[i] /* = qq_old[i] = ql_old[i] */ = Real_t(0.);
293.       F                 }
294.       F                 else {
295.       F                   Real_t ssc = (pbvc[i] * e_new[i]
296.       F                            + vhalf * vhalf * bvc[i] * pHalfStep[i]) / rho0;
297.       F
298.       F                   if ( ssc <= Real_t(.1111111e-36) ) {
299.       F                     ssc = Real_t(.3333333e-18);
300.       F                   } else {
301.       F I                   ssc = SQRT(ssc);
302.       F                   }
303.       F                   q_new[i] = (ssc*ql_old[i] + qq_old[i]);
304.       F                 }
305.       F                 e_new[i] = e_new[i] + Real_t(0.5) * delvc[i] *
306.       F                   ( Real_t(3.0)*(p_old[i]+q_old[i]) -
307.       F                     Real_t(4.0)*(pHalfStep[i]+q_new[i]) );
308.       F------>       }

CC-6339 CC: VECTOR CalcEnergyForElems, File = lulesh_kernel.cc, Line = 288
  Loop was not vectorized because of a potential hazard in conditional code on line 295.
```

Excerpt 10.3.1 No vectors with LULESH loop in AVX2 without predication.

TABLE 10.3.2 Detailed description of CC-6339 message from explain command.

```
VECTOR:  A loop was not vectorized because of a potential hazard in
conditional code on line num.

Vectorization of conditional code on architectures that lack
predicated execution requires careful manipulation to avoid
triggering floating point exceptions or memory segmentation faults.
The line identified by this message has one such hazard, and at
present it inhibits vectorization of the entire loop.  This
restriction may be removed in a future compiler release.
```

Robust support for predication, gather/scatter, and other vector ISA features like horizontal reductions, make SVE a relatively easy target for compilers when it comes to vectorizing real-world codes.

```
287.                     #pragma ivdep
288.  +     VFr2----<     for (long i = 0 ; i < length ; ++i) {
289.        VFr2            Real_t vhalf = Real_t(1.) / (Real_t(1.) + compHalfStep[i]);
290.        VFr2
291.        VFr2            if ( delvc[i] > Real_t(0.) ) {
292.        VFr2              q_new[i] /* = qq_old[i] = ql_old[i] */ = Real_t(0.);
293.        VFr2            }
294.        VFr2            else {
295.        VFr2              Real_t ssc = (pbvc[i] * e_new[i]
296.        VFr2                      + vhalf * vhalf * bvc[i] * pHalfStep[i]) / rho0;
297.        VFr2
298.        VFr2              if ( ssc <= Real_t(.1111111e-36) ) {
299.        VFr2                ssc = Real_t(.3333333e-18);
300.        VFr2              } else {
301.        VFr2 I              ssc = SQRT(ssc);
302.        VFr2              }
303.        VFr2              q_new[i] = (ssc*ql_old[i] + qq_old[i]);
304.        VFr2            }
305.        VFr2            e_new[i] = e_new[i] + Real_t(0.5) * delvc[i] *
306.        VFr2              ( Real_t(3.0)*(p_old[i]+q_old[i]) -
307.        VFr2                Real_t(4.0)*(pHalfStep[i]+q_new[i]) );
308.        VFr2---->     }

CC-6204 CC: VECTOR CalcEnergyForElems, File=lulesh_kernel.cc, Line=288
  A loop was vectorized.
```

Excerpt 10.3.2 LULESH loop vectorizes with predication with AVX-512.

10.3.2 Broadcom Vulcan

Broadcom has been talking about creating a new ARMv8 core and SoC as early as the 2013 Linley Tech Processor Conference [4], [5]. These new "Vulcan" cores are to be quad-issue, have 4-way SMT, and support deep superscalar out-of-order execution. The new cores are expected to be manufactured on a 16 nm process; if the expected 2.5 GHz to 3 GHz is able to be reached, Vulcan would be quite competitive from a performance perspective. Broadcom intends for Vulcan to have stronger cores than many other ARM designs, with something like 90% the performance of Intel's Haswell core. If expected performance levels are met, it would mean one of the most common criticisms of ARM CPUs in the HPC space, low per-core performance, has been successfully addressed. That said, there is some concern about the future of this product. Rumors started appearing in the trade rags during the second half of 2016 claiming that Broadcom decided to drop the product line after the company was acquired by Avago [6], [7], [30]. While there isn't much more information available at this time, industry rumors suggest Vulcan has found a buyer and will see the light of day after all; Chris Williams from The Register has stated that some suggest, "Cavium is interested in buying the Vulcan blueprints from Broadcom" [30]. Unsupported rumors aside, this kind of market churn is the current reality for the HPC ARM ecosystem. The market is transitioning out of the very early initial stages, where there are lots of players and the winners have not yet been chosen. Presently, and into the near future, there will be a consolidation of the players in this market as it matures.

10.3.3 Cavium Thunder X

Cavium is one of the early entrants to the emerging ARM server and HPC markets. Cavium reworked an existing MIPS CPU design, adapting it to the ARM ISA. Their 48-core ARMv8 Thunder X SoC series features a relatively high number of relatively weak, dual-issue cores with limited out-of-order support. All cores share a centralized 16 MB level-2 cache running at full core speed along with up to four channels of DDR4. Each Thunder X core can generally be thought of as being a good bit weaker than an ARM Cortex-A57 reference design. While Cavium's Thunder X line deserves some amount of coverage, it is not expected to be the highest-performing or most popular ARM solution in the HPC space. Note that the initial Thunder X architecture appears to be markedly different than the recently announced high-end Thunder X2 architecture from Cavium, which reportedly has a stronger out-of-order core design.

10.3.4 Fujitsu Post-K

One of the ARM designs announced most recently is the new HPC chip introduced by Fujitsu at HotChips28 [32]. The RIKEN Advanced Institute for Computational Science in Japan selected Fujitsu to construct the Post-K supercomputer, expected in 2022 with a performance around 1,000 PFLOPS. Fujitsu announced their intention to design custom ARMv8 CPUs for this machine and is the first company to publicly announce plans to support the new SVE SIMD ISA. Use of the ARM ISA is a change for Fujitsu, as the previous K computer utilized SPARC64 VIIIfx CPUs. The primary reason given for the ISA change is the richness of the ARM software ecosystem. Indeed, there are many more chips supporting ARM sold than those supporting SPARC. This should allow the new Post-K computer to run a wider array of software applications and tools than would have been possible with the SPARC ISA. Along with the new ISA, more cores and wider vectors are expected.

10.3.5 Qualcomm Centriq

In December 2016, Qualcomm gave a demonstration of its new 48-core Centriq 2400 64-bit ARM SoC based on their new core code-named Falkor [26]. There aren't too many details to be found about this SoC or core yet, but some limited information is available. The Centriq 2400-series family of microprocessors is expected to be fabricated on a 10 nm FinFET process, which is actually a strong position to be in when compared to Intel's currently shipping 14 nm products. However, transistor size and other manufacturing process features are only part of the performance equation.

10.4 FUTURE MEMORY TECHNOLOGIES

As SoCs put increasing pressure on the memory system with ever-growing vector lengths as well as core counts, many application codes are becoming increasingly memory bound. Data movement is slow and power-hungry, to the extent that it is actually the largest obstacle to reaching exascale: it is estimated that using DDR to provide all the memory bandwidth needed for an exascale system would consume the entire power budget expected to be available, about 20 MW, and then some. New, efficient memory technologies are strongly needed. To improve performance and efficiency, the memory hierarchy is expanding in complexity, often with the addition of more cache levels and/or multiple tiers of main memory. It will not be uncommon to see small, fast high-bandwidth memory combined with large, relatively slow "capacity" memory. These new high-bandwidth memories will be almost exclusively based on die-stacking technologies.

10.4.1 Die-Stacking Technologies

Die-stacked memory consists of multiple silicon dies stacked on top of one another with through-silicon vias (TSV) and micro-bumps connecting them. Each such stack is generally composed of some number of memory dies stacked on top of a logic die. As with many things related to integrated circuit technologies, distance is the enemy of speed and energy efficiency. By placing the memory dies close together, faster and more efficient components can be created.

The two primary die-stacked memories applicable to the HPC market are HBM and HMC. HBM or "high-bandwidth memory" has a wide, parallel bus with a relatively low clockspeed. These HBM stacks need to be kept very close to the CPU cores due to the memory interface used. This is depicted in Figure 10.4.1 which shows not only a die-stacked memory, but also the silicon interposer used to tightly couple the SoC to the memory stack.

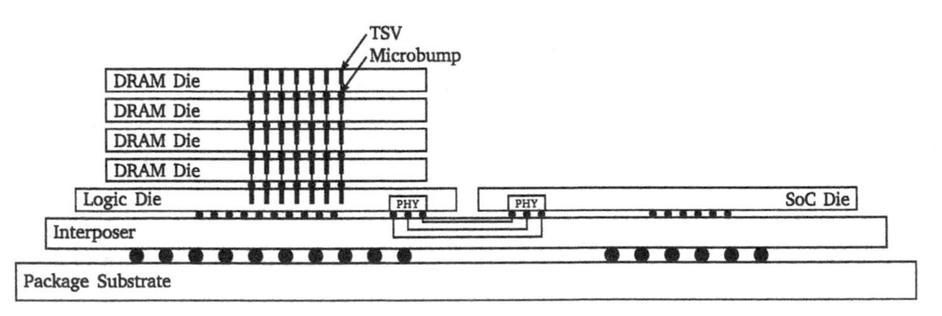

FIGURE 10.4.1 HBM die stack utilizing silicon interposer.

Note that this figure is not to scale. While the memory stacks may be four to eight dies high, the physical height of the stacks is still very small. In fact, the entire memory die stack may have a similar height as the SoC die itself. HMC or "hybrid memory cube" has a narrower, serial bus with a faster clock. The serial bus comes with a SerDes (Serializer/Deserializer) requirement and implies higher latency, but supports placing the memory stacks farther away from the CPU as well as the chaining of memory stacks. These new memory technologies have already been introduced with mainstream GPU and accelerator products from companies such as AMD, NVIDIA, and Intel. General purpose HPC and server CPUs are next in line, with products utilizing stacked memory technologies expected in the near future. Such faster, denser, and more energy-efficient memory technologies are needed to overcome the limitations of standard DDR. These two die-stacked memories are compared to traditional DDR4 in Table 10.4.1.

TABLE 10.4.1 Key performance numbers for DDR4, HBM2, and HMC3.

	Bandwidth (GB/s)	Latency (ns)	Energy (pJ/bit)	Cost
DDR4	26	85	60	1x
HBM2	256	75	7	2x
HMC3	320	130	20	6x

10.4.2 Compute Near Data

The introduction of a logic die on the bottom of memory die stacks provides an opportunity for the addition of compute near data (CND) features. Such capabilities are also commonly referred to as process in memory (PIM) and process near memory (PNM). The high-level idea behind CND is to send operations to the data, in or close to memory, rather than sending data to the operations, traditionally processed in the core. CND can help with memory clearing and initialization, computational reductions, streaming computation, and data rearrangement such as fine-grained and coarse-grained gathers and scatters. Data rearrangement is of particular interest here, as a memory system with CND support could present a reordered view of data to the rest of the chip, making non-contiguous data appear contiguous to the cache hierarchy and vector units. This can significantly improve cache utilization and vector performance, as things which would otherwise have required strided or sparse accesses are now able to be done contiguously. There are exciting possibilities in this area, but work is in early stages; CND technologies are still a reasonably long way away from becoming mainstream. For further reading on compute near data, consider: *Computational RAM: A Memory-SIMD Hybrid and Its Application to DSP* [11], *Processing in Memory: The Terasys Massively Parallel PIM Array* [14], *The Architecture of the DIVA Processing-In-Memory Chip* [9], and *TOP-PIM: Throughput-Oriented Programmable Processing In Memory* [33].

10.5 FUTURE HARDWARE CONCLUSIONS

While there is a wide range of new technologies coming in the future, these technologies are largely designed to address similar performance issues and highlight three primary trends:

1. More cores per SoC and more SMT threads per core.
2. Longer vector lengths with more flexible SIMD instructions.
3. Increasingly complex memory hierarchies.

10.5.1 Increased Thread Counts

The first primary trend that can be seen from looking at future hardware designs shouldn't seem all that surprising or new. As the number of transistors able to fit in a given area of silicon is still increasing (albeit at a slower pace than in the recent past) there is a need to find something productive to do with the additional transistors. While there may be some room for instruction level parallelism (ILP) increases in some of the smaller ARM cores, this opportunity isn't always available. Generally, codes have an inherent limit to the amount of ILP made available to the cores, and the performance per watt tradeoff from making larger cores to extract additional ILP is not a clear win. Instead, the additional transistors from a die shrink are increasingly used to add more cores. In fact, even in the cases when a core is enlarged in order to be able to extract more ILP, SMT is frequently added as well to increase efficiency. SMT helps in the case of a big out-of-order core, as one thread may be able to run on the core while another thread is stalled due to an inherent computational or memory dependency of the code. Most future HPC CPUs are expected to feature SMT to some degree, around 2- to 8-way SMT. In order to be prepared for these future hardware designs, application developers should work to ensure their codes are as multithreaded as is reasonably possible.

10.5.2 Wider Vectors

"Vose's Law – The vector width of general-purpose CPUs will double roughly every five years."

Aaron Vose
Cray's ORNL Center of Excellence

The second primary trend found in future hardware designs is presented in Figure 10.5.1. The figure shows that supported vector lengths of a number of common CPU microarchitectures have been increasing over time in a manner more or less consistent with Vose's Law.

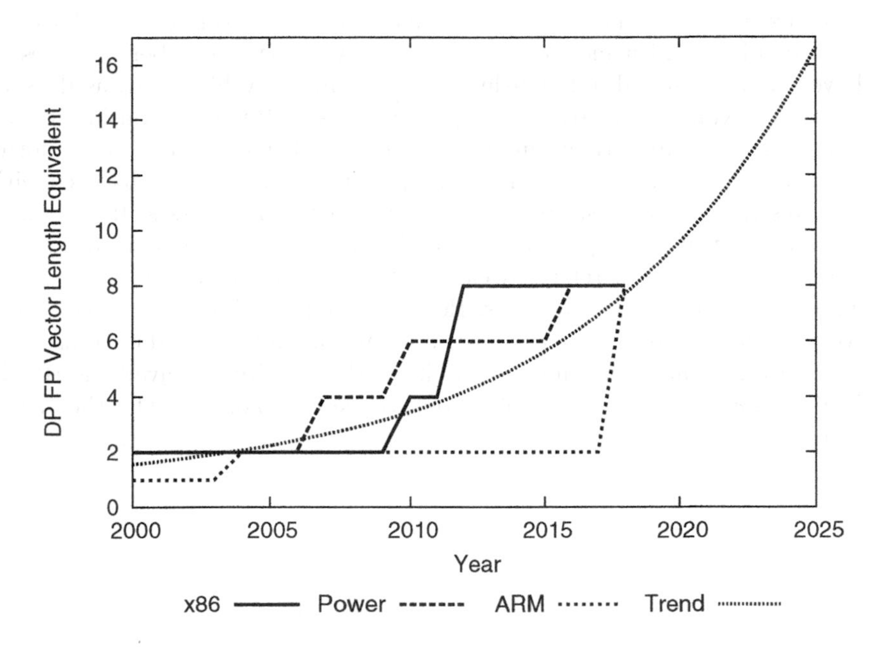

FIGURE 10.5.1 Twenty-five years of vector width increases for three common classes of CPU architecture implementations.

Note that this trend is describing the combined vector unit width (parallel double-precision floating-point equivalent) of physical CPUs, not the ISA itself – e.g., two 512-bit SIMD units are counted as a CPU vector width of 1024 bits. It might be tempting to credit ARM with a vector length of 2048 bits seeing as the SVE ISA supports this width. However, initial SVE hardware implementations are expected to arrive with combined SIMD unit widths smaller than the max the ISA supports.

Consider looking for opportunities to vectorize outer loops with longer trip counts. While it may be a little easier and more traditional to vectorize small inner loops, the trip counts of inner loops are often too small to make effective use of the vector units. Look for loops with a large number of iterations, which often means performing outer-loop vectorization. The growth of CPU vector widths is expected to continue into the foreseeable future, so optimizing application codes to be amenable to long vector lengths, where possible, is highly recommended.

10.5.3 Increasingly Complex Memory Hierarchies

Finally, the third trend found in future hardware designs is the increasing complexity of the memory hierarchy, as additional levels such as L4 caches and high-bandwidth memories are added. A growing number of codes will have a make-or-break relationship with this memory hierarchy, as data motion is becoming an increasingly significant issue. Begin by considering your applications computational intensity – the number of operations performed per array access. Also consider your application's data reuse distance, either in bytes between accesses or time in cycles between accesses. If the computational intensity is high enough, and the reuse distance is low enough, new hardware may provide relatively easy performance gains through longer vector widths and increased cache sizes. However, many codes have neither a large computational intensity nor a large amount of exploitable data reuse. Such codes are not cache-friendly and will instead be limited by available HBM or DDR bandwidth. For these applications, the key is to optimize for the memory hierarchy.

APPENDIX A

Supercomputer Cache Architectures

CONTENTS

A.1 ASSOCIATIVITY

A cache is a highly structured, expensive collection of memory banks. There are many different types of caches. A very important characteristic of the cache is associativity. In order to discuss the operation of a cache we need to envision the following memory structure. Envision that memory is a two-dimensional grid, where the width of each box in the grid is a cache line. The size of a row of memory is the same as one associativity class of the cache. Each associativity class also has boxes the size of a cache line. When a system has a direct-mapped cache, this means that there is only one associativity class. On the more recent x86 systems, the associativity of the level-1 cache is 8-way. This means that there are eight rows of associativity classes in level-1 cache. Now consider a column of memory. Any cache line in a given column of memory must be fetched to the corresponding column of the cache. The following discussion addresses a 2-way associative cache. While we are targeting an 8-way associative cache, it is easier to display a 2-way cache. In a 2-way associative cache, there are only two locations in level-1 cache for any of the cache lines in a given column of memory. Figure A.1.1 depicts the concept.

Figure A.1.1 depicts a 2-way associative level-1 cache. There are only two cache lines in level-1 cache that can contain any of the cache lines in the Nth column of memory. If a third cache line from the same column is required from memory, one of the cache lines contained in associativity class 1 or associativity class 2 must be flushed out of cache, typically to level-2 cache. Consider the loop in Excerpt A.1.1.

Let's assume that `A(1)` through `A(8)` are contained in the Nth column in memory. What is contained in the second row of the Nth column? The width of the cache is the size of the cache divided by its associativity. The size of this level-1 cache is 65,536 bytes and each associativity class has 32,768 bytes

Level-1 Cache:

$(N,1)$	$(N+1,1)$	$(N+2,1)$	$(N+3,1)$	$\cdots$
$(N,2)$	$(N+1,2)$	$(N+2,2)$	$(N+3,2)$	$\cdots$

Main Memory:

$(N,1)$	$(N+1,1)$	$(N+2,1)$	$(N+3,1)$	$\cdots$
$(N,2)$	$(N+1,2)$	$(N+2,2)$	$(N+3,2)$	$\cdots$
$(N,3)$	$(N+1,3)$	$(N+2,3)$	$(N+3,3)$	$\cdots$
$(N,\cdots)$	$(N+1,\cdots)$	$(N+2,\cdots)$	$(N+3,\cdots)$	$\cdots$

FIGURE A.1.1 A 2-way associative level-1 cache (column, associativity) and a main memory (column, row).

```
REAL*8 A(65536),B(65536),C(65536)

DO I=1,65536
  C(I) = A(I)+SCALAR*B(I)
ENDDO
```

Excerpt A.1.1 Example code showing cache associativity collision.

of locations. Since the array A contains 8 bytes per word, the length of the A array is 65536*8 = 131,072 bytes. The second row of the Nth column will contain A(4097-8192), the third A(8193-12288), and the sixteenth row will contain the last part of the array A. The seventeenth row of the Nth column will contain B(1)-B(4096) since the compiler will store the array B right after the array A, and C(1)-C(4096) will be contained in thirty-third row of the Nth column. Given the size of the dimensions of the arrays A, B, and C, the first cache line required from each array will be in the exact same column. We are not concerned with the operand SCALAR – this will be fetched to a register for the duration of the execution of the loop.

When the compiler generates the fetch for A(1), the cache line containing A(1) will be fetched to either associativity 1 or associativity 2 in the Nth column of level-1 cache. Then the compiler fetches B(1), and the cache line containing that element will go into the other slot in the Nth column of level-1 cache, with either associativity 1 or associativity 2. Figure A.1.2 illustrates the contents of level-1 cache after the fetch of A(1) and B(1).

The add is then generated. To store the result into C(1), the cache line containing C(1) must be fetched to level-1 cache. Since there are only two slots available for this cache line, either the cache line containing B(1) or the cache line containing A(1) will be flushed out of level-1 cache into level-2 cache. Figure A.1.3 depicts the state of level-1 cache once C(1) is fetched to cache.

Level-1 Cache:

$(N,1)$ `A(1-8)`	$(N+1,1)$	$(N+2,1)$	$(N+3,1)$	...
$(N,2)$ `B(1-8)`	$(N+1,2)$	$(N+2,2)$	$(N+3,2)$	...

FIGURE A.1.2 Contents of 2-way associative level-1 cache (column, associativity) after loading A and B.

Level-1 Cache:

$(N,1)$ `C(1-8)`	$(N+1,1)$	$(N+2,1)$	$(N+3,1)$	...
$(N,2)$ `B(1-8)`	$(N+1,2)$	$(N+2,2)$	$(N+3,2)$	...

FIGURE A.1.3 Contents of 2-way associative level-1 cache (column, associativity) after loading A, B, and C.

During the second pass through the loop, the cache line containing `A(2)` or `B(2)` will have to be fetched from level-2 cache. However, the access will over-write one of the other two slots, and we end up thrashing cache with the execution of this loop. Excerpt A.1.2 presents similar code which avoids this problem.

```
REAL*8 A(65544),B(65544),C(65544)

DO I=1,65536
  C(I) = A(I)+SCALAR*B(I)
ENDDO
```

Excerpt A.1.2 Example code avoiding cache associativity collisions.

The `A` array now occupies 16 full rows of memory plus one cache line. This causes `B(1)` through `B(8)` to be stored in the N+1 column of the seventeenth row and `C(1)-C(8)` is stored in the N+2 column of the thirty-third row. We have offset the arrays so they do not conflict in the cache. This storage therefore results in more efficient execution than the previous version. The next example investigates the performance impact of this rewrite. Figure A.1.4 depicts the status of level-1 cache given this new storage scheme.

Figure A.1.5 shows the performance of the code presented in Excerpt A.1.3 for various vector lengths on a Haswell processor with an eight-way associative cache. In these tests, the loop has more operands to illustrate an issue on the eight-way cache when a large number of operands are used. Once again the test is run at various vector lengths 1000 times. This is done to measure the performance when all the arrays are fully contained within the caches. If the trip count is greater than 910 (Equation A.1), the nine arrays cannot be contained within level-1 cache and we have some overflow into level-2 cache.

Level-1 Cache:

$(N, 1)$ A(1-8)	$(N+1, 1)$ B(1-8)	$(N+2, 1)$ C(1-8)	$(N+3, 1)$	$\cdots$
$(N, 2)$	$(N+1, 2)$	$(N+2, 2)$	$(N+3, 2)$	$\cdots$

FIGURE A.1.4 Contents of 2-way associative level-1 cache (column, associativity) after loading A, B, and C with restructured code.

```
REAL*8 A(65536),B(65536),C(65536)
REAL*8 D(65536),E(65536),F(65536)
REAL*8 G(65536),H(65536),I(65536)

DO I=1,65536
  A(i) = B(i) + 3.14 * C(i) +D(i)*E(i)+F(i)*G(i)+H(i)*P(i)
ENDDO
```

Excerpt A.1.3 Example code showing more cache associativity collisions.

$$\frac{65536 \text{ bytes in level-1 cache}}{8 \text{ bytes} * 9 \text{ operands}} = 910 \tag{A.1}$$

When the trip count is greater than 3640 (Equation A.2), then the nine arrays will not fit in level-2 cache, and the arrays will spill over into level-3 cache.

$$\frac{524588 \text{ bytes in level-2 cache}}{8 \text{ bytes} * 9 \text{ operands}} = 3640 \tag{A.2}$$

In Figure A.1.5, there is a degradation of performance as `N` increases, and this is due to where the operands reside prior to being fetched.

More variation exists than that attributed to increasing the vector length. There is a significant variation of performance of this kernel due to the size of each of the arrays. Each individual series indicates the dimension of the three arrays in memory. Notice that the series with the worse performance is dimensioned by 65536 `REAL(8)` words. The performance of this memory alignment is extremely bad because the three arrays are over-writing each other in level-1 cache as explained earlier. The next series which represents the case where we add a cache line to the dimension of the arrays gives slightly better performance. However, it is still poor. The third series once again gives slightly better performance and so on. The best performing series is when the arrays are dimensioned by 65536 plus a page (512 8-byte words).

The reason for this significant difference in performance is due to the memory banks in level-1 cache. When the arrays are dimensioned to be a large power of two, they are aligned in cache. As the three arrays are accessed, the accesses pass over the same memory banks and the fetching of operands stalls until the banks can refresh. When the arrays are offset by a full page of 4096 bytes, the banks have time to recycle before another operand is accessed.

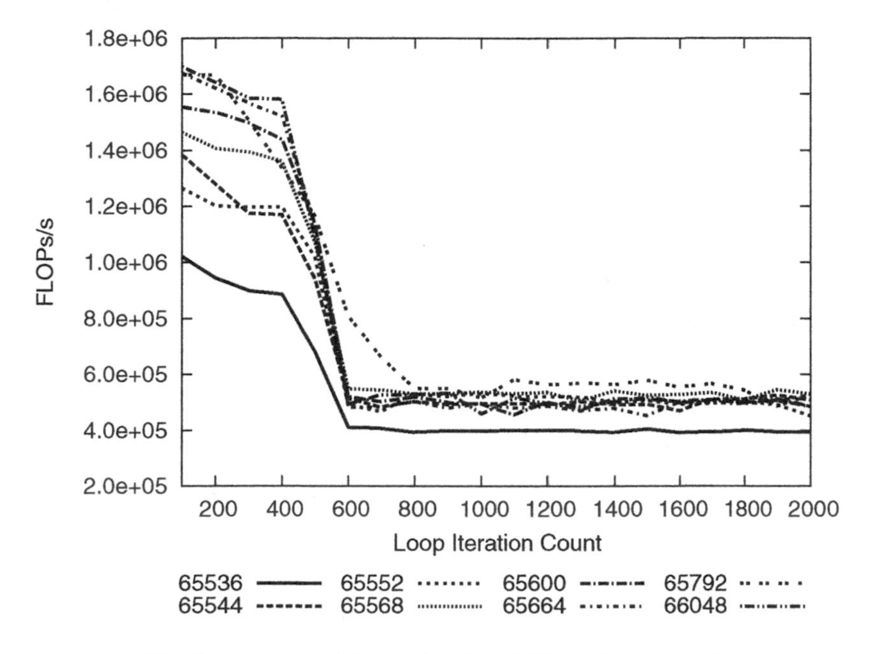

FIGURE A.1.5 Performance of kernel with different array sizes.

The lesson from this example is to pay attention to the alignment of your arrays. As will be discussed in the compiler section, the compiler can only do so much to "pad" arrays to avoid these alignment issues. The application programmer needs to understand how best to organize their arrays to achieve the best possible cache utilization. Memory alignment plays a very important role in effective cache utilization which is an important lesson a programmer must master when writing efficient applications.

APPENDIX B

The Translation Look-Aside Buffer

CONTENTS

B.1 INTRODUCTION TO THE TLB

The mechanics of the translation look-aside buffer (TLB) is the first important lesson in how to effectively utilize memory. When the processor issues an address for a particular operand, that address is a logical (virtual) address. Logical addresses are what the application references, and in the applications view, consecutive logical addresses are believed to be contiguous in memory. In practice this is not the case. Logical memory is mapped onto physical memory with the use of the TLB. The TLB contains entries that give the translation from the logical memory pages to pages within physical memory, as seen in Figure B.1.1. The default page size on most Linux systems is 4096 bytes. If one is storing 4-byte operands, there will be 1024 4-byte operands in a page, and the page holds 512 8-byte operands.

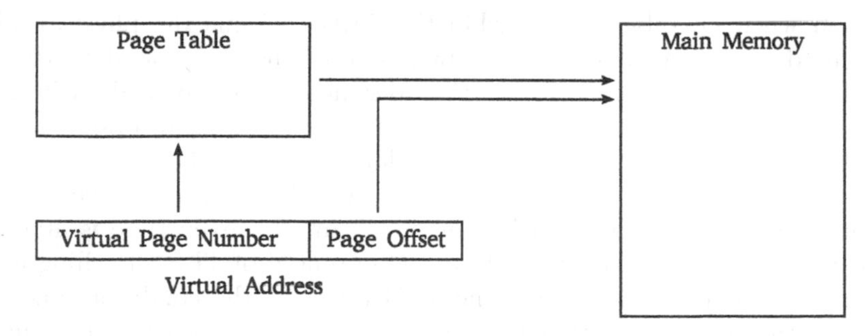

FIGURE B.1.1 Virtual to physical address translation with page table.

Physical memory can be fragmented, and two pages adjacent in logical memory may not be adjacent in physical memory, as shown in Figure B.1.2.

The mapping between the two, logical and physical, is performed by an entry in the TLB. The first time an operand is fetched to the processor, the page table entry must be fetched and the logical address translated to a physical address. Then the cache line that contains the operand is fetched to level-1 cache. To access a single element of memory, two memory loads are issued, one each for the table entry and the cache line, and 64 bytes are transferred to level-1 cache – all this for supplying a single operand. If the next operand is contiguous to the previous one in logical memory, the likelihood of it being in the same page and same cache line is very high. The only exception would be if the first operand was the last operand in the cache line which happens to be the last cache line in the page.

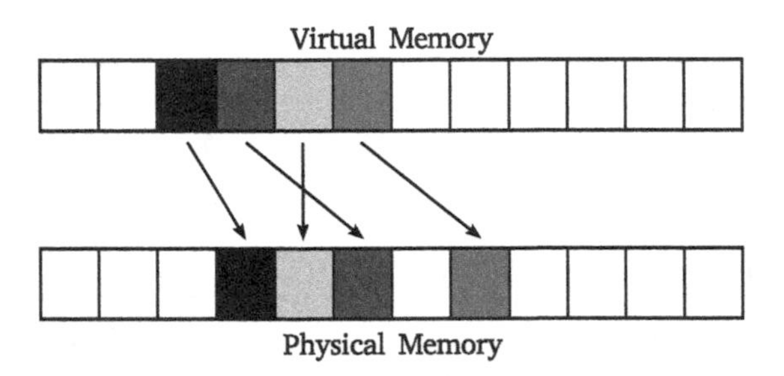

FIGURE B.1.2 Contiguous pages in virtual address space may be fragmented in physical memory.

A TLB table entry allows the processor to access all of the 4096 bytes within the page. As the processor accesses additional operands, either the operand will reside within a page whose address resides in the TLB, or another table entry must be fetched to access the physical page containing the operand. A very important hardware statistic that can be measured for a section of code is the effective TLB miss ratio. A TLB miss is the term used to describe the action when no table entry within the TLB contains the physical address required – then a miss occurs, a page entry must be loaded to the TLB, and the cache line that contains the operands can be fetched. A TLB miss is very expensive since it requires two memory accesses to obtain the data. Unfortunately, the size of the TLB is relatively small. On some AMD microprocessors the TLB only has 48 entries, which means that the maximum amount of memory that can be "mapped" at any given time is 4096*48 = 196,608 bytes. Given this limitation, the potential of TLB thrashing is likely in many situations. Thrashing the TLB refers to the condition where very few of the operands within the page are referenced before the page table entry is flushed from the TLB. In subsequent chapters, examples where the TLB performance is poor will be discussed in more detail. The programmer should always obtain hardware counters for the important sections of their application to determine if their TLB utilization is a problem.

APPENDIX C

Command Line Options and Compiler Directives

CONTENTS

C.1 COMMAND LINE OPTIONS AND COMPILER DIRECTIVES

Frequently used command line arguments and compiler directives for Cray's CCE compiler and Intel's ICC compiler are shown in Tables C.1.1 and C.1.2.

TABLE C.1.1 Command line options for Cray and Intel compilers.

Purpose	Cray CCE	Intel ICC
Obtain messages concerning optimization of loops.	-hlist=m	-qopt_report=[n]
Specify if compiler should replace certain sequences of code with library calls.	-h[no]pattern	--qopt-matmul
Specify highest relatively safe optimization.	-Ofp3	-fast
Instrument looping structures in program.	-hprofile_generate	
Specify agressiveness of compiler's inlining.	-hipa[n]	-inline

TABLE C.1.2 Compiler directives for Cray and Intel compilers.

Purpose	Cray CCE	Intel ICC
Tell compiler to assume array refs are aligned.	VECTOR ALIGNED	ASSUME_ALIGNED addr:n
Specify which arrays to place in MCDRAM.	MEMORY (BANDWIDTH)	ATTRIBUTES FASTMEM
Control cache blocking on nested loops.	BLOCKABLE NOBLOCKING	(NO)BLOCK_LOOP
Instructs compiler to place variable in cache.	CACHE var	
Instructs the compiler to collapse nested loops.	(NO)COLLAPSE	
Instructs the compiler not to split loop.	NOFISSION	
Instructs the compiler (not) to fuse loops.	(NO)FUSION	NOFUSION
Force or inhibit inlining for next routine call.	(NO)INLINE	(NO)INLINE
Instructs the compiler to interchange nested loops.	(NO)INTERCHANGE	
Instructs the compiler to ignore assumed data dependencies.	IVDEP	IVDEP
Gives compiler information about the next loop.	LOOP_INFO min_trips(c) est_trips(c) max_trips(c)	LOOP COUNT
Instruct compiler to use certain optimization level on next loop.	OPTIMIZE (option)	(NO)OPTIMIZE[:lvl]
Instruct compiler to replace certain code sequences with library calls.	(NO)PATTERN	
Integer arrays used as index has no repeating values.	PERMUTATION	IVDEP
Indicates which loop from loop-nest should be vectorized.	PREFERVECTOR	VECTOR
Instructs the compiler to prefetch certain variables within looping structure.	PREFETCH	(NO)PREFETCH
Memory references can safely be made within conditionals.	SAFE_ADDRESS	
Memory references can safely be made within conditionals and computations are safe.	SAFE_CONDITIONAL	
Instructs the compiler to unroll following loop.	(NO)UNROLL	(NO)UNROLL
Instructs the compiler on vectorization of next loop.	(NO)VECTOR	(NO)VECTOR

APPENDIX D

Previously Used Optimizations

CONTENTS

D.1 LOOP REORDERING

This is a very powerful restructuring technique that is occasionally used by the compiler – the Cray compiler indicates this with the `i` in the loop listing as indicated in the following listing:

Many times, the compiler cannot do this due to potential data dependency issues. Loop recording can be used to restructure a loop nest to get better array accessing, to vectorize on the longer index, and so on. Whenever these are statements between the loops being switched, those loops must be split out from the subsequent multicasted loop.

1. Excerpt 7.5.2

2. Excerpt 7.8.2

D.2 INDEX REORDERING

Index reordering is used to remove strides on arrays. This must be used with caution because it is a global change. When the indices of an array are reordered, the restructuring has global implications. If the array is passed as an argument to a subroutine, both the caller and callee must be modified. If the array is in a module or a common block, all the routines that access that module or common block must be rewritten.

1. Excerpt 7.5.2

D.3 LOOP UNROLLING

Loop unrolling is a very powerful technique to increase computational intensity in a loop. Additionally, loop unrolling gives the compiler more operations to shuffle around to get more overlap. This technique is often employed by the compiler. However, the user is better equipped with the knowledge of the size of the loop and should not be afraid to use it manually.

1. Excerpt 6.5.4
2. Excerpt 7.4.3
3. Excerpt 7.9.2

D.4 LOOP FISSION

Loop splitting or loop fissure is used to remove recursion, subroutine calls, and other non-vectorizable components from large computational loops, as well as for loop splitting of non-tightly nested loops. Exercise caution, as loop splitting tends to increase data motion and should not be used when a computational loop is memory bandwidth limited.

1. Excerpt 7.7.3

D.5 SCALAR PROMOTION

Scalar promotion is needed when a loop is split, and a scalar contains information for each value of the loop index. In this case, the scalar must be promoted to an array. Scalar promotion is used extensively in restructuring (example 47020). When scalar promotion is used, it is important to keep the array sizes small enough to fit into the cache. If the promoted scalar cannot fit in cache, then increase memory traffic results and will degrade any performance gain.

1. Excerpt 7.6.2

D.6 REMOVAL OF LOOP-INDEPENDENT IFS

In (example 47020), the inner `I` loop was split into numerous loops on `I` and the loop-independent `IF` statements removed from the inner `DO` loops. This is a very powerful technique and not one that is employed by the compiler. As discussed earlier, this may lead to scalar promotion and one must be careful to minimize memory traffic.

1. Excerpt 7.10.2
2. Excerpt 7.10.4

D.7 USE OF INTRINSICS TO REMOVE IFS

Whenever `MAX`, `MIN`, `SIGN`, and `ABS` can be used to replace a conditional statement, a significant performance gain can be achieved. The compiler replaces these intrinsics with very efficient machine operations.

D.8 STRIP MINING

In the rewrite of Excerpt 7.10.10, strip mining was used to reduce the number of operations performed in each pass through the inner `DO` loop. In this case, strip mining was used to reduce the number of unnecessary operations. However, another very powerful strip mining technique is used for reducing the amount of memory used when promoting scalars to arrays. By having a strip-mining loop around a set of loops where scalar temporaries must be introduced, the total amount of data allocated can be reduced to fit into cache. Additionally, numerous examples in the book use strip mining to optimize cache usage.

1. Excerpt 7.10.11

D.9 SUBROUTINE INLINING

Subroutine inlining is used extensively by the compiler when it can find the called routine, and often this restructuring is controlled by compiler options and/or directives in the source code. This is an extremely important optimization for C and C++ applications that tend to call small routines. By inlining a set of routines, the compiler has a much better chance of resolving potential data dependencies. Often the user is better equipped to employ this restructuring.

D.10 PULLING LOOPS INTO SUBROUTINES

Often an application will call subroutines within looping structures, and one would want to vectorize the loop containing the callee. Subroutine inlining is best when applied to smaller routines. When the routines are larger, it would be better to pull the loop and/or a chunk of the loop down into the subroutine.

1. Excerpt 7.11.2
2. Excerpt 7.11.6
3. Excerpt 7.11.8

D.11 CACHE BLOCKING

Blocking for effective cache reuse is extremely important on today's bandwidth-limited systems. Without good cache reuse, performance will suffer on all current and future systems. Numerous examples are used to illustrate various ways of obtaining better cache reuse in the book.

1. Excerpt 6.5.6
2. Excerpt 6.5.8
3. Excerpt 6.5.9

D.12 LOOP FUSION

The restructuring performed in Chapter 4's Excerpt 4.8.2 in rewriting the Fortran array syntax to a `DO` loop is an example of loop fusion or loop jamming. This results in better cache utilization by eliminating loads and stores.

1. Excerpt 4.8.2
2. Excerpt 6.6.2
3. Excerpt 6.6.3

D.13 OUTER LOOP VECTORIZATION

Sometimes excellent performance gains can be obtained by vectorizing outer loops. In particular, the compiler can do great register allocation when it has a lot of invariants within the outer loop.

1. Excerpt 7.13.2
2. Excerpt 8.4.4
3. Excerpt 8.4.6

APPENDIX E

I/O Optimization

CONTENTS

E.1 INTRODUCTION

I/O is extremely important in high performance computing. Unfortunately, large scientific applications do not run for hours and then print out YES or NO. When using numerous nodes, one needs to consider how to utilize parallel I/O. To perform parallel I/O efficiently, one needs a good parallel filesystem. Over the past 10 years Lustre has evolved to be the most popular parallel filesystem. This chapter will discuss the details behind Lustre and some tips on the most effective way of using a Lustre filesystem. We start with common I/O strategies and then move into how the Lustre filesystem can be utilized most efficiently. This appendix contains discussion of a presentation developed by several Cray Inc. application specialists – our thanks to Stefan Andersen, Harvey Richardson, Frank Kampe, and others.

E.2 I/O STRATEGIES

E.2.1 Spokesperson

In this method, one process controls all of the I/O. All the processors performing computation send their data to a single processor, who usually is also performing computation, to write out important data to be saved. The spokesperson also would read all data and then broadcast that data to the other processors. This method is easy to program, but it does not scale. First the interconnect is burdened with the traffic to aggregate data onto the spokesperson (memory also needs to be available on the spokesperson for the aggregated data), and then the data is transferred out across the interconnect from the spokesperson to the filesystem. This is also referred to as All-to-One I/O and should be avoided when scaling to a larger number of processors.

E.2.2 Multiple Writers – Multiple Files

In this method, everyone writes their data to a separate file. This also is very easy to write and is only limited by the size of the file system. This approach has two drawbacks. First if it is used when running on thousands of processors you will have thousands of files. The second drawback is how this might interface to Lustre which will be discussed later. The first drawback might be to use a random access file where each processor writes records into a large file. Random file access is supported by MPI-I/O and many languages/libraries. It can be extremely efficient. This approach is frequently called Multiple Writers–Single File. To obtain the best performance, data layout within the shared file is very important.

E.2.3 Collective I/O to Single or Multiple Files

In this approach, there are designated I/O processors, which collect data from the other processors, aggregate that information, and write it out to a file or perhaps several files. The I/O itself can be parallel I/O or to independent files. This approach may be best for interfacing to Lustre (discussed in the next section), since it decreases the number of writers communicating with Lustre.

E.3 LUSTRE MECHANICS

Lustre is advertised as a scalable cluster filesystem for Linux. Lustre is an object oriented storage system with Metadata Servers (MDS), Object Storage Servers (OSS), and Object Storage Targets (OST). These are the three major functional units.

1. One or more metadata servers (MDS) owns one or more of the metadata targets on each Lustre filesystem. The MDS stores namespace metadata, such as filenames, directories, access permissions, and file layout.

2. One or more object storage servers (OSS) store file data on one or more object storage targets (OST) with each OST managing a single local disk filesystem. The capacity of a Lustre filesystem is the sum of the capacities provided by the OSTs.

3. Clients that access and use the data. Lustre presents all clients with a unified namespace for all of the files and data in the filesystem using standard POSIX semantics and allows concurrent and coherent read and write access to the files in the filesystem.

4. The OST is a dedicated filesystem that exports an interface to byte ranges of objects for read/write operations. An MDT is a dedicated filesystem that controls file access and tells clients the layout of the object(s) that make up each file.

5. The OSS is responsible for modifying objects on the OST filesystem.

6. Lustre has a distributed lock manager in the OpenVMS style to protect the integrity of each file's data and metadata. Access and modification of a Lustre file is completely cache coherent among the clients.

7. File data locks are managed by the OST on which each object of the file is striped, using byte-range extent locks.

8. Client requests can be granted both overlapping read extent lock for part of all of the file, allowing multiple concurrent readers of the same file, and/or non-overlapping write extent locks for regions of the file. This allows many Lustre clients to access a single file concurrently for both read and write, avoiding bottlenecks during file I/O.

Figure E.3.1 illustrates the configuration of a typical Lustre file system. Note that this architecture can vary both on the network and storage. The backend storage can vary from OSS server internal disks, to software RAID to external SAS-attach hardware raid controllers fronting a large number of disks. In some cases an OSS could be providing 1 GB/s bandwidth to backend storage.

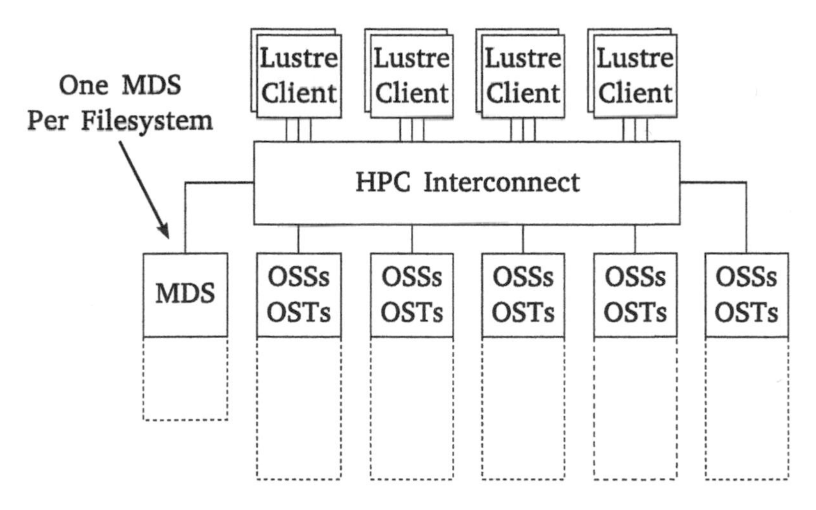

FIGURE E.3.1 Typical high-level configuration of Lustre filesystem.

The obvious bottleneck of this I/O architecture is the MDS. There is only one that is servicing the Lustre file system. The MDS is a shared resource that provides access to each file system's MDTs, it is involved in creating, opening, closing, and getting attributes of the file. The MDS is a shared resource and it can be stressed in large systems by a lot of simple requests.

As a user of many Lustre file systems this has far-reaching impact across all users of the system. If numerous requests are coming into the MDS for reads/writes, even directory information, the system can bog down. Some installations employ a Lustre file system for their home file system and when users are performing operations like "`ls -ltr`" on their directories, this can have a diverse impact on the file system.

When using Lustre for large parallel I/O, it is important to stripe the files across several OST to maximize the bandwidth to the disk. As the number of stripes increases, the potential bandwidth to the disks will increase. Of course efficient I/O also requires large chunks of contiguous data to be transferred, so there is a happy medium between the number of stripes and the size of the data transfers. Lustre has a set of commands for examining, setting, and obtaining stripe characteristics of the file. There are default settings for each file system which can be controlled with the use of these commands.

Lustre is a high-performance, high-bandwidth parallel file system; however, it requires multiple writers to multiple stripes to achieve the best performance. Users must match the size and number of Lustre stripes to the way the files are accessed. Typically, one wants to use large stripes and counts for big files and small stripes and counts for small files.

APPENDIX F

Terminology

CONTENTS

F.1 SELECTED DEFINITIONS

alignment – Alignment of arrays relates to how the starting of the array references are aligned with a cache line.

ARM – An ARM processor is one of a family of CPUs based on the RISC (reduced instruction set computer) architecture developed by Advanced RISC Machines (ARM). ARM makes 32-bit and 64-bit RISC multicore processors.

cache associativity – Desribed in Appendix A.

cache blocking – Technique for reducing the working set of a multinested looping structure to fit into lower level caches.

cache line – Desribed in Appendix A.

chip – An integrated circuit, or chip, is a set of electronic circuits on a small piece of semiconductor material. A chip can contain one or more cores and memory.

clock cycle – Provides the basic timing unit of the CPU. Effecitvely, all operations in the CPU are synchronized and coordincated on clock cycle boundaries, with all operations requiring at least one clock cycle.

cores – Today's HPC CPUs are multicore processors which contain more than one independent processing unit, referred to as cores.

DDR – Double Data Rate memory is used for main memory on today's HPC systems.

GPU/GPGPU – Graphics Processing Unit/General Purpose Graphics Processing Unit; a highly parallel processing unit originally employed for graphics processing. With the advent of error correction and higher precision computation GPU's became viable for HPC.

hardware prefetching – To amortize the latency to memory on a high performance system, the hardware detects patterns on a stream of memory accesses and actually fetches ahead of the on-demand fetch instructions. This can be very powerful in a memory limited looping structure.

hybridization – The process of refactoring an application to employ MPI and OpenMP within the same application.

hyper-threading – Hyper-threads are logical processors that share an actual core. The hyper-threads have their own register sets and control logic to enable sharing of the physical core in a single clock-cycle. Also called simultaneous multi-threading (SMT).

MCDRAM – Multi-channel DRAM. A recent memory technology that employs a 3D stacked memory for achieving very high memory transfer rates.

memory bandwidth – Memory bandwidth relates to the maximum rate that operands can be fetched from the memory on the node, typically in units of bytes/second.

memory page size – Described in Appendix B.

MPI – Message Passing Interface, the de facto standard for communicating within an application that uses messages rather than communicating via shared memory.

MPI tasks per node – Number of MPI tasks/processes used on each node. Typically a constant for each node.

multicore – Machine architecture with more than one processor/node. Typically 2 to 32 cores.

manycore – Machine architecture employed by Intel on Knights series hardware with 60 to 80 cores/node.

node – A node is a compute unit that contains multiple cores sharing memory as well as circuitry to connect to a network for communicating between nodes.

NUMA – Non-Uniform Memory Access. Indicates that the shared memory on the node is composed of multiple memories which have different latency and bandwidth to a given core.

NVLink – NVIDIA's NVLink is a high-bandwidth interconnect which enables fast cache-coherent communication between the CPU and GPUs, as well as between GPUs.

OpenMP – Standard language extension for Fortran, C, and C++ for spreading the computational work across shared memory threads.

peak performance – The maximum processing rate of the core, or the collection of cores on the node, or the collection of nodes in the system. Today the KNL system can generate 8 double precision add/multiplies each clock cycle on each core.

predicated execution – Predicated execution is the process of using a mask register to control a SIMD instruction.

register set – Each hyper-thread has a set of registers for feeding the computational units. There are different registers for floating-point values, integers, and sometimes addresses. Newer systems have vector registers and mask registers for handling vector operands. On KNL and Skylake the width of the vector registers is 8 double-precision floating-point elements.

shared memory across cores – Indicates that the cores are all accessing the same memory place and can communicate through that memory.

shared memory parallelization – Technique for using multiple threads which are running across cores and/or hyper-threads that divide the work in a parallel region.

SIMD – Single Instruction Multiple Data instructions are powerful instructions that can be issued in one clock cycle and generate many results. Today's SIMD instructions are referred to as vector instructions. However, they are not traditional vector instructions which employ segmented units.

socket – The connector on the motherboard that interfaces the CPU to the rest of the system.

software prefetching – Many compilers also identify operands that could be fetched before they are needed, and issue the prefetch so that the operand is in the cache earlier.

stride – A stride on a memory access is the distance between consecutive elements being used in the computation. A stride can be encountered by accessing an array on an outer index in Fortran or an inner index in C. It can also be encountered by accessing an array where its index is not contiguous.

strong scaling – Increasing the number of MPI tasks while keeping the problem size constant.

thread affinity – The binding of a thread to a specifc set of processors/cores.

threading – Employing OpenMP and/or pthreads to manage shared memory processes.

threads per MPI task – The number of threads employed for shared memory parallelism for each MPI task/process.

vector – Traditionally referred to a single instruction that employed a segmented functional unit, an assembly line for processing an operation. For example, the vector add unit on the Cray 1 had six segments. It took 6 clock cycles to get the first result, and then each additional result took a single clock cycle. On current hardware the so-called vector instructions are actually SIMD instructions.

weak scaling – Increasing the problem size proportional to the increase in MPI tasks.

X86 – X86 is a family of backward-compatible instruction set architectures based on the Intel 8086 CPU and its Intel 8088 variant.

APPENDIX G

12-Step Process

CONTENTS

G.1 INTRODUCTION

This appendix outlines a 12-step process that can be used for porting/optimizing an existing application to a multi/manycore architecture.

G.2 PROCESS

1. **Step One** – Understand the hardware. (Chapter 3)
2. **Step Two** – Select a problem to be solved on target hardware. (Chapter 2)
3. **Step Three** – Profile a real problem that needs to be solved on target hardware. (Chapter 5)
4. **Step Four** – Plan of attack. What and where is the bottleneck? In remembrance of Ricky Kendall, we use the whack-a-mole approach. Deal with the top user of time and then reprofile and address the next, etc.
5. **Step Five** – Computation. (Chapter 7 and Chapter 8)
   (a) Load imbalance.
   (b) Vectorization.
   (c) Threading.
6. **Step Six** – Memory access patterns. (Chapter 6)
   (a) Contiguous or strided?
   (b) Alignment.
   (c) Indirect addressing.
   (d) Where is the data coming from?

7. **Step Seven** – Messaging. (Chapter 9)
   (a) Load imbalance.
   (b) Type of message passing – all to one, all to all, pair-wise.
   (c) Sizes of transfers.
8. **Step Eight** – Input/output. (Appendix E)
   (a) Type of I/O – one writer, all writers, group writers.
   (b) Sizes of transfers.
   (c) Asynchronous I/O.
9. **Step Nine** – Check timings and verification of answers.
10. **Step Ten** – Reprofile the refactored application.
    (a) Has the refactoring changed the bottleneck(s)?
    (b) Still Computation? Go to Step 5.
    (c) Still Memory Accessing? Go to Step 6.
    (d) Still Messaging? Go to Step 7.
    (e) Still I/O? Go to Step 8.
11. **Step Eleven** – Run on other hardware to examine performance portability.
12. **Step Twelve** – Get acceptance of code developers/users and update code repository.

Developing one-off versions of the application is not the goal. It is very important to gain acceptance of other code developers and update the code repository with refactored code.

Bibliography

[1] David Bailey, D Browning, R Carter, L Dagum, R Fatoohi, S Fineberg, P Frederickson, T Lasinski, R Schreiber, H Simon, et al. The nas parallel benchmarks. *NASA Ames Research Center: Moffett Field, CA*, March 1994.

[2] Hesheng Bao, Jacobo Bielak, Omar Ghattas, Loukas F Kallivokas, David R O'Hallaron, Jonathan R Shewchuk, and Jifeng Xu. Large-scale simulation of elastic wave propagation in heterogeneous media on parallel computers. *Computer Methods in Applied Mechanics and Engineering*, 152(1):85–102, 1998.

[3] E Barszcz, R Fatoohi, V Venkatakrishnan, and S Weeratunga. Solution of regular, sparse triangular linear systems on vector and distributed-memory multiprocessors. *NASA Rept. RNR-93-007*, 1993.

[4] Broadcom. Broadcom announces server-class armv8-a multi-core processor architecture. http://www.broadcom.com/press/release.php?id=s797235, October 2013. Accessed: 2016-09-12.

[5] Broadcom. Broadcom announces server-class armv8-a multi-core processor architecture. https://web.archive.org/web/20160304072150/http://www.broadcom.com/press/release.php?id=s797235, October 2013. Accessed: 2016-12-17.

[6] Charlie Demerjian. Anyone want to buy a high performance arm server soc? http://semiaccurate.com/2016/10/31/anyone-want-buy-high-performance-arm-server-soc/. Accessed: 2016-12-17.

[7] Charlie Demerjian. Someone did want to buy a high performance arm server soc. https://semiaccurate.com/2016/12/13/someone-want-buy-high-performance-arm-server-soc/. Accessed: 2016-12-17.

[8] Maximillian J Domeika, Charles W Roberson, Edward W Page, and Gene A Tagliarini. Adaptive resonance theory 2 neural network approach to star field recognition. In *Aerospace/Defense Sensing and Controls*, pages 589–596. International Society for Optics and Photonics, 1996.

[9] Jeff Draper, Jacqueline Chame, Mary Hall, Craig Steele, Tim Barrett, Jeff LaCoss, John Granacki, Jaewook Shin, Chun Chen, Chang Woo Kang,

et al. The architecture of the diva processing-in-memory chip. In *Proceedings of the 16th International Conference on Supercomputing*, pages 14–25. ACM, 2002.

[10] Eduardo F DAzevedo, Mark R Fahey, and Richard T Mills. Vectorized sparse matrix multiply for compressed row storage format. In *International Conference on Computational Science*, pages 99–106. Springer, 2005.

[11] Duncan G Elliott, W Martin Snelgrove, and Michael Stumm. Computational ram: A memory-simd hybrid and its application to dsp. In *Custom Integrated Circuits Conference*, volume 30, pages 1–30, 1992.

[12] Oliver Fuhrer, Carlos Osuna, Xavier Lapillonne, Tobias Gysi, Ben Cumming, Mauro Bianco, Andrea Arteaga, and Thomas Christoph Schulthess. Towards a performance portable, architecture agnostic implementation strategy for weather and climate models. *Supercomputing Frontiers and Innovations*, 1(1):45–62, 2014.

[13] A Yu Gelfgat, PZ Bar-Yoseph, and A Solan. Stability of confined swirling flow with and without vortex breakdown. *Journal of Fluid Mechanics*, 311:1–36, 1996.

[14] Maya Gokhale, Bill Holmes, and Ken Iobst. Processing in memory: The terasys massively parallel pim array. *Computer*, 28(4):23–31, 1995.

[15] Intel. Intel 64 and ia-32 architectures optimization reference manual. http://www.intel.com/content/dam/www/public/us/en/documents/manuals/64-ia-32-architectures-optimization-manual.pdf, June 2016. Accessed: 2016-09-21.

[16] Jim Jeffers and James Reinders. *High Performance Parallelism Pearls Volume Two: Multicore and Many-core Programming Approaches*. Morgan Kaufmann, 2015.

[17] Ian Karlin, Abhinav Bhatele, Bradford L. Chamberlain, Jonathan Cohen, Zachary Devito, Maya Gokhale, et al. Riyaz Haque. Lulesh programming model and performance ports overview. Technical Report LLNL-TR-608824, December 2012.

[18] Samuel W Key and Claus C Hoff. An improved constant membrane and bending stress shell element for explicit transient dynamics. *Computer Methods in Applied Mechanics and Engineering*, 124(1):33–47, 1995.

[19] John Levesque and Gene Wagenbreth. *High Performance Computing: Programming and Applications*. CRC Press, 2010.

[20] John M Levesque and Joel W Williamson. *A Guidebook to FORTRAN on Supercomputers*. Academic Press, 1989.

[21] P. J. Mendygral, N. Radcliffe, K. Kandalla, D. Porter, B. J. O Neill, C. Nolting, P. Edmon, J. M. F. Donnert, and T. W. Jones. WOMBAT: A Scalable and High-performance Astrophysical Magnetohydrodynamics Code. *The Astrophysical Journal Supplement Series*, 228:23, February 2017.

[22] J Michalakes, J Dudhia, D Gill, T Henderson, J Klemp, W Skamarock, and W Wang. The weather research and forecast model: software architecture and performance. In *Proceedings of the Eleventh ECMWF Workshop on the Use of High Performance Computing in Meteorology*, pages 156–168. World Scientific: Singapore, 2005.

[23] Matthew Otten, Jing Gong, Azamat Mametjanov, Aaron Vose, John Levesque, Paul Fischer, and Misun Min. An mpi/openacc implementation of a high-order electromagnetics solver with gpudirect communication. *The International Journal of High Performance Computing Applications*, 30(3):320–334, 2016.

[24] RIKEN. Himeno benchmark. http://accc.riken.jp/en/supercom/ himenobmt/. Accessed: 2016-09-27.

[25] Hideki Saito, Greg Gaertner, Wesley Jones, Rudolf Eigenmann, Hidetoshi Iwashita, Ron Lieberman, Matthijs van Waveren, Brian Whitney, et al. Large system performance of spec omp2001 benchmarks. In *International Symposium on High Performance Computing*, pages 370–379. Springer, 2002.

[26] Anton Shilov. Qualcomm demos 48-core centriq 2400 server soc in action, begins sampling. http://www.anandtech.com/show/10918/qualcomm-demos-48core-centriq-2400-server-soc-in-action-begins-sampling. Accessed: 2016-12-17.

[27] RD Smith and PR Gent. Reference manual for the parallel ocean program (pop), ocean component of the community climate system model (ccsm2. 0 and 3.0). Technical report, Technical Report LA-UR-02-2484, Los Alamos National Laboratory, Los Alamos, NM, http://www.ccsm.ucar.edu/models/ccsm3.0/pop, 2002.

[28] Nigel Stephens. Armv8-a next generation vector architecture for hpc. http://www.hotchips.org/wp-content/uploads/hc_archives/hc28/HC28.22-Monday-Epub/HC28.22.10-GPU-HPC-Epub/HC28.22.131-ARMv8-vector-Stephens-Yoshida-ARM-v8-23_51-v11.pdf . HotChips28, August 2016.

[29] A Wallcraft, H Hurlburt, EJ Metzger, E Chassignet, J Cummings, and Ole Martin Smedstad. Global ocean prediction using hycom. In *DoD High Performance Computing Modernization Program Users Group Conference, 2007*, pages 259–262. IEEE, 2007.

[30] Chris Williams. Broadcom quietly dismantles its 'vulcan' arm server chip project. http://www.theregister.co.uk/2016/12/07/broadcom_arm_processor_vulcan/. Accessed: 2016-12-17.

[31] Samuel Williams, Leonid Oliker, Richard Vuduc, John Shalf, Katherine Yelick, and James Demmel. Optimization of sparse matrix–vector multiplication on emerging multicore platforms. *Parallel Computing*, 35(3):178–194, 2009.

[32] Toshio Yoshida. Introduction of fujitsu's hpc processor for the post-k computer. http://www.hotchips.org/wp-content/uploads/hc_archives/hc28/HC28.22-Monday-Epub/HC28.22.10-GPU-HPC-Epub/HC28.22.131-ARMv8-vector-Stephens-Yoshida-ARM-v8-23_51-v11.pdf . HotChips28, August 2016.

[33] Dongping Zhang, Nuwan Jayasena, Alexander Lyashevsky, Joseph L Greathouse, Lifan Xu, and Michael Ignatowski. Top-pim: throughput-oriented programmable processing in memory. In *Proceedings of the 23rd International Symposium on High-Performance Parallel and Distributed Computing*, pages 85–98. ACM, 2014.

Crypto

BTC:

1DW8hyB4LgXu4kSdwJ8dYedGNvv49Pudsr (₿)

GPG:


```
-----BEGIN PGP PUBLIC KEY BLOCK-----
Version: GnuPG v1

mQINBFhqk6sBEACiKHQVV33a/PVa+bEviBH45Fca8IdCbNym9W6DNVgs3zle2KaP
YVdHUThq8Rc3VCNZnep4m5s7b2MSx0RSWc034IsK/Vk5+QzjR2ztvjW1LRkXuJhd
zPgwSRx3MKZtpFuqx0hzxT73s0cJr8kU1HmMdFQMbPEnUhLzPMA7t2cs7YmIb0Pi
8nR2U6JJpG+iSfs9Bw0fikluFLx7bS7uamwNt3SB6vBX8pP1G6ZuZUjqdFoJbyii
8WWEQjcTGGLvVxQCC+dz6C2Y3pDpx90ZZ935WfMboTcT21BkikvsAPUZiV+0QZt1
9ymjgXAE1HO26x0FjTNiyg7jhbZsuwmoeJaPbMCdaXdFZXRH1dfdX3GiuYAaNLZj
z2jW+QNebDcjmzhcNa0Ch93bFBg3nykazt4SFGqV+TzpJDTd8d91zIr0j6WfNVdd
6PRTWy9HJXtPSc0gu2wfkVC1z7qBWDy1741OUMpRHvAbuEl8rLtEs/xB52r5m0ZZ
rYHKTl7pWct4po0tx9xXQ0Bzmr+tfK/+y0bJ3q0fdg/UfmBRbCQ6XoIGXTkFpUpf
8++0mjswW3SEstykNWpMX2BT21RfSELm4Gqn8atEofp0hEgQXE3YcgFWdzK8w7vt
naI29EJmN+mPw/5YDcuW9S+J1K2zXwiT+LHu905Vz/E12U/f0yPkXg2/0wARAQAB
tAlITU1DIEJvb2uJAjgEEwECACIFAlhqk6sCGwMGCwkIBwMCBhUIAgkKCwQWAgMB
Ah4BAheAAAoJEMpAcmM0ykKH3yMP+g0rlhT0U3gdM3xFJbg10QkkQbHgcVXjQyCV
00LVBDbKG4uSZJd0t79BjA1hyzNzhS0Zga9IzEw1Gux1AQnQtfJi0Sy438YdXyIe
8H/2U6EY6Q40ypPnKZ07723MSHF7TIfKfzS7JemnfXhwsRYWzeiahU3VG0r25zxJ
C2P6rjp/8v5JMdDWrknGb3iXAyVZ6uQbnRpYnmJmbW85LKSlspveCfpeGD9U0Dz2
NXRB9arWrM7FGtoGPQsXzMI/KXu0I5zG+qD8/2c7EaF8KA2JXgQ5nXMpHGHnAb1p
7RO5sTqeP1f7GcLyaYB6TR38Vh/50BHBc9lbjdISfLTwARJ5X+EP4ePg6EUi/FpN
sh0SueryTVPxSQH/3zzQ4ia/g3ivGKp1xK+7ZmzfvgN51mjQNqswjAlGMKZUJhZb
ivWF4PRnyT4jMMziEAtVapxJ2WQlXdHY0NMcB/mfIlkU39p0131NcGn+gT56wdzo
6UHxcFY/vjcvxuC7te9UEBv4RvEGy3QLjKrTTMn2zK0/fW7uWG7MKPq0SR3yhAAD
4sZ6grhcESaT0KPkljYiyBMDPo/voKfpJqTQDW/SI/n0si0BpkwmDo313LlVdzLW
bWHsKUAxypU50Uii+IknTGsY0X4jDv/UJK0bwfQGzLsEgMzvkYSFzkdgGlLL331M
AFqgVsjPuQINBFhqk6sBEAC0EN8/ZDZbxUFBPUQguEqoiDH/r3LdG6yb4CLdjApA
kmGVVVbDITgydYG/17ihtXplAQmJ3PgEbnzYST07K0nfeJ0fRtTGTNMwokY8yaCB
2LAGkrN8zJ+gw2XeekCZAp0GhiD4dL6esHNalG20FKbepxzk5P8w1md0UAYRMClI
tKY7T3xCpbqe+DLIjlU7NddofP2TVFCkiu1I1KYp1M2TTL2kn6QRidbL0q10mriz
V1kP2Zuz5DyE+nKnWvQp+EC2Pu1tUvV6u7QN/36PPsVrRt7M32f6VjvL6+zXCaTJ
ptZTb6DHPqzzlXCmjv9xbFleUE9uj03kJ0W4D456LZATX/8VeSPlUQR0z/kq0KIz
L7k3jZbiVRdyP3Wo2oxJ/Cz6J3iX0cfXq4EU825+mTrBC5CjCiMy7CJpWlEQDwa4
DfniXbU64gECPhzw1F9UPr6a2Rwoxn0TiG/sq+ex8W8vFKsBEkF5ijDc1mZ9RPf4
nIZAvG3BqbaXQdboR/Nn2/WfaYxIPTrfLzwRdNKdIgdsXjvhm0tJiEQQz48nrW08
C/tmJWtfWD+mYYolVUOSXd5ZM95kg9P9CZT5+1IXjANTP4iKUpJ0DnfGH845jvWZ
/4Lsf1fb9g4oYGw4qsubUHfcJBShrnXPFvqRtUGz1z0j1UBsT8DiZnBZ0oLpEbBU
5wARAQABiQIfBBgBAgAJBQJYap0rAhsMAAoJEMpAcmM0ykKHaAsP/R8VPGVQzl8L
0pDXbgM3TyTVDvWkSYF2MnM1Xu1hZreuS2V9rRE9DFxJ8La/rX9Mh2CM+N7W6CIa
npoxQWv1+tm1naGJdTGFDCy2FbZraGvvjd1TcTHPqMDCb+Kf3PVVqcZN9P8hgX0s
Wl3mMcdSnrth0B7i0bggUwu8B/7Lkp+R3r3hjHL8/kUYdgDqf5d0zL/jeFiyhzWR
8ebPlBkaApBcZE3JICygSps+CTY0ev+LPMM0hNSAbPMl5grKY6InqqP+NVm+11zS
Y0+htRxbV7Uao5shw0R2ZJgCwjKxzMUXd+0/6FUJQWA1SbtS0jDL0JbB94WxRc/X
iEw0PtIyFfoao/iVoBV0lKRTVKZsJI3pHWMTbQwpLVjvYtRLmo140qY9S1iMEvF0
rkPvkkGdYl3SugmQnQN2XiF+sgXY4kaZKusse+Fo+HIVwUJPp00k0QPUTDKGMBbf
Rzbgh7vl/Q7ENFVWTanPIYd9sy2KNgIvDtxfcq7ZcyesVSjvl7mf5dmAw98o/kP5
Tgr5NXr3ynYIknNnPx9Q0Akt0Zrmi4pIkKZinoKUAVqiQWRoqlFNvc6eloM8h/wn
J7sPkx1xWuMlZB0Y+p3CE9MpWS7jZmi3CJEt1IkZjrcppghZHP0EBJFxB+s7NIfj
ztAfYpxUoT2u/RlrAoPgieCg0j0qTs0k
=E6A2
-----END PGP PUBLIC KEY BLOCK-----
```

Index

D

G

M

Q